VOLUME SEVENTY SIX

Advances in
ORGANOMETALLIC CHEMISTRY

3rd Symposium in Carbene and Nitrene Chemistry

VOLUME SEVENTY SIX

ADVANCES IN ORGANOMETALLIC CHEMISTRY

3rd Symposium in Carbene and Nitrene Chemistry

Edited by

PEDRO J. PÉREZ

*Laboratorio de Catálisis Homogénea
CIQSO-Centro de Investigación en Química Sostenible and
Departamento de Química
Universidad de Huelva - Huelva
Spain*

Founding Editors

F. GORDON A. STONE

ROBERT WEST

ELSEVIER

ACADEMIC PRESS
An imprint of Elsevier

Academic Press is an imprint of Elsevier
50 Hampshire Street, 5th Floor, Cambridge, MA 02139, United States
525 B Street, Suite 1650, San Diego, CA 92101, United States
The Boulevard, Langford Lane, Kidlington, Oxford OX5 1GB, United Kingdom
125 London Wall, London, EC2Y 5AS, United Kingdom

First edition 2021

Copyright © 2021 Elsevier Inc. All rights reserved.

No part of this publication may be reproduced or transmitted in any form or by any means, electronic or mechanical, including photocopying, recording, or any information storage and retrieval system, without permission in writing from the publisher. Details on how to seek permission, further information about the Publisher's permissions policies and our arrangements with organizations such as the Copyright Clearance Center and the Copyright Licensing Agency, can be found at our website: www.elsevier.com/permissions.

This book and the individual contributions contained in it are protected under copyright by the Publisher (other than as may be noted herein).

Notices
Knowledge and best practice in this field are constantly changing. As new research and experience broaden our understanding, changes in research methods, professional practices, or medical treatment may become necessary.

Practitioners and researchers must always rely on their own experience and knowledge in evaluating and using any information, methods, compounds, or experiments described herein. In using such information or methods they should be mindful of their own safety and the safety of others, including parties for whom they have a professional responsibility.

To the fullest extent of the law, neither the Publisher nor the authors, contributors, or editors, assume any liability for any injury and/or damage to persons or property as a matter of products liability, negligence or otherwise, or from any use or operation of any methods, products, instructions, or ideas contained in the material herein.

ISBN: 978-0-12-824582-8
ISSN: 0065-3055

For information on all Academic Press publications
visit our website at https://www.elsevier.com/books-and-journals

Publisher: Zoe Kruze
Acquisitions Editor: Sam Mahfoudh
Developmental Editor: Tara Nadera
Production Project Manager: Denny Mansingh
Cover Designer: Alan Studholme

Typeset by SPi Global, India

Contents

Contributors	vii
Preface	ix

1. Additions of N, O, and S heteroatoms to metal-supported carbenes: Mechanism and synthetic applications in modern organic chemistry   **1**

Hillary J. Dequina, Kate A. Nicastri, and Jennifer M. Schomaker

1. Introduction	2
2. Ammonium ylides	3
3. Oxonium ylides	44
4. Sulfur ylides	71
Acknowledgment	88
References	88

2. π-Alkene/alkyne and carbene complexes of gold(I) stabilized by chelating ligands   **101**

Miquel Navarro and Didier Bourissou

1. Introduction	101
2. Gold(I) π-complexes	104
3. Gold(I) carbene complexes	123
4. Concluding remarks	134
Acknowledgments	135
References	135

3. Imido complexes of groups 8–10 active in nitrene transfer reactions   **145**

Caterina Damiano, Paolo Sonzini, Alessandro Caselli, and Emma Gallo

1. Introduction: Involvement of imido complexes in the formation of C—N bonds	145
2. Imido complexes of group 8	148
3. Imido complexes of groups 9 and 10	172
4. Summary and outlook	179
References	180

4. **Recent progress on group 10 metal complexes of pincer ligands: From synthesis to activities and catalysis**   **185**

Krishna K. Manar and Peng Ren

1. Introduction	185
2. Well-defined group 10 (nickel, palladium, and platinum) pincer complexes	188
3. General synthetic routes to group 10 pincer complexes	200
4. Catalyzed cross-coupling reactions	237
5. Miscellaneous reactions	243
6. Conclusions and perspectives	250
Acknowledgments	250
References	250

Contributors

Didier Bourissou
CNRS/Université Paul Sabatier, Laboratoire Hétérochimie Fondamentale et Appliquée (LHFA, UMR 5069), Toulouse, France

Alessandro Caselli
Department of Chemistry, University of Milan, Milan, Italy

Caterina Damiano
Department of Chemistry, University of Milan, Milan, Italy

Hillary J. Dequina
Department of Chemistry, University of Wisconsin-Madison, Madison, WI, United States

Emma Gallo
Department of Chemistry, University of Milan, Milan, Italy

Krishna K. Manar
School of Science, Harbin Institute of Technology (Shenzhen), Shenzhen, China

Miquel Navarro
CNRS/Université Paul Sabatier, Laboratoire Hétérochimie Fondamentale et Appliquée (LHFA, UMR 5069), Toulouse, France

Kate A. Nicastri
Department of Chemistry, University of Wisconsin-Madison, Madison, WI, United States

Peng Ren
School of Science, Harbin Institute of Technology (Shenzhen), Shenzhen, China

Jennifer M. Schomaker
Department of Chemistry, University of Wisconsin-Madison, Madison, WI, United States

Paolo Sonzini
Department of Chemistry, University of Milan, Milan, Italy

Preface

This is a special volume of *Advances in Organometallic Chemistry* on occasion of the 3rd International Symposium on Carbene and Nitrene Chemistry, chaired by Michael Doyle, Chi-Ming Che and Jianbo Wang and celebrated in San Antonio, Texas, in February 2020. Representative areas of research within this field are presented in this volume, where contributors participated as invited speakers at such forum.

In Chapter 1, Schomaker provides an update on the advances regarding the transformations of carbene-generated ylide intermediates, focusing on ammonium, oxonium, and sulfonium ylides. Chapter 2 is devoted to gold-carbene chemistry where Borissou has reviewed the particular area of chelating ligands-containing compounds and their effect in the ligation to π-alkenes and alkynes. Chapter 3 focuses on nitrene chemistry, where Gallo provides an overview of imido derivatives of groups 8–10 relevant to transfer reactions. A final chapter, not related to the aforementioned Symposium, completes the volume: the recent developments on the chemistry of group 10 complexes bearing pincer ligands constitute the topic reviewed by Peng.

I very much appreciate the participation and commitment of all the authors, particularly in these difficult times. My appreciation is also extended to the editorial team Tara Nodera and Denny Mansingh for making possible that this volume reaches the readers.

Pedro J. Pérez

CHAPTER ONE

Additions of N, O, and S heteroatoms to metal-supported carbenes: Mechanism and synthetic applications in modern organic chemistry

Hillary J. Dequina[†], Kate A. Nicastri[†], and Jennifer M. Schomaker*

Department of Chemistry, University of Wisconsin-Madison, Madison, WI, United States
*Corresponding author: e-mail address: schomakerj@chem.wisc.edu

Contents

1. Introduction	2
2. Ammonium ylides	3
2.1 Intramolecular [2,3]-sigmatropic rearrangements	6
2.2 Intermolecular [2,3]-sigmatropic rearrangement	11
2.3 Intramolecular 1,2-Stevens rearrangements	15
2.4 Intermolecular 1,2-Stevens rearrangements	23
2.5 Miscellaneous	25
2.6 N-H insertion	28
2.7 Intermolecular electrophilic trapping of protic ammonium ylides: Multicomponent reactions	35
2.8 Intramolecular electrophilic trapping of protic ammonium ylides	38
2.9 Concluding remarks	44
3. Oxonium ylides	44
3.1 [2,3]-Sigmatropic rearrangements	46
3.2 Intramolecular [2,3]-sigmatropic rearrangements	47
3.3 Intermolecular [2,3]-sigmatropic rearrangements	51
3.4 1,2-Shifts	56
3.5 O-H insertion	60
3.6 Electrophilic trapping	65
3.7 Multicomponent reactions (intermolecular electrophile trapping)	65
3.8 Intramolecular electrophile trapping	69
3.9 Concluding remarks	71

[†] These authors contributed equally.

4. Sulfur ylides 71
   4.1 Introduction to sulfur ylides and their reactivity 71
   4.2 Generation of sulfur ylides from metal carbenes 72
   4.3 [2,3]-Sigmatropic rearrangements 73
   4.4 Intermolecular [2,3]-sigmatropic rearrangements 74
   4.5 1,2-Stevens (thia-Stevens) rearrangement 79
   4.6 Intermolecular 1,2-Stevens (thia-Stevens) rearrangements 79
   4.7 Intramolecular 1,2-Stevens rearrangements 84
   4.8 Applications in total synthesis 86
   4.9 Concluding remarks 88
Acknowledgment 88
References 88

1. Introduction

Carbene transfer reactions comprise a useful class of transformations for the formation of new C–C or C–X bonds, particularly in the construction of complex carbo- and heterocyclic motifs present in diverse bioactive small molecules and natural products. These reactions are typically conducted in the presence of transition metal catalysts to facilitate access to the desired metal carbene intermediates from bench-stable precursors and to achieve some level of control over carbene reactivity *via* the choice of metal and ligand. Metal carbenes derived from α-diazocarbonyl compounds are among the most common, as they display high electrophilicity, allowing them to readily react with nucleophiles to generate versatile ylide intermediates.[1] An ylide is a molecular species containing a negatively charged carbon atom adjacent to a positively charged heteroatom (referred to as an "onium" group).[2] Synthetic applications of ylides were not widely explored until after the pioneering work of Wittig in 1953, where phosphonium ylides were employed for the preparation of alkenes from aldehydes and ketones.[3]

The general process of ylide formation between a metal carbene complex and a Lewis base begins with a double umpolung of the carbenic carbon (Fig. 1).[2] The metal complex, generally derived from a copper(I) or dirhodium(II) catalyst, behaves as a Lewis acid and accepts electrons from the diazo carbon at its vacant coordination site; since the turn of the century, the breadth of metal complexes aiding in productive carbene transfer has been extended to include those of ruthenium,[4–11] iron,[12–17] palladium,[18–20] silver,[21–24] and cobalt.[25–33] Electron back-donation from the metal complex

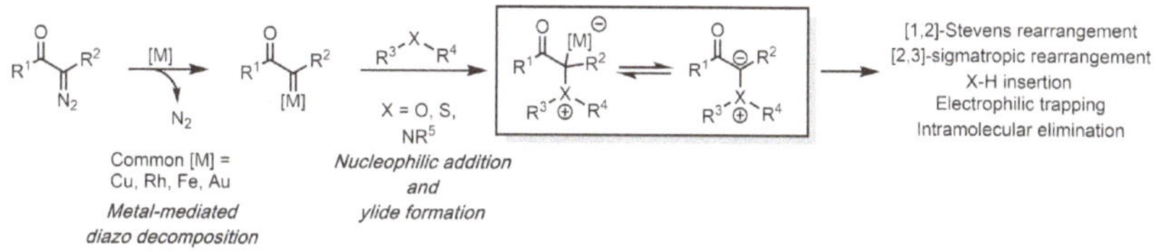

Fig. 1 Metallocarbene-derived ylide formation and subsequent transformations.

to the carbenic carbon and the concomitant loss of dinitrogen generates the metal carbene species. In its capacity as a Lewis acid, the metal carbene intermediate can then accept electrons from the heteroatom of a Lewis base to generate the metal associated ylide, which can subsequently dissociate to furnish the free ylide species. Common Lewis bases reported to add to metal-supported carbenes to furnish ylides include amines, ethers, and sulfides; ylide formation can occur in either an inter- or intramolecular fashion. These versatile ylide intermediates can undergo a host of subsequent transformations, including rearrangements ([1,2]-Stevens and [2,3]-sigmatropic rearrangements),[34,35] X–H insertions, and can participate in multicomponent reactions (MCRs)[36] to provide a powerful synthetic approach toward the rapid construction of structurally complex products.

This chapter will discuss examples of the applications of metal carbenoid-derived ammonium, oxonium, and sulfonium ylides to organic synthesis. It is not intended to be an exhaustive discussion of the primary literature on this topic, but rather serve as an overview to highlight significant advances from the past two decades in the hopes of stimulating continued exploration in harnessing carbenoid-derived ylide intermediates for asymmetric synthesis.

2. Ammonium ylides

Nitrogen ylides are compounds that contain a positively charged nitrogen atom adjacent to a carbanion (Fig. 2A). The hybridization of the

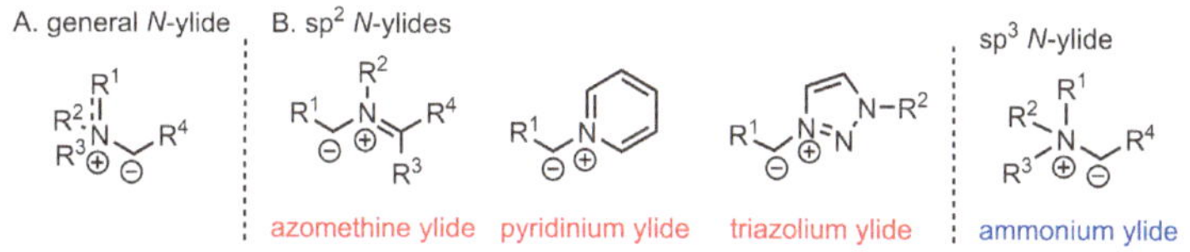

Fig. 2 (A) General representation of ammonium ylides. (B) General structures of both sp^2 and sp^3 nitrogen ylides.

nitrogen atom plays a significant role in the resultant reactivity[37–39]; as such, this general structure can be further defined by subdividing the ylides into N-sp^2 and N-sp^3 ylides. The N-sp^2 ylides include azomethine, pyridinium, and triazolium ylides (Fig. 2B, left).[40–42] In contrast, ammonium ylides contain a sp^3-hybridized nitrogen atom adjacent to a carbanionic site (Fig. 2B, right). This section discusses only ammonium ylides, as there are several other timely reviews that cover the application of sp^2 N-ylides to organic synthesis.

Ammonium ylides are commonly generated in two ways. The first approach involves base-induced deprotonation of a quaternary ammonium salt *in situ* to deliver the corresponding ylide (Fig. 3A). An extensive body of work has been reported in this area in recent years and its application in asymmetric cyclization reactions has been recently reviewed[39]; thus, base-induced ammonium ylide formation will not be covered in this section. A second approach to generate ammonium ylides involves reaction between an amine and a metal-supported carbene (Fig. 3B). These reactions are less well-explored than base-induced ammonium ylide formations but do offer unique opportunities to expand novel chemical space and generate complex molecules in relatively rapid fashion.

Reactions of ammonium ylides generated *via* reaction of an amine with a metal-supported carbene can be divided into four categories. One class of reactions involves rearrangement processes that include [2,3]-sigmatropic and [1,2]-Stevens rearrangements. A [2,3]-sigmatropic rearrangement is a pericyclic reaction in which an amine substituted with an allyl group reacts with a metal-supported carbene to furnish an ammonium ylide, where the initial N-allyl group is transposed to the anionic ylide carbon (Fig. 4A). In contrast, [1,2]-Stevens rearrangements are symmetry-forbidden and are hypothesized to occur through stepwise radical pathways (Fig. 4B). For this reason, the [1,2]-Stevens rearrangement can occur on many different

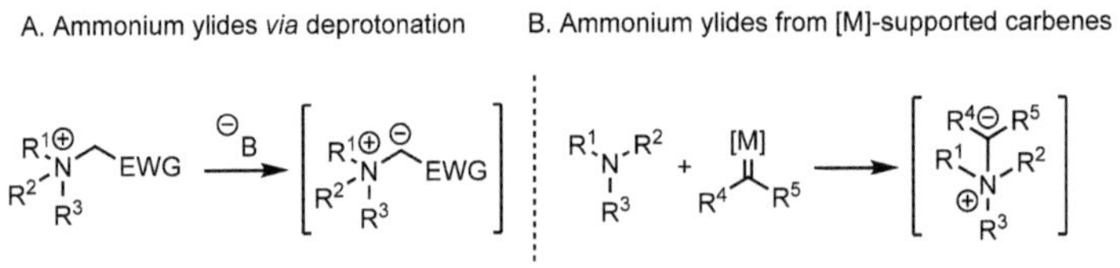

Fig. 3 (A) Base-mediated generation of ammonium ylides from the quaternary ammonium salts. (B) Generation of ammonium ylides *via* reaction with a metal-supported carbene.

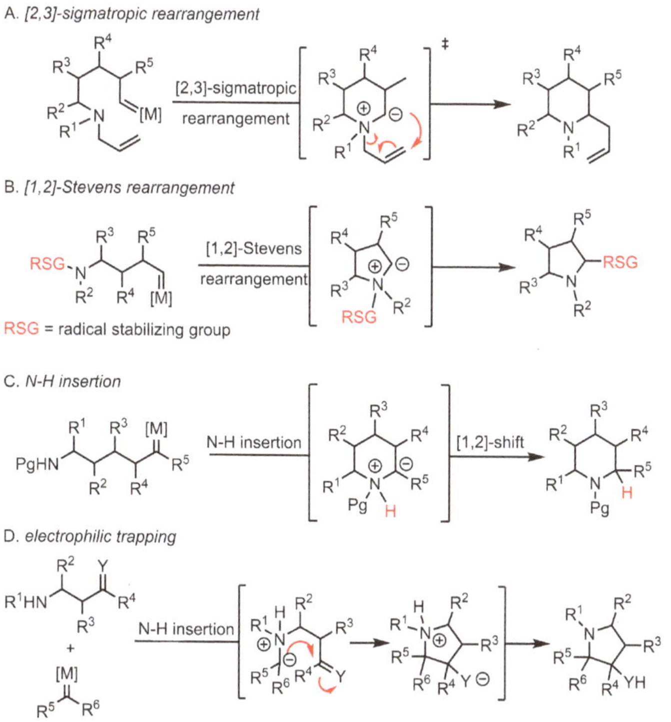

Fig. 4 (A) General [2,3]-sigmatropic rearrangement of allylic amines. (B) General scheme for [1,2]-Stevens shift. (C) General scheme for N-H insertion reaction *via* protic ammonium ylides. (D) General scheme for electrophilic trapping of a protic ammonium ylide (cyclization is shown but can present as a multi-component reaction).

types of amines as long as the reaction proceeds in the presence of a radical-stabilizing group. Reaction of metal-supported carbenes with secondary amines is another common way to generate protic ammonium ylides. These intermediates can undergo two divergent reaction pathways that involve either N-H insertion or electrophilic trapping. N-H insertion occurs when the ammonium ylide intermediate undergoes a [1,2]-shift to deliver a new amine product (Fig. 4C). In contrast, if a protic ammonium ylide is intercepted with an electrophile prior to the [1,2]-shift, a delayed proton transfer can yield products resulting from intramolecular cyclizations or novel multi-component reactions (Fig. 4D, intramolecular cyclization pictured). The following section is subdivided into reactions of each type, with particular emphasis on examples that have been reported in the past 20 years. Older historical examples will be included when relevant.

2.1 Intramolecular [2,3]-sigmatropic rearrangements

The first example of a [2,3]-sigmatropic rearrangement of an ammonium ylide generated by metal catalysis was reported by Doyle in 1981.[43] In this report, the authors evaluated a variety of [2,3]-sigmatropic rearrangements, including those of sulfur and halonium ylides. The Doyle group found that the [2,3]-sigmatropic rearrangement of dimethyl allyl amine **1** could take place in the presence of $Rh_2(OAc)_4$ or $Rh_6(CO)_{16}$ under milder reaction conditions than those usually employed with Cu catalysis (Scheme 1). It was noted that these transformations furnish approximately the same results as the analogous base-promoted processes, implying that the [2,3]-sigmatropic rearrangement likely proceeds without the aid of the metal. This first report stimulated further efforts to harness ammonium ylide-mediated [2,3]-sigmatropic rearrangements for applications in synthesis, as described in more detail below.

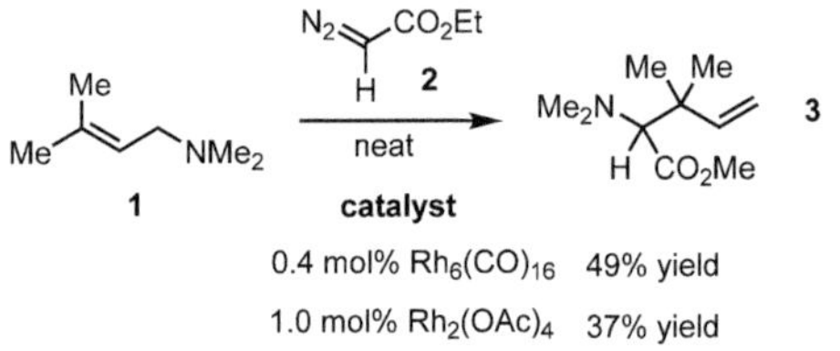

Scheme 1 Rh-catalyzed [2,3]-sigmatropic rearrangement of allylamines *via* a proposed ammonium ylide species.

In 2001, Clark and coworkers described intramolecular [2,3]-sigmatropic rearrangements of ammonium ylides to generate mono- and bicyclic amines.[44] The use of α-diazo precursors with modified tether lengths ($n=0-2$, Scheme 2) enabled the corresponding [2,3]-sigmatropic rearrangements to be achieved in the presence of a Cu catalyst. It is important to note that no reaction was observed at room temperature, likely due to coordination of the amine to the copper catalyst. Rh catalysis was also successful in these reactions, but Cu was selected for further scope studies due to higher overall yields. A variety of pyrrolidinones **8**, piperidinones **9**, azepinones **10**, and azocinones **11** could be accessed *via* this methodology (Scheme 2A). The same group later applied these conditions to the generation of bicyclic amines to furnish pyrrolizidine, indolizidine and quinolizidine alkaloids **16–19** in good yields and as single stereoisomers (Scheme 2B).[45]

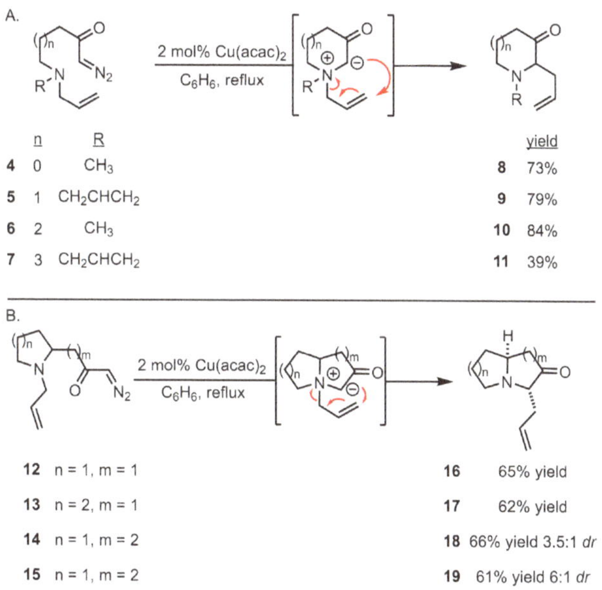

A.

n	R		yield
4	0	CH₃	8 73%
5	1	CH₂CHCH₂	9 79%
6	2	CH₃	10 84%
7	3	CH₂CHCH₂	11 39%

B.

12	n = 1, m = 1	16	65% yield
13	n = 2, m = 1	17	62% yield
14	n = 1, m = 2	18	66% yield 3.5:1 dr
15	n = 1, m = 2	19	61% yield 6:1 dr

Scheme 2 (A) Application of [2,3]-sigmatropic rearrangement of ammonium ylides to generate monocyclic amines. (B) Extension of Cu-catalyzed [2,3]-sigmatropic rearrangements to the synthesis of bicyclic amines with N-ring junctions.

In 2002, Clark and coworkers reported the syntheses of novel α-substituted and α,α-disubstituted non-natural amino acids using a [2,3]-sigmatropic rearrangement[46]; these motifs can be readily incorporated into polypeptides. This strategy enables the facile exchange of groups alpha to the diazo moiety or to the substituents on the allylamine. An aromatic template was specifically chosen to tether the diazocarbonyl and allylamine functionalities, as it can both function as a protecting group and enable direct coupling to other amino acids by activating the carboxyl group. Diazoallylamine substrates, such as **20** and **23**, were exposed to the Cu catalyst to deliver azalactones **21** and **24** (Scheme 3A). Cyclic amino acids could also be accessed under these conditions by performing a tandem ylide generation and rearrangement to give a diene product that undergoes sequential ring closing metathesis to deliver **24**. These products were afforded in excellent yields and could be deprotected to furnish the cyclic amino acid derivatives (**25**) (Scheme 3B).

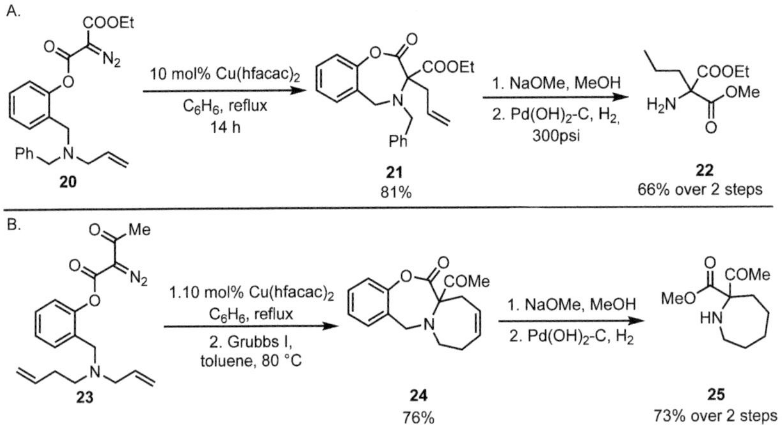

Scheme 3 (A) [2,3]-Sigmatropic rearrangement to generate azalactones and conversion to α-aminoesters. (B) [2,3]-Sigmatropic rearrangement to generate azadienes that can undergo RCM to yield unnatural cyclic α-aminoesters.

In 2003, Sweeney and coworkers presented an interesting intramolecular [2,3]-sigmatropic rearrangement for the synthesis of proline derivatives that involves a ring contraction.[47] The group began their investigations by exploring the reaction of **2** with tetrahydropyridine **26** in the presence of a Rh catalyst. Unfortunately, even at high temperatures (>80 °C), the highest yield obtained was only ~20%. Switching to a Cu catalyst greatly enhanced the efficiency of the reaction, giving up to 59% yield with trace amounts of the corresponding elimination product **31** (Scheme 4A). Interestingly, a correlation between the amount of elimination byproduct and the use of more electronegative ligands on the Cu catalyst was noted. The group extended their initial results to other carbene precursors, such as **27**, to furnish the corresponding pyrrole products. The change to acceptor/acceptor-type carbene precursors resulted in a significant increase in the yield of the overall transformation. In addition, it was found that treatment of tetrahydropyridine **26** under the same conditions with Meldrum's acid-derived diazo **32** gave spiropyrrolidines of the form **33** (Scheme 4B).[48]

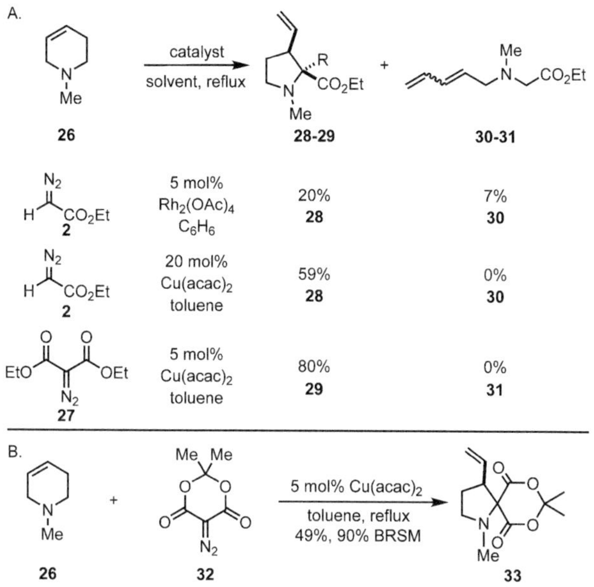

Scheme 4 (A) Development of intramolecular [2,3]-sigmatropic rearrangement to yield ring-contracted proline derivatives. (B) Application to the synthesis of spiropyrrolidines from reaction of **26** and diazoesters derived from Meldrum's acid.

The [2,3]-sigmatropic rearrangement was expanded to include rearrangements of ene-*endo*-spirocyclic tetrahydropyridine ammonium ylides (Scheme 5).[49] Sweeney and coworkers hypothesized that the rearrangement of spirocyclic 6,7-ylides would enable access to the pyrroloazepine scaffold, a common core in the *Stemona* alkaloids. The authors considered

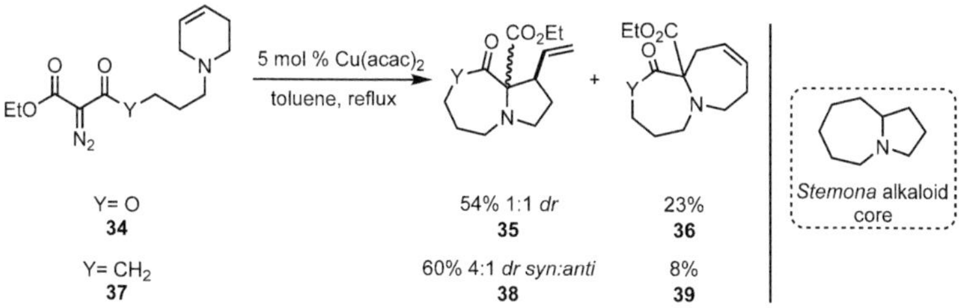

Scheme 5 Cu-catalyzed reaction of spirocyclic ylides to furnish all-carbon fused frameworks (left). General structure of the *Stemona* alkaloid core (right).

that [1,2]-rearrangement might be competitive, as the geometric constraints of the spirocyclic ylides might disfavor a symmetry-allowed [2,3]-sigmatropic rearrangement. Exposure of the diazoester precursor **34** to Cu catalysis resulted in the desired [2,3]-sigmatropic rearrangement product **35** in 54% yield, along with 23% yield of [1,2]-Stevens rearrangement product **36** (Scheme 5, entry 1). Bulking up the oxygen atom of **37** to a methylene group led to a higher selectivity for [2,3]-sigmatropic rearrangement by suppressing the [1,2]-shift product **39** to only 8% yield and increasing the yield of the [2,3]-sigmatropic rearrangement product **38** to 60% (Scheme 5, entry 2). When the chemistry was expanded to include [5,5]-spirocyclic ylides, a 1:1 ratio of [1,2]-Stevens:[2,3]-sigmatropic rearrangement resulted, further supporting the hypothesis that increased rigidity enhances geometric constraints that disfavor the [2,3]-sigmatropic rearrangement. Ultimately, this stereoselective reaction enables rapid access to the *Stemona* alkaloid family and highlights the potential for synthetic applications of spirocyclic ammonium ylides.

In 2015, Xu and coworkers reported one-pot syntheses of benzopyrrolizidinyl sulfonamides from Cu- and Rh-catalyzed reactions of 4-*N*-allylarylpropylamino-1-sulfonyl triazoles.[50] Rh-mediated decomposition of the triazole tether led to four potential reaction sites for the azavinyl carbene **41** (Scheme 6): insertion into the π-bond (i), C-H insertion (ii), attack by the heteroatom electron pair (iii), or attack on the aromatic ring (iv). Experimental observations provided evidence that nucleophilic

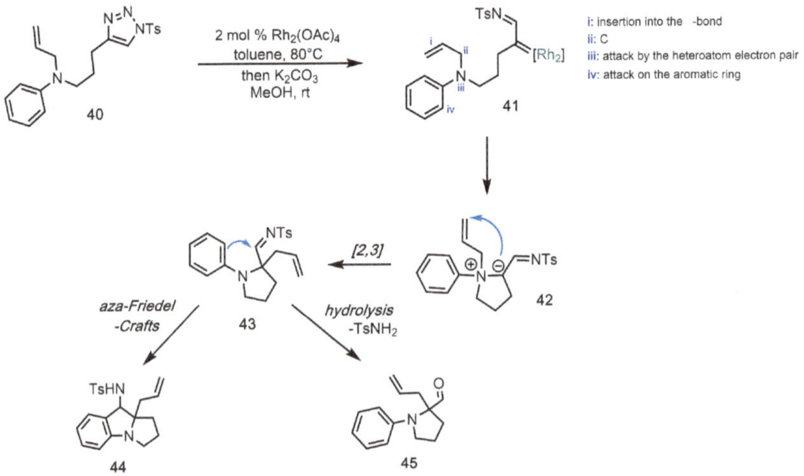

Scheme 6 Reaction pathways for the formation of benzopyrrolizidine **44** and 2-allyl-1-phenylpyrrolidine-2-carbaldehyde **45** from Rh-carbene complex **41**.

addition outcompetes the other three possibilities for *in situ* generation of the rhodium carbene in the azavinyl carbene complex **41**. The proposed mechanism involves an intramolecular nucleophilic addition of the arylamino nitrogen to the rhodium carbene center to generate ammonium ylide species **42**. A [2,3]-sigmatropic rearrangement and C–N bond cleavage give intermediate **43**, which undergoes a subsequent Cu(OTf)$_2$-catalyzed aza-Friedel–Crafts cyclization to generate benzopyrrolizidinyl sulfonamide product **44**. The benzopyrrolizidinyl sulfonamide products were obtained from this cascade reaction in moderate yields; this one-pot, two-relay catalysis protocol represents an efficient method for the construction of polycyclic alkaloid scaffolds.

2.2 Intermolecular [2,3]-sigmatropic rearrangement

In 2008, Inoue and coworkers reported a Cu-catalyzed, intermolecular [2,3]-sigmatropic rearrangement of allylic amine **46** with **2** to deliver homoallylic amine **48**.[51] The use of Cu(acac)$_2$ and Cu(hfacac)$_2$ furnished the products in modest yields of 35% and 31%, respectively (Scheme 7). These conditions were applied to the synthesis of trisubstituted alkenes, giving the corresponding trisubstituted homoallylic amines **49** in up to 50% yield. This was one of the first successful examples of ammonium ylide intermolecular [2,3]-sigmatropic rearrangement and stimulated more recent applications, as discussed below.

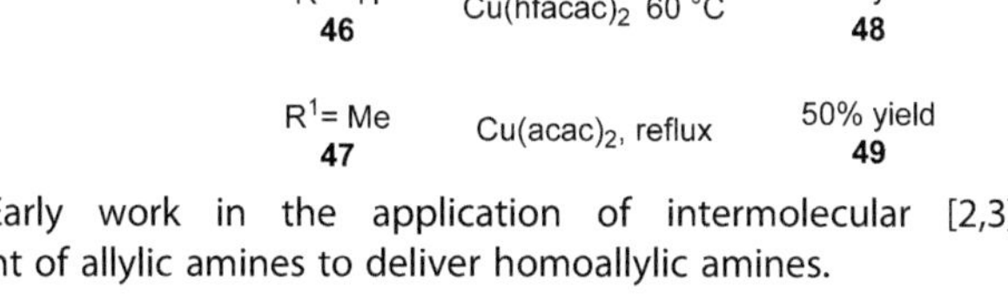

Scheme 7 Early work in the application of intermolecular [2,3]-sigmatropic rearrangement of allylic amines to deliver homoallylic amines.

Pyrrolidine scaffolds are present in many biologically active natural products and pharmaceuticals[52]; despite the multitude of methods for constructing this privileged scaffold, few catalytic approaches using diazo compounds have been developed.[53–60] In 2015, Nikolaev and coworkers secured functionalized N-arylpyrrolidines via metal carbene-mediated ylide formation, followed by intramolecular Michael addition.[61] Initial studies were conducted on amino ester **50** and diazomalonate **51** in the presence of $Rh_2(OAc)_4$ in dichloromethane at room temperature (Scheme 8A). Two products were isolated in a combined yield of 56% (60% based on recovered starting material, BRSM). While the reaction was believed to proceed via an insertion of the Rh carbene intermediate into the N–H bond of the amino group, further structural investigation revealed that the expected N–H insertion products were not obtained. Instead, the products were identified as pyrrolidines **52** (mixture of two diastereomers) and **53**, the latter resulting from the C–H-insertion of the carbene into the p-C–H-bond of the N–phenyl group.

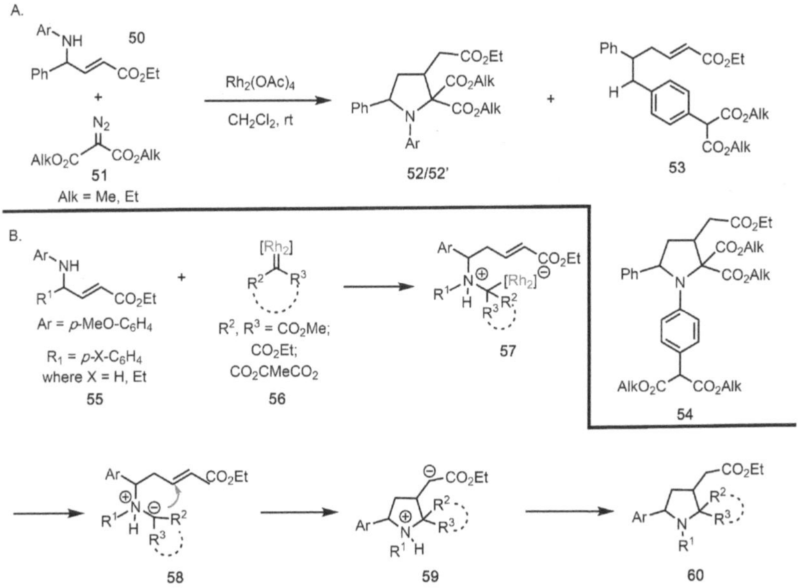

Scheme 8 (A) General scheme for the $Rh_2(OAc)_4$-catalyzed reaction of diazomalonates **51** with amino esters **50**. (B) Proposed mechanistic rationale behind the synthesis of pyrrolidines **60** from amino esters **55** and Rh carbene complexes **56**.

Promotion of the formation of the desired product **52** was attempted by increasing the equivalents of diazomalonate **51**. However, a new product was observed of the form of **54**, which represents a 1:2-adduct of the starting amino ester **50** with di(alkoxycarbonyl)carbene (Scheme 8A). These results prompted adjustment of the experimental design to prevent the formation of side products from the attack of the Rh carbene on the *p*-C–H bond. Switching to *p*-MeO-phenyl amino esters **55** increased the yields of pyrrolidines **52** and **52′** up to 70% and prevented the competing C–H insertion from occurring. Formation of ylide intermediate **57** occurs by a well-established nucleophilic addition process; in this case, the electron-rich amino group of **55** adds to the metal carbene complex **56** (Scheme 8B). Intramolecular conjugate addition of ylide **58** to the enoate tether forms zwitterion **59**, which ultimately undergoes ring closure and 1,4-*H*-migration to give pyrrolidine **60**.

In 2017, the Schomaker group employed several unusual features of bicyclic methyleneaziridines[62–69] (MAs) to achieve a formal [3 + 1] ring expansion to methyleneazetidines (Scheme 9A).[70] The authors attributed successful inhibition of nitrogen inversion to the unique geometry and ring strain of the MAs, which ultimately renders the nitrogen lone pair unusually nucleophilic and sterically accessible. The key aziridinium ylide intermediate was generated by nucleophilic addition of the aziridine nitrogen to a rhodium-bound carbene; as previously noted, successful ylide formation benefited from the bicyclic nature of the aziridine and the good *E:Z* ratios in the MA.

In terms of scope, the electronics of the aryl substituents on the diazoester carbene precursors did not have a significant effect on the reaction outcome (Scheme 9A). Steric hindrance prevented formation of **62** using a MA with a substituent *cis* to the aziridine nitrogen; however, fully substituted methyleneazetidine **63** was accessed upon switching to a less-hindered styrenyl diazoester partner. Methyleneazetidine **64**, containing challenging adjacent quaternary stereocenters, was obtained as a mixture of diastereomers in 89% yield and 3:1 *dr*; the *syn*-Me/CO$_2$Me isomer was obtained in 54% yield and 15:1 *dr* following separation of the diastereomers.

Experimental and computational results were evaluated to elucidate possible pathways for the [3 + 1] ring expansion. An efficient transfer of chirality from enantioenriched (*S*)-**65** to methyleneazetidine (*S*,*S*)-**67** was observed (Scheme 9B); this result suggests ablation of the stereochemical information present in the aziridine precursor does not occur in the intermediates of the ring expansion. Based on these results and DFT computational studies, the most likely pathway was concluded to involve a highly asynchronous, concerted [2,3]-Stevens rearrangement. Overall, the strained bicyclic framework of MAs and the unique reactivity of aziridinium ylides allowed for an efficient stereospecific transformation to forge two new C–C and C–N bonds, as well as two adjacent stereocenters.

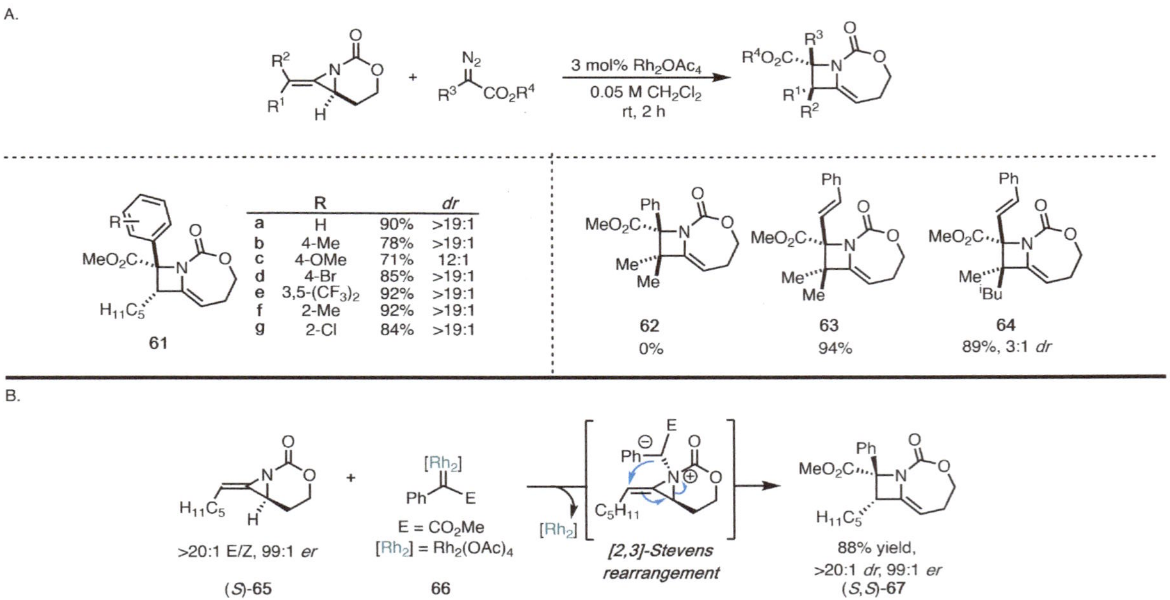

	R		*dr*
a	H	90%	>19:1
b	4-Me	78%	>19:1
c	4-OMe	71%	12:1
d	4-Br	85%	>19:1
e	3,5-$(CF_3)_2$	92%	>19:1
f	2-Me	92%	>19:1
g	2-Cl	84%	>19:1

Scheme 9 (A) General scheme for the [3 + 1] ring expansion of bicyclic methyleneaziridines and selected substrates from the diazoester and aziridine scopes. (B) Chirality transfer experiment using enantiopure methyleneaziridine (*S*)-**65**.

2.3 Intramolecular 1,2-Stevens rearrangements

Early reports of [1,2]-Stevens rearrangements proposed to proceed through the intermediacy of ammonium ylides were reported by West and coworkers in a series of papers in 1993; these reports covered both intramolecular [1,2]-Stevens rearrangements[71] to furnish cyclic amines (Scheme 10A) and intermolecular [1,2] rearrangements[72] to furnish linear amines (Scheme 10B). The intramolecular version of this transformation achieved the syntheses of 2-substituted piperidin-3-ones **71**, **72**, and **74** from the corresponding diazoketones **68–70** using Rh catalysis. The authors found a variety of substituted amines were successful in this chemistry, provided the migrating group was capable of stabilizing the radical intermediate. A radical pathway is supported by the isolation of **73**. Challenges arose in the attempted synthesis of benzoannulated derivatives (Scheme 10A, entry 3); use of an aromatic linker led to a 3-aminoindanone byproduct **75**, which was hypothesized to arise from the enforced proximity between the benzylic methylene and the carbenoid.

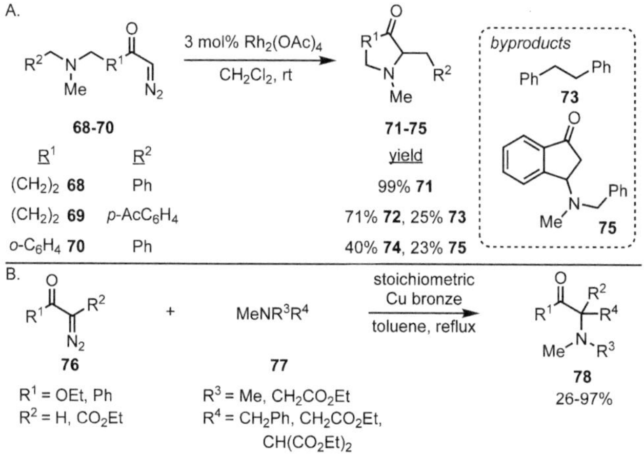

Scheme 10 (A) Rh-catalyzed intramolecular [1,2]-Stevens rearrangements of allylamines to form 2-substituted piperidin-3-ones. (B) Intermolecular [1,2]-Stevens rearrangements to α-aminoesters with stoichiometric Cu.

Shortly after this initial report, the West group reported an intermolecular variant to generate α-amino esters from the reaction of **76** with tertiary amines of the form **77**. Rh catalysis was initially investigated for this transformation but proved challenging, as amines can impede diazo decomposition. Switching to stoichiometric amounts of Cu-bronze in refluxing toluene ultimately furnished [1,2]-shift products in good-to-excellent yields (Scheme 10B). The Cu could be reduced to sub-stoichiometric amounts, albeit with much slower reaction rates. These two reports demonstrate the first metal-mediated carbene [1,2]-Stevens shifts, opening the door to further useful developments in the past 30 years. The remaining section covers some of these advancements, with an emphasis on efficient, stereoselective transformations that lead to highly functionalized molecules.

In 2001, Padwa and coworkers published their investigations of ammonium ylide cascade sequences to deliver 5,7-fused N-heterocyclic frameworks found in the cephalotaxine and lennoxamine natural products (Scheme 11).[73] Diazo derivatives **79–81** were subjected to Cu(acac)$_2$-catalyzed conditions to deliver the corresponding isoindolobenzazepine derivatives **82–84** in good to excellent yields as 1:1 mixture of diastereomers (Scheme 11A). This reaction was later attempted on a substrate resembling the lennoxamine scaffold (**85**); Rh catalysis afforded the desired isoindolobenzazepine **86** in 83% yield, while avoiding the formation of carbonyl ylide-derived isobenzofuran byproducts (Scheme 11B). Unfortunately, the origin of this selectivity remains unclear. This method represents an early application of the [1,2]-Stevens rearrangement in synthesis and demonstrates the sensitivity of ammonium ylide reactions to the identities of the substrate and catalyst.

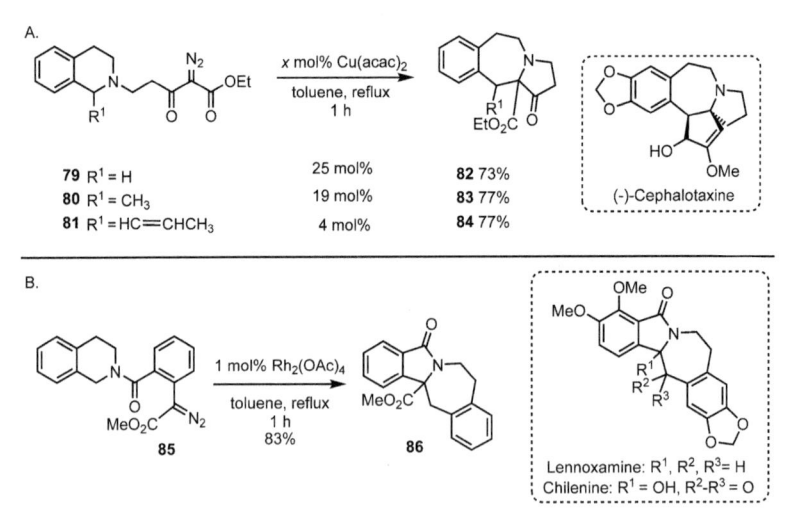

Scheme 11 (A) Synthesis of isoindolobenzazepine derivatives via Cu catalysis (left) and natural products bearing this scaffold (right). (B) Application of this method to a substrate mimicking the core of lennoxamine.

The intramolecular [1,2]-Stevens rearrangement was expanded by the West group for the synthesis of enantiopure quinolizidine scaffolds from the corresponding 2-silylpyrrolidine compounds (Scheme 12).[74] This unique transformation sought to identify a suitable hydroxyl surrogate that was capable of stabilizing the intermediate radical of the [1,2]-Stevens rearrangement and could later be converted into polyhydroxylated skeletons of indolizidine or quinolizidine alkaloids. The West group selected a silyl group to explore its compatibility with the [1,2]-Stevens rearrangement, given the ease of synthesizing 2-silylpyrrolidines combined with the robustness of the Fleming-Tamao oxidation which would be used to convert [1,2]-rearrangement products to their hydroxylated derivatives. Initial attempts utilized (S)-N-Boc-2-trimethylsilyl pyrrolidine, which was converted to the diazoketone **87**. Treatment with Cu(acac)$_2$ at elevated temperatures furnished the desired quinolizidine products **88–89** in 58% yield in 2:1 dr, favoring the *syn* diastereomer (Scheme 12A). The stereoselectivity is hypothesized to arise from preferential attack of the intermediate metal-mediated carbene *cis* to the larger non-H substituent at C-2, followed by migration with retention. NMR analysis of the products using a chiral shift reagent revealed the reaction proceeds with stereochemical retention, suggesting that radical recombination is fast compared to bond rotation. The West group further explored a phenyldimethylsilyl derivative that could be easily converted to a hydroxyl group by Fleming-Tamao oxidation. Conversion of the enantiopure diazoketone **90** to the corresponding

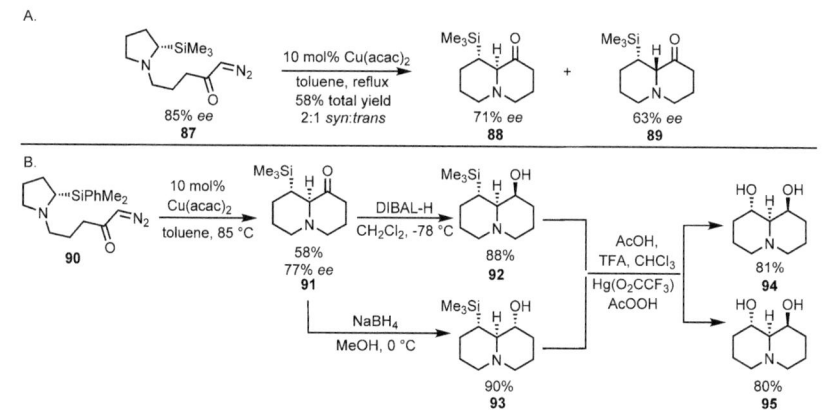

Scheme 12 (A) Studies of the silyl-[1,2]-Stevens rearrangement using a TMS group. (B) Application to a phenyldimethylsilyl-substituted pyrrolidine to yield a quinolizidine that undergoes a Tamao-Fleming oxidation.

quinolizidine **91** was achieved *via* [1,2]-Stevens rearrangement to give a single diastereomer. The high diastereoselectivity of this transformation likely arises from the increased steric bulk of the silyl group, which rigidifies the pyrrolidine prior to carbene addition; comparison with a chiral shift reagent indicated retention of chirality. Finally, the quinolizidine derivative underwent Fleming-Tamao oxidation to yield quinolizidinediols **93–94**, highlighting the potential applications of this method toward accessing privileged natural product scaffolds (Scheme 12B).

The West group expanded the utility of the [1,2]-Stevens rearrangement for the synthesis of pyrrolizidine-type alkaloids by exploring rearrangements of azetidinium ylides.[75] It was hypothesized that the release of strain in the azetidinium ylide intermediate would lower the activation energy barrier for the initial homolysis step and drive the [1,2]-Stevens rearrangement to furnish pyrrolizidine type alkaloids. Upon exposure to Rh or Cu catalysts, the diazoketone **96** underwent the desired [1,2]-rearrangement to give the corresponding pyrrolizidine **97** in 82% yield and 4:1 *dr* (Scheme 13A). Following reduction of the corresponding ester groups, the syntheses of

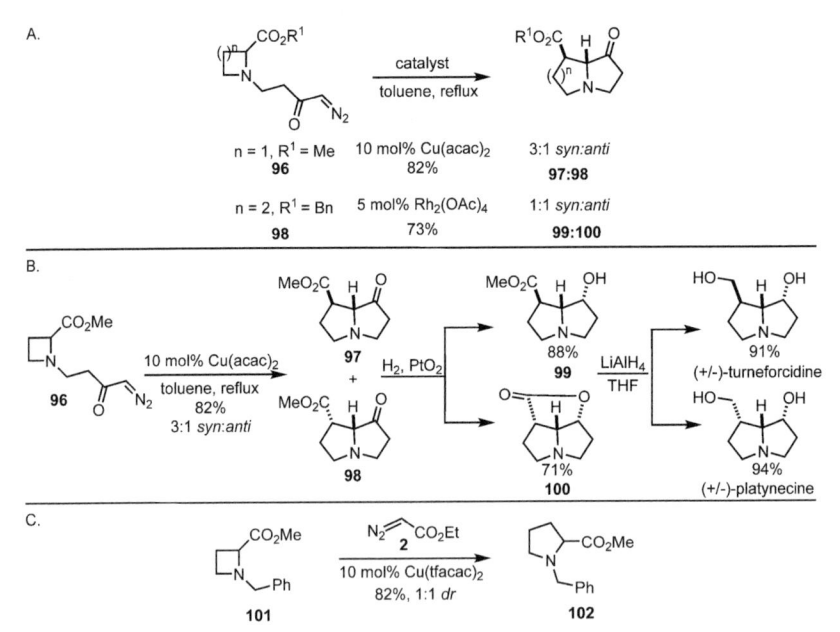

Scheme 13 (A) [1,2]-Stevens rearrangement of azetidinium and pyrrolidinium ylides to pyrrolizidine and indolizidine-type alcohols. (B) Application of this methodology to the synthesis of (+/−)-turneforcidine and (+/−)-platynecine. (C) Extension of the method to an intermolecular [1,2]-Stevens rearrangement.

(+/−)-turneforcidine and (+/−)-platynecine was achieved in five steps from the starting benzyl azetidine compound (Scheme 13B). Excitingly, the West group found the ring expansion could be conducted in an intermolecular fashion using azetidine **101** in the presence of **2** under Cu catalysis. The pyrrolidine diester **102** was obtained as a 1:1 mixture of diastereomers, supporting the possibility of future intermolecular [1,2]-Stevens shifts of ammonium ylides (Scheme 13C).

The [1,2]-Stevens rearrangement of spirocyclic ylides was extended to include the synthesis of pyrrolo[1-2-*a*]benzodiazepinone derivatives (Scheme 14).[76] This Rh-catalyzed transformation was readily achieved with a variety of substrates containing differential substitution of the ester and aromatic moieties to deliver compounds such as **104** in 45–60% yields as single diastereomers. The high diastereoselectivity was attributed to homolytic cleavage of the carbon–nitrogen bond to generate a radical pair held tightly together by a solvent cage. Following homolysis, solvent cage/rapid recombination of the radical center with the neighboring ylide carbon generates the rearranged product as the *trans*-diastereomer.

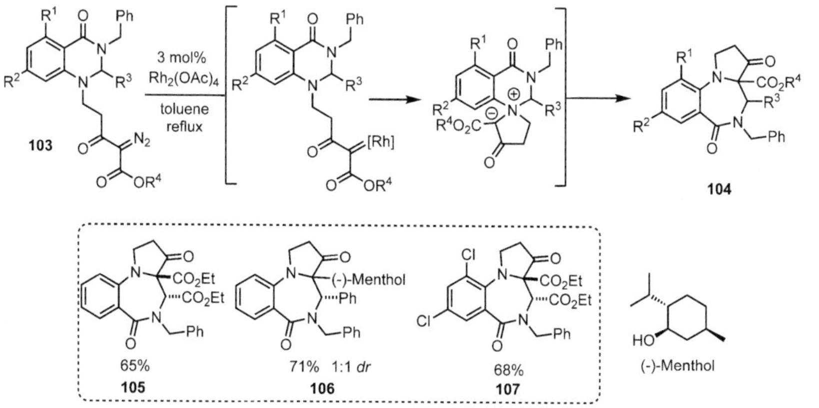

Scheme 14 Rh-catalyzed synthesis of pyrrolo[1-2-*a*]-benzodiazepinone derivatives (top) and examples (bottom).

In 2008, the Iwasawa group reported the use of a W catalyst for the synthesis of N-fused tricyclic indoles by a tandem [1,2]-Stevens-type rearrangement/1,2-alkyl migration of ammonium ylides.[77] It was hypothesized that treatment of an *o*-alkynylphenyl pyrrolidine or piperidine with a metal catalyst capable of activating the alkyne moiety could lead to nucleophilic attack by the nitrogen atom to generate a metal-containing ammonium ylide. This ylide could then undergo a [1,2]-Stevens-type rearrangement and subsequent [1,2]-alkyl migration to deliver a *N*-fused tricyclic indole (Scheme 15A). Use of [W(CO)$_6$] or [ReBr(CO)$_5$] in combination with photoirradiation generated the desired compound in 65% yield. Catalysts frequently utilized in alkyne activation chemistry, such as Au, were inactive in this transformation under both thermal and photoirradiation conditions, likely due to deactivation by coordination of the amine to the metal center. The lower affinity of W for hard bases and its ability to selectively activate the alkyne was proposed as key to the success of this chemistry to deliver *N*-fused bicyclic and tricyclic indole skeletons with good functional group tolerance, including electronically diverse indoles such as **111**. In contrast to the pyrrolidine products, the application of [W(CO)$_6$] for the

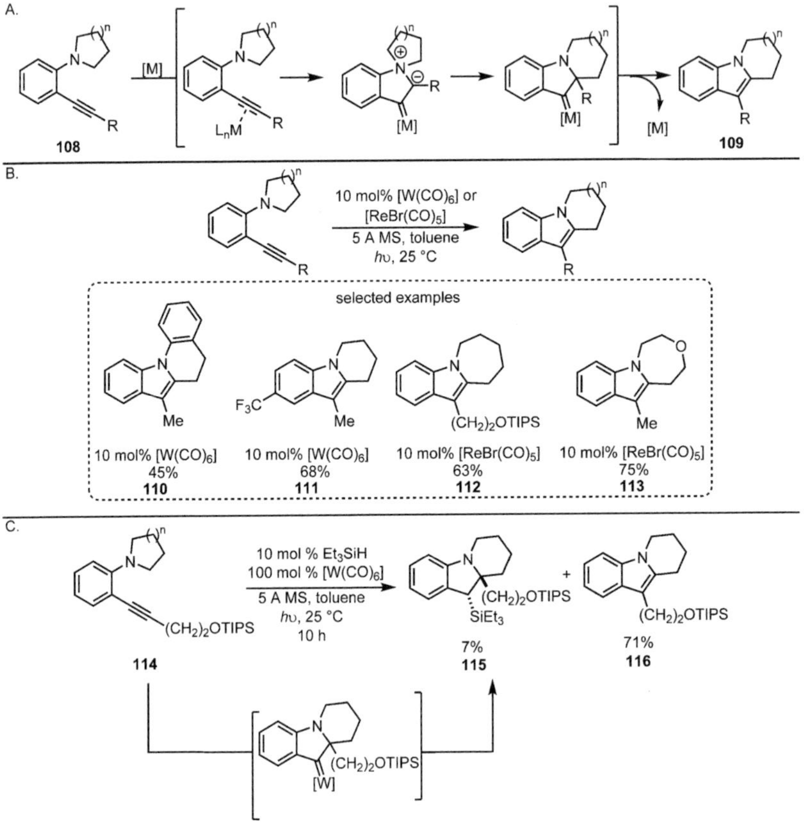

Scheme 15 (A) Proposed mechanism for a [1,2]-Stevens-like migration to deliver *N*-fused tricyclic indoles. (B) Optimized reaction conditions and selected results. (C) Metallocarbene trapping experiment.

synthesis of six-membered azacycles such as **112** and **113** was ineffective, even with increased catalyst loading. The synthesis of protected ethers **112**, and heterocycles **113** (Scheme 15B) was achieved in the presence of [ReBr(CO)$_5$] but unfortunately the origin of this restored reactivity is not commented upon any further. Importantly, the intermediacy of metallocarbene was verified *via* a trapping experiment utilizing excess triethylsilane to furnish the tricyclic indoline derivative **115** (Scheme 15C).

In their endeavor to synthesize (±)-Preussin from decanal, Burtoloso and coworkers explored a Stevens rearrangement of an ammonium ylide to access *cis*-2,5-disubstituted pyrrolidinones.[78] A stereoselective Cu-catalyzed ylide formation and subsequent [1,2]-Stevens rearrangement were the key steps of their concise route (Scheme 16). Formation of the ylide **119b** occurs after intramolecular addition of the tertiary amine to its metal carbene tether in **118** and dissociation of the copper species. The subsequent [1,2]-Stevens rearrangement is postulated to proceed through a radical mechanism in which ammonium ylide radical intermediates **120** and **121** are accessed from homolytic dissociation of the ylide. Radical coupling completes the stereocontrolled benzyl migration to yield the thermodynamically stable *cis*-2,5-isomer of **122**. This key intermediate was afforded in moderate yield and ultimately reduced to furnish the desired (±)-Preussin. The authors noted this methodology could be extended to other Preussin analogs and employed to construct other all-*cis*-substituted pyrrolidine alkaloids.

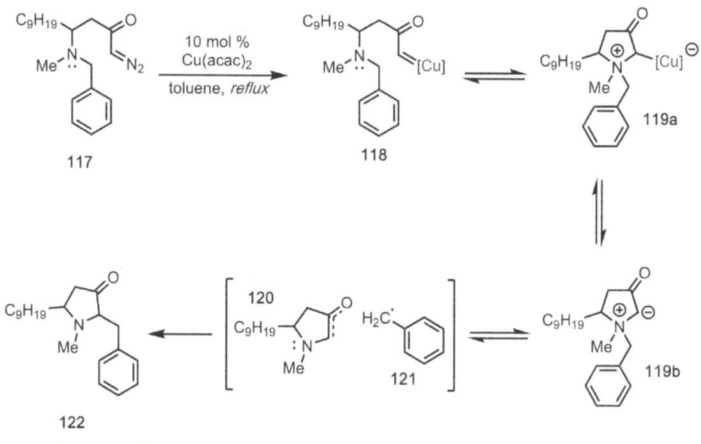

Scheme 16 Proposed mechanism for the Cu-catalyzed formation of ammonium ylide **119** and the subsequent [1,2]-Stevens rearrangement to access (±)-Preussin intermediate **122** from β-amino diazoketone **117**.

Over the years, significant attention has been directed toward developing efficient methods to construct nitrogen-bridged bicyclic structures, due to their prevalence in the framework of bioactive alkaloids.[79] Nemoto and coworkers serendipitously developed a rhodium-catalyzed formal carbene insertion into amide C–N bonds during their initial studies aimed at the synthesis of the tricyclic framework of (−)-agelastatin A (Scheme 17A).[80] From an analysis of the product structure, the authors noted that the formation of the Rh-supported ammonium ylide **124** was favored over a five-membered ring formation from the intended C–H insertion.

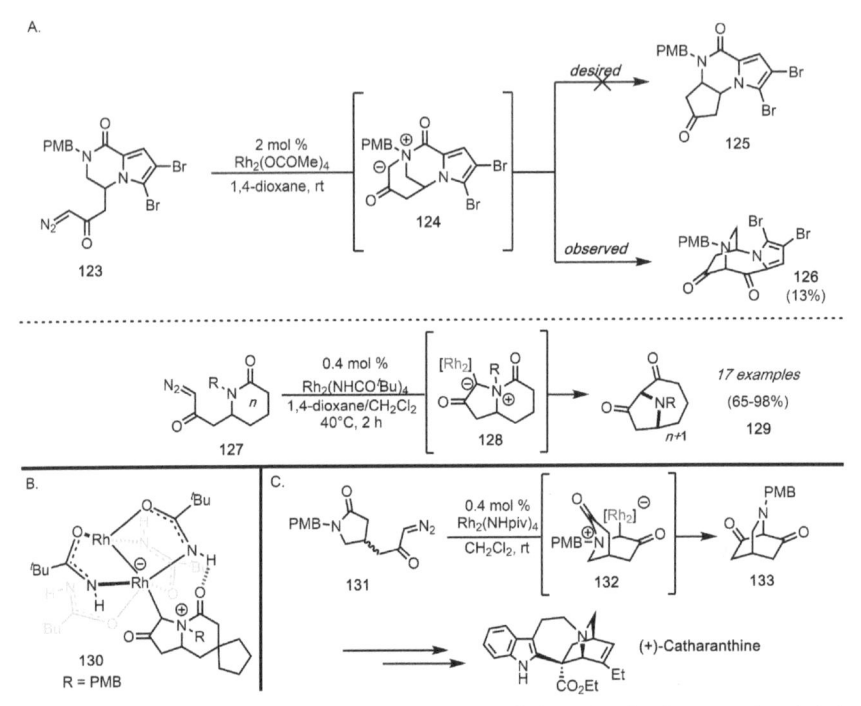

Scheme 17 (A) Nemoto's attempted synthesis of the tricyclic framework of (−)-agelastatin A and the unexpected azabicyclic product formation. (B) Proposed transition state model for the hydrogen bonding between the amide carbonyl group and the hydrogen in the carboxamidate ligand. (C) Key amide insertion step to access a meso isoquinuclidine intermediate toward the asymmetric synthesis of (+)-Catharanthine.

A proposed transition state model **130** for this Rh-catalyzed formal carbene insertion into an amide C–N bond suggested a hydrogen bonding interaction between the amide carbonyl group and the hydrogen in the carboxamidate ligand triggers the [1,2]-Stevens shift of the acyl group

(Scheme 17B). DFT calculations and reaction coordinate analysis of the acyl transfer ultimately favored a concerted, asynchronous process over a stepwise Stevens rearrangement. The azabicyclo[$x.y.z$]alkane derivatives **129** were obtained in good-to-excellent yields; it was envisioned these unexpected products could be utilized for the construction of diverse nitrogen-bridged bicyclic systems.

The prevalence of the 2-azabicyclo[2.2.2]octane, or isoquinuclidine, scaffold in bioactive alkaloids stimulated interest in methods for its asymmetric synthesis.[81–83] A 2019 study by Nemoto and coworkers described an asymmetric formal synthesis of (+)-Catharanthine that proceeds through desymmetrization of a key *meso*-isoquinuclidine intermediate.[84] The formal insertion of a rhodium carbene into an amide C–N bond was employed to obtain the *meso*-isoquinuclidine. The Rh-bound ylide **132** was obtained following an intramolecular nucleophilic addition of the PMB-nitrogen of the lactam **131** to its Rh-carbene tether. A subsequent Stevens rearrangement of the ylide produced isoquinuclidine **133**, which was carried forward en route to (+)-Catharanthine (Scheme 17C).

2.4 Intermolecular 1,2-Stevens rearrangements

Numerous reports have shown *N*-sulfonyl-1,2,3-triazoles function as practical diazo surrogates for carbene precursors, allowing them to engage in classical transformations, such as transannulation, cyclopropanation, and C–H functionalization.[85–88] Metal-catalyzed decomposition of *N*-sulfonyl-1,2,3-triazoles and subsequent ylide formation upon nucleophilic addition has become a useful synthetic strategy for the construction of highly functionalized heterocycles.

In 2018, the Lacour group shared an efficient synthesis of highly functionalized benzodiazepines **136** that proceeds through ammonium ylide intermediates **140**, accessed by nucleophilic addition of Tröger bases **134** to *N*-sulfonyl 1,2,3-triazole-derived α-imino carbenes **135** (Scheme 18A).[89] It was observed that reactions involving Tröger bases with electron-donating substituents occurred more effectively than those with electron-withdrawing groups; promotion of the initial step of the mechanism was believed to be facilitated by the higher nucleophilic character of the nitrogen atoms. An inverse electronic demand was observed in the examination of the *N*-sulfonyl triazole scope. Moderate yields (34–55%) were attained with triazoles substituted with electron-donating groups, but higher yields (74–83%) were achieved with precursors containing electron-withdrawing groups.

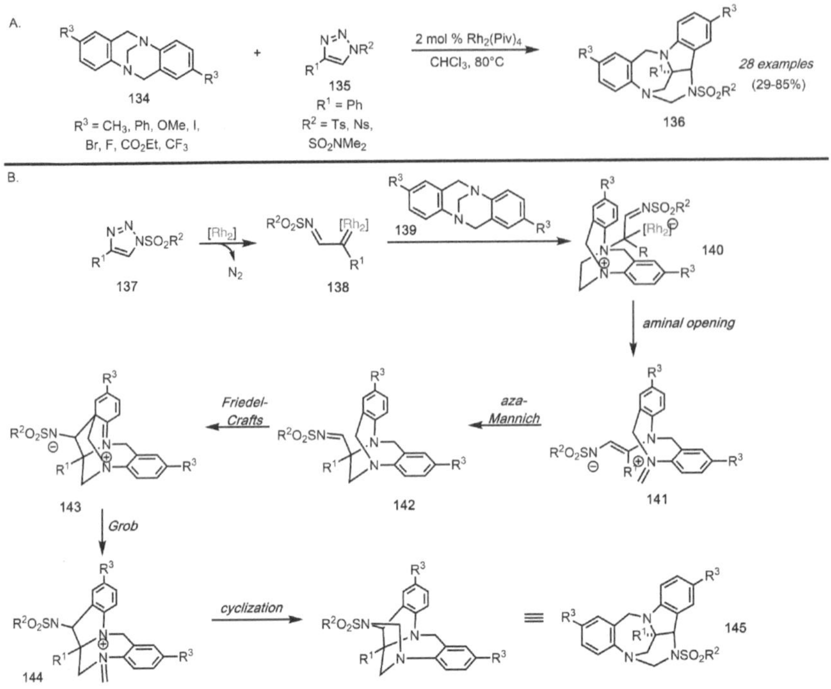

Scheme 18 (A) General scheme for Lacour's electrophilic addition of α-imino carbenes to Tröger bases to access polycyclic indoline-benzodiazepines. (B) Proposed mechanistic rationale for the observed reactivity.

Ylide formation was followed by a cascade of transformations (Scheme 18B), beginning with an aminal opening to provide **141**, followed by an intramolecular aza-Mannich reaction to afford **142**. A subsequent Friedel–Crafts cyclization of the sulfonyl imine **142** was envisaged, due to the spatial proximity between the electrophilic imine and the electron-rich aniline. The authors note the overall transformation from **137** to **142** is considered a formal [1,2]-Stevens rearrangement. An alignment between the nitrogen lone pair and the σ^* orbital of the adjacent C–C bond triggers a penultimate Grob fragmentation of zwitterionic **143** to give **144**. Cyclization of this intermediate through intramolecular trapping of the iminium by the sulfonamide ultimately produces the diazepine product **145**. This overall reaction cascade afforded polycyclic indoline-benzodiazepines as single diastereomers containing four stereocenters, including two located at bridgehead nitrogen atoms.

2.5 Miscellaneous

Further development of their aziridinium ylide chemistry was explored by the Schomaker group with their investigation of aziridine precursors lacking the exocyclic alkene present in MAs.[90] However, formation of imine **148** was observed following the reaction between bicyclic aziridine **146** and diazoester **147** (Scheme 19A). This unexpected result has been attributed to the fragmentation of the aziridinium ylide intermediate *via* cheletropic extrusion.

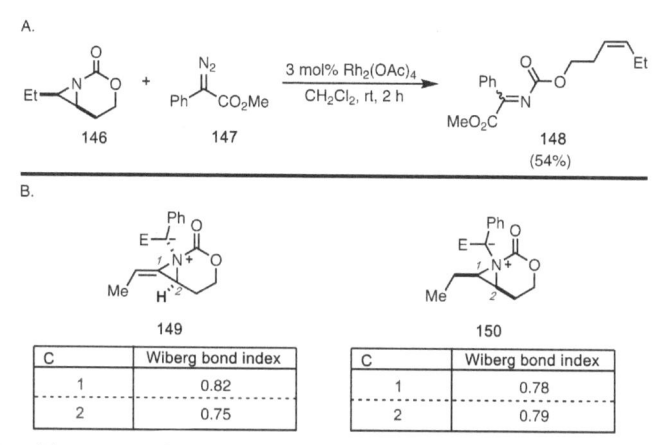

Scheme 19 (A) Attempted expansion of aziridinium ylide reactivity to bicyclic aziridines and the resultant cheletropic extrusion from the Rh-mediated carbene transfer between cis-**146** and diazoacetate **147**. (B) Computed Wiberg bond indices for the C–N bonds of methyleneaziridine- **149** and bicyclic aziridine-derived ylides **150**.

Computational analysis was used to evaluate the features of the respective aziridinium ylides arising from a methyleneaziridine **149** and a bicyclic aziridine **150** (Scheme 19B). The differing vinylic and allylic C–N bond strengths (0.82 and 0.75 kcal/mol, respectively) in **149** favor initial rupture of the allylic C–N bond toward ring expansion. The C–N bonds of **150** possess nearly equivalent bond strengths (0.78 and 0.79 kcal/mol, respectively), which favors a concerted extrusion over ring expansion. Computational analysis suggested the absence of the exocyclic alkene in the bicyclic aziridines contributes to the observed cheletropic extrusion in the unbiased system.

In efforts to better understand the reactivities of these unusual aziridinium ylides, the Schomaker group explored accessing the reactive intermediate from the reaction of the bicyclic aziridines with vinyl diazoacetate-derived rhodium carbenes. In 2020, the group reported a formal [3+3] ring expansion of bicyclic aziridines **151** to highly substituted dehydropiperidines **153** (Scheme 20A).[91] Competitive cheletropic extrusion was inhibited by delocalization of the negative charge through the vinyl group of the diazo precursor **152**. Productive carbene transfer was demonstrated with a variety of *cis*-substituted bicyclic aziridines; however, no reaction was observed with the *trans*-aziridine isomers, mostly likely due to hindrance of ylide formation by steric congestion at the nitrogen lone pair. A survey of aryl-substituted diazoesters with varying electronic features revealed the electronics of the styrene in the carbene precursor do not heavily affect the reaction outcome.

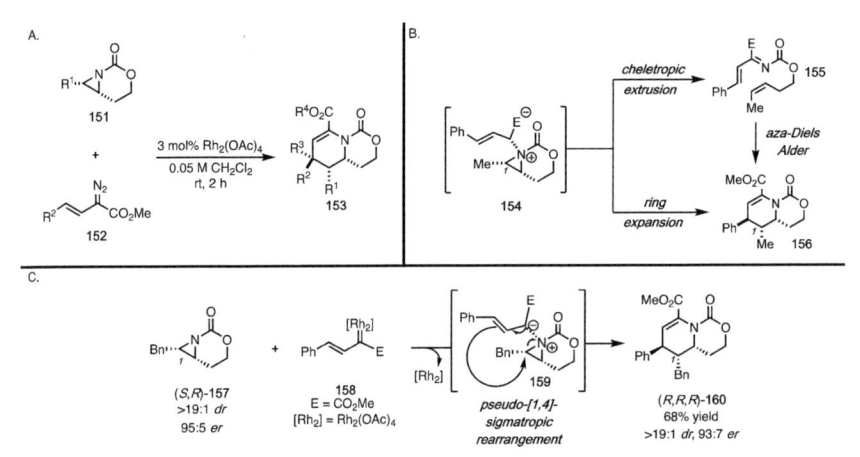

Scheme 20 (A) General scheme for the Rh-catalyzed [3+3] ring expansion of bicyclic aziridines toward the synthesis of dehydropiperidines. (B) Potential reaction pathways for aziridinium ylide intermediate **154**. (C) Chirality transfer experiment using enantiopure aziridine (*S,R*)-**157**.

According to DFT computations, aziridinium ylide **154** can undergo either a cheletropic extrusion pathway or ring expansion through a ring-opening/ring-closing cascade (Scheme 20B). In the former case, cheletropic extrusion of the ylide would yield azadiene **155**. The dehydropiperidine product **156** would then be accessed following a subsequent aza-Diels Alder cycloaddition. The alternative ring expansion pathway predicted a direct formation of dehydropiperidine **156** from ylide **154**. Chirality from enantioenriched aziridine (*S,R*)-**157** was transferred to (*R,R,R*)-**160** with excellent retention at **C1** (Scheme 20C), thus supporting the latter pathway. The key transformation of this pathway was ultimately considered to be a pseudo-[1,4]-sigmatropic rearrangement.

Regarding the rearrangement step, further insight was obtained from the retention of stereochemical information. Inversion of stereochemistry at **C1** was not supported by the chirality retention experiment, thus invalidating a proposed intramolecular S_N2 attack of the benzylic carbon. Instead, it was proposed that the rearrangement proceeds through a stereoretentive S_N1-like mechanism. Ultimately, this study represents the first account of preclusion of cheletropic extrusion in favor of the productive ring expansion of aziridinium ylides derived from unbiased aziridines.

The Schomaker group continued to explore different carbene precursors to further develop aziridinium ylide chemistry toward the synthesis of complex *N*-heterocycles.[92] In the presence of a transition metal catalyst, pyridotriazoles can undergo tautomerization and decomposition to yield pyridyl carbene precursors. Instead of the expected fused dehydropiperazine product **167**, the reaction of pyridotriazole **161** with bicyclic aziridine **164** afforded ketimine product **166** through an apparent cheletropic extrusion of aziridinium ylide **165** (Scheme 21). DFT calculations were conducted to understand the unexpected fate of the aziridinium ylide intermediate of this transformation. To afford the desired tricyclic ring expansion product **167**, the ketimine would have to undergo a subsequent aza-Diels-Alder reaction. However, the calculations indicated this would be kinetically unfeasible; a high energy barrier of >50 kcal/mol complements the required loss of aromaticity of the pyridyl substituent necessary for this transformation.

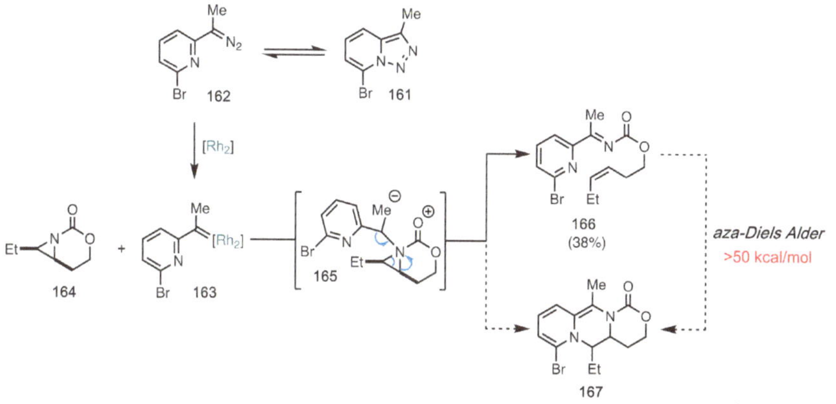

Scheme 21 Schomaker's attempted carbene transfer using pyridotriazole-derived α-imino metal carbene **163** to access fused piperazines **167**.

To achieve the desired ring expansion and access dehydropiperazines, the Schomaker group altered the identity of the carbene precursor. As their utility as carbene precursors has been extensively documented, the reactivity of N-sulfonyl-1,2,3-triazoles was investigated using the group's bicyclic aziridine substrates. Highly functionalized dehydropiperazines **173** were afforded in good yields and with excellent diastereoselectivity from a variety of substituted bicyclic aziridines **171** and tosyl-protected aryl N-sulfonyl-1, 2,3-triazoles **168**. DFT calculations support ylide formation following the nucleophilic addition of the bicyclic aziridine **171** to the electrophilic center of the rhodium-supported carbene **170** (Scheme 22). A direct formation of dehydropiperazine **173** from ylide **172** through a sigmatropic rearrangement is kinetically favored; this transformation involves a concomitant, yet highly asynchronous breaking of the aziridine C–N bond and formation of a new C–N bond.

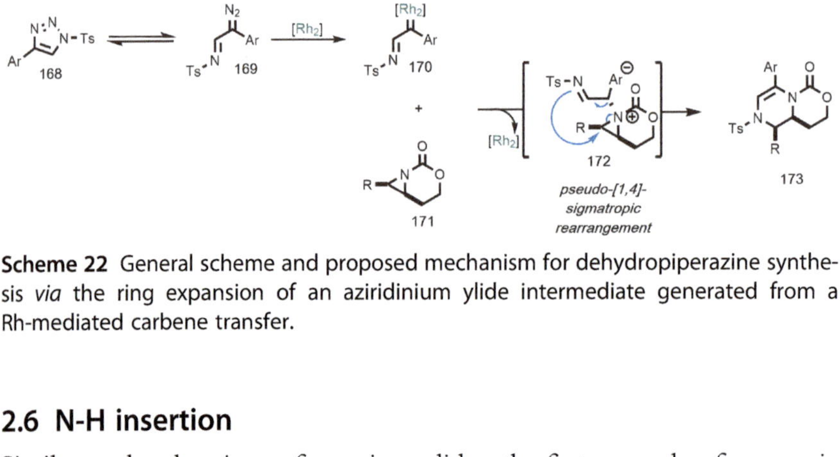

Scheme 22 General scheme and proposed mechanism for dehydropiperazine synthesis *via* the ring expansion of an aziridinium ylide intermediate generated from a Rh-mediated carbene transfer.

2.6 N-H insertion

Similar to the chemistry of oxonium ylides, the first example of a racemic N-H insertion was discovered by Yates in 1952 *via* the reaction of α-acetophenone **175** with aniline **174** to deliver α-aniline acetophenone **176** (Scheme 23A).[93] As shown above (Fig. 4C) racemic and asymmetric N-H insertion processes are hypothesized to proceed through protic ammonium ylides that can undergo a subsequent [1,2]-H shift to deliver the functionalized products. Since the initial discovery by Yates, there have been many examples of racemic N-H insertion in organic synthesis, including an application to a 1980 total synthesis of (+)-Thienamycin by Merck (Scheme 23B).[94] As many racemic N-H insertion reactions exist and have been covered elsewhere,[95] this section will focus on recent developments in synthetically valuable enantioselective methodologies that serve as an overview of the current state of the field.

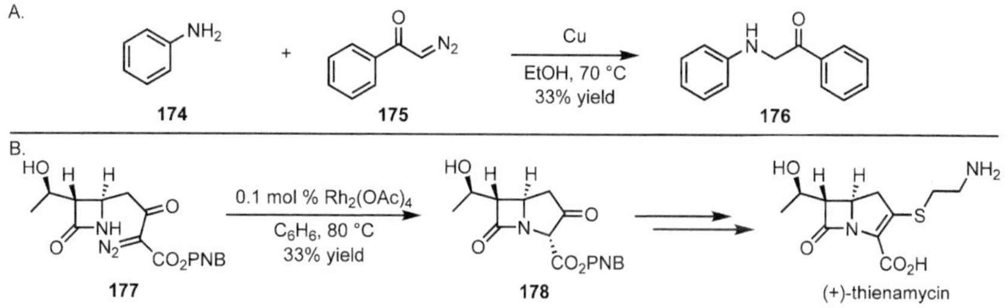

Scheme 23 (A) First example of racemic insertion of a carbene into a N–H bond. (B) Application of N-H carbene insertion to the synthesis of (+)-Thienamycin.

The first examples of enantioselective N-H insertion were reported by McKervey in 1996 (45% *ee*) and Moody (9% *ee*) in 2002 using asymmetric Rh catalysts.[96,97] McKervey achieved a greater degree of success by utilizing Cbz-protected amine **179** in intramolecular N-H insertion to deliver cyclic amino acid **180**. There was a significant dependence of both catalyst identity and reaction temperature on the chemo- and enantioselectivity of the reaction, with optimal conditions employing Rh$_2$(*S*)-[mandelate]$_4$ (Scheme 24).

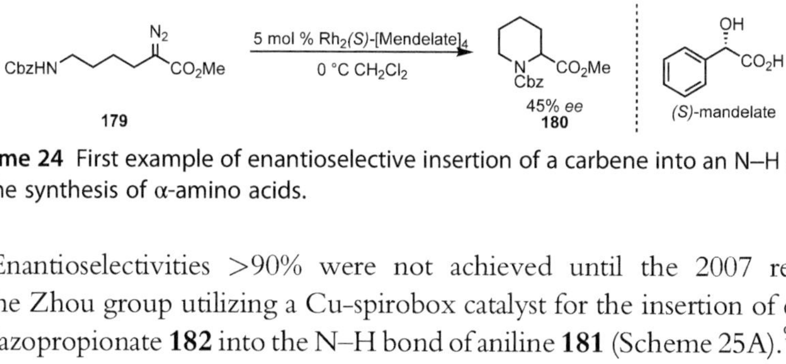

Scheme 24 First example of enantioselective insertion of a carbene into an N–H bond for the synthesis of α-amino acids.

Enantioselectivities >90% were not achieved until the 2007 report by the Zhou group utilizing a Cu-spirobox catalyst for the insertion of ethyl α-diazopropionate **182** into the N–H bond of aniline **181** (Scheme 25A).[98a,99] Interestingly, the group found the enantioselectivity of the process was directly correlated to the identity of the copper anion from the copper pre-catalyst. Their results show that smaller, more coordinating anions lead to lower *ee*. Cu(BArF$_4$)$^-$ ultimately proved to be the best pre-catalyst, due to the bulk of this non-coordinating anion. The (*S*,*S*)-SpiroBox complex formed from this copper salt (also used in O-H insertion chemistry, Scheme 58) furnished aniline N-H insertion products with >98% *ee*.

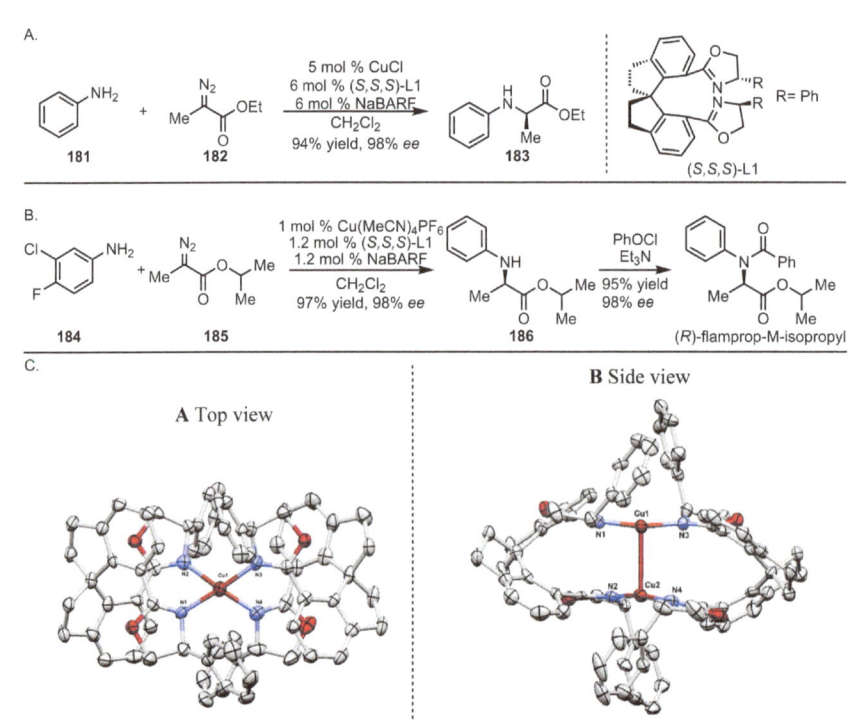

Scheme 25 (A) Optimal conditions for enantioselective N-H insertion of aniline derivatives with Cu-SpiroBox catalyst. (B) Application to the enantioselective synthesis of herbicide (R)-flamprop-M-isopropyl. (C) Crystal structure of the dimer Cu/SpiroBox complex. Reprinted (adapted) with permission from Zhu SF, Zhou Q-L. Transition-metal-catalyzed enantioselective heteroatom–hydrogen bond insertion reactions. Acc Chem Res. 2012;45:1365–1377. Copyright 2012 American Chemical Society.

The Zhou group obtained a single crystal structure of this complex, which displayed a unique binuclear structure with two Cu(I) atoms coordinated by two nitrogen atoms that belong to two spirobisoxazoline ligands in a *trans* orientation (Scheme 25C). This structure enables the phenyl groups to form a C2-symmetric chiral pocket around the Cu center to promote the observed remarkable enantioselectivity. Optimized conditions were used to convert a variety of primary anilines and α-alkyl-alpha-diazoacetates/α-diazopropionates to the corresponding products in good yields (62–98%) and with excellent enantioselectivities (85–98%). The Zhou group demonstrated the utility of this method by applying it to the synthesis of herbicide (R)-flamprop-M-isopropyl (Scheme 25B).[98b]

Soon after Zhou's initial publication, the Fu group published a similar transformation for the synthesis of α-amino acids using a Cu/bpy* system with an expanded range of amine and diazoacetate sources for enantioselective N-H insertion (Scheme 26).[100] The Cu catalyst system previously developed by the

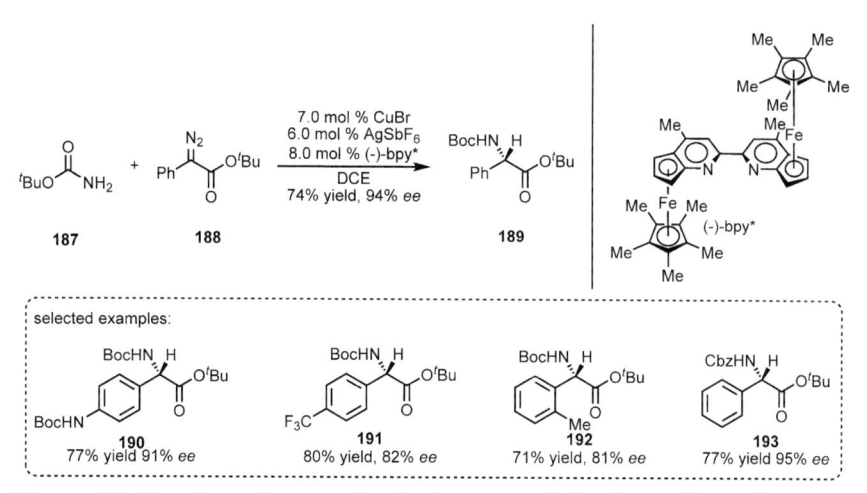

Scheme 26 Complementary enantioselective approach for the synthesis of α-amino acids using a Cu/bpy* system and selected examples.

Fu group for the insertion of α-diazo esters into O–H bonds (Scheme 57) furnished only a modest 50% ee. After extensive optimization, a bulkier (−)-bpy* catalyst was found to furnish better yields and enantioselectivities (**189**, 74% yield and a 94% ee) using CuBr as the precatalyst. In contrast to Zhou's initial studies, the aromatic moiety in Fu's chemistry can be modified and tolerates various electronic and steric modifications, including heteroatom substitution. The Boc amine could also be altered to a Cbz amine for easier protecting group removal and furnished the desired products in similar yields and ee. Overall, this method complements Zhou's initial report.

The Zhou group followed up Fu's report of α-amino esters with an enantioselective approach to α-aminoketones.[101] Previously, the only diazo derivatives amenable to enantioselective N-H insertion were diazoesters. The Zhou group hypothesized that diazoketones or other highly electron-withdrawing carbene precursors were inaccessible using current catalyst systems, given their propensity to form metal-free ylides and the corresponding enols, which precludes any role of the metal in the chirality-determining proton-transfer step. The Zhou group circumvented this issue by employing a chiral spiro phosphoric acid (CSPA) into their catalyst to enable an enantiodetermining proton transfer step. This strategy is similar to the Rh/CPA strategy applied by Hu in their oxonium ylide mediated MCRs (Schemes 64 and 65). The Zhou group found that the highly Lewis acidic catalyst $Rh_2(TFA)_4$, in combination with substituted **CSPA1**, furnished the corresponding α-aminoketone **195** in 80% yield and with 96% ee (Scheme 27A). It is important to note that the asymmetric insertion reaction was first attempted with a variety of chiral Rh catalysts, as well as the Cu/SpiroBox catalyst discussed above. None of these conditions gave any *ee*, supporting the hypothesis that this transformation proceeds through a metal-free ylide. The reaction tolerated variations in the steric and electronic groups on the aromatic diazoketone, as well as functional groups that include alkenes **196** and *N*-methyl pyrroles **197**. This chemistry was applied to the total synthesis of (L)-(−)-733,061, an antagonist of the NK1 receptor (Scheme 27B).

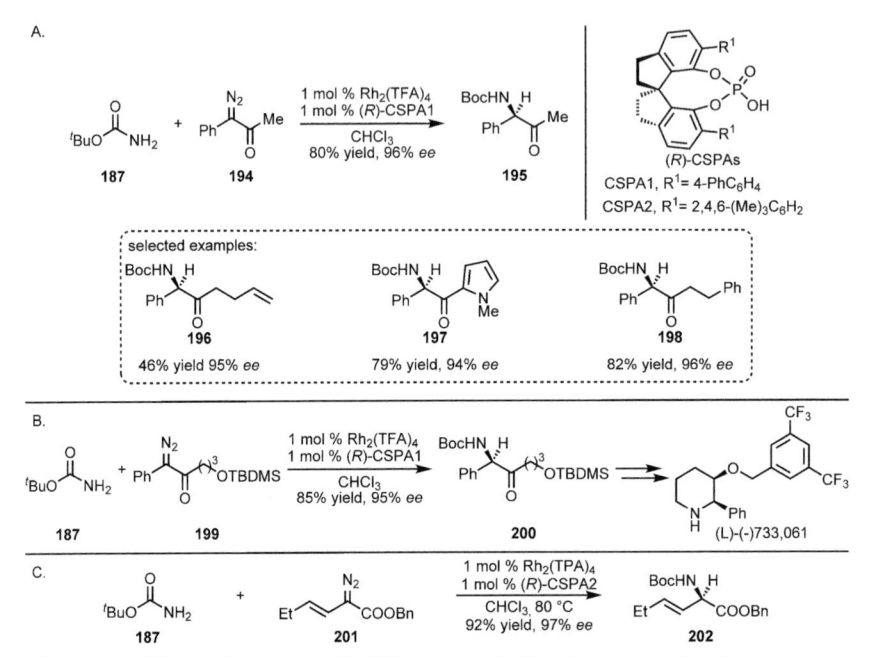

Scheme 27 (A) Application of Rh/CSPA catalysis for the synthesis of enantiopure α-amino acid derivatives. (B) Application to the formal total synthesis of (L)-(−)733,061. (C) Modification of the diazo precursor for the synthesis of α-alkenyl-α-amino acid derivatives.

The Zhou group followed this initial report by extending the Rh/CSPA dual catalytic system to the synthesis of α-alkenyl-α-amino acids, such as **202**.[102] The combination of a bulkier CSPA with Rh$_2$(TPA)$_4$ was optimal, yielding enantioselectivities up to 96% *ee* (Scheme 27C). The fact that the identity of the Rh catalyst had an impact on the reaction lead the authors to suggest the metal may be involved in the enantio-determining proton transfer step; however, no comment was provided as to why both a bulky Rh catalyst and CSPA are needed to give high enantiopurity. A variety of functional groups and substrate with diverse steric and electronic profiles performed well in this reaction.

In 2019, the Zhou group expanded on their work in enantioselective N–H insertion by reporting the first example of the enantioselective insertion of carbenes into aliphatic amine N–H bonds.[103] This reaction had largely eluded the synthetic community, due to the ready coordination of aliphatic amines to metal catalysts, which inhibits carbenoid formation in the absence of high temperatures. In addition, aliphatic amines can readily displace metal catalysts from the ylide intermediates before asymmetric induction can occur. The Zhou group overcame these challenges by employing a homoscorpionate-coordinated copper complex CuTp* (Scheme 28A) in combination with a chiral amino-thiourea (CAT) to

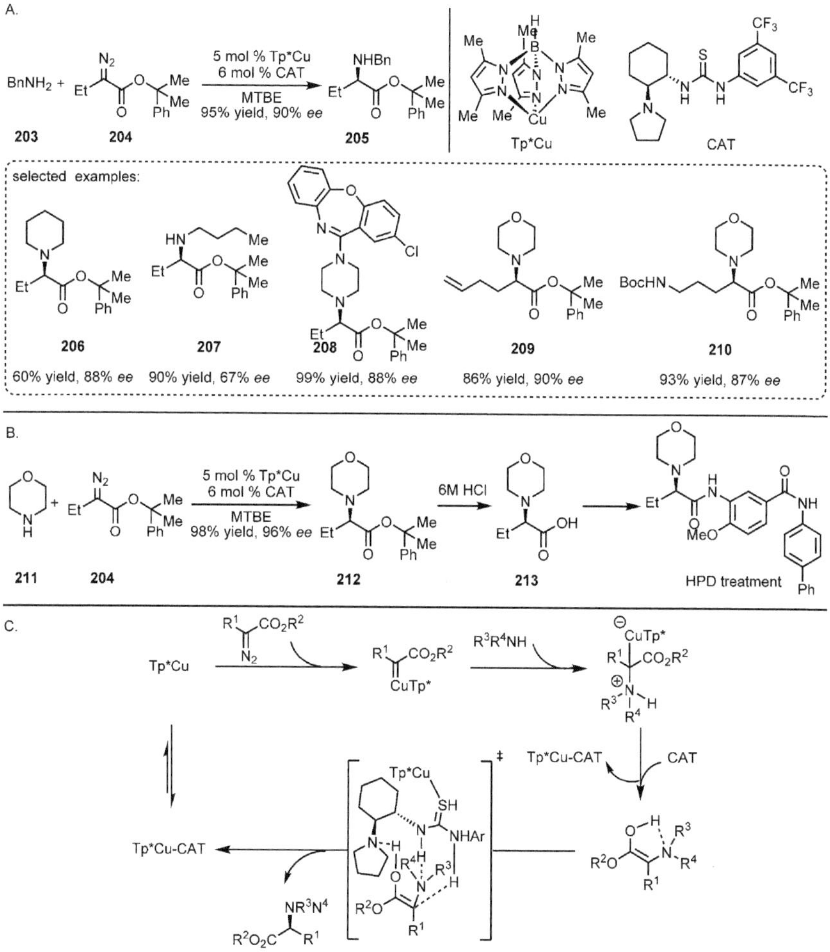

Scheme 28 (A) Application of Rh/CSPA for the synthesis of α-amino acids. (B) Application to an antiproliferative compound. (C) Proposed mechanism of enantioselective benzylic and aliphatic N-H insertion.

facilitate enantioselective proton transfer in the presence of alkyl/aryl diazoacetates and benzyl/aliphatic amines (Scheme 28A). The scope of this reaction was broad; a variety of secondary **207** and tertiary amines **206** were accessed with a wide range of functionalities, including alkenes **209** and protected amino acids **210**. The method was applied to the syntheses of several biologically significant molecules, including compounds for the treatment of hyper-proliferative disorders (Scheme 28B).

Interestingly, initial screening experiments identified several other Tp* ligands with differential substitution that were able to promote the transformation in inferior yields, but with consistent *ee* between 80% and 90%. In contrast, tuning the electronic properties of the arene ring on the CAT gave consistent yields, but drastically decreased *ee*. Based on this experimental evidence, the Zhou group hypothesized that Cu coordination enhances the Brϕnsted acidity of the CAT but does not influence enantioinduction. To further the understanding of this unique mechanism, the Zhou group undertook an extensive kinetic analysis using *in situ* IR spectroscopy. The reaction rate showed a first-order dependence on the concentrations of the Cu catalyst and the diazo ester, suggesting that metal mediated carbene formation is the rate-limiting step. *In situ* IR experiments also indicated a negative first-order dependence on the CAT, suggesting a pre-equilibrium formation of the resting-state complex between the CAT and the CuTp*. Importantly, the benzylamine **203** showed a zero-order kinetic dependence, arguing against pre-coordination between the amine and copper complex, a common issue in ammonium ylide formation. The group proposed that the negative Tp* ligand may generate a Cu catalyst that is a softer Lewis acid and favors interactions with the softer sulfur of the CAT, as compared to the amine. Based on this experimental data, a mechanism was proposed (Scheme 28C), where the Tp*Cu-CAT complex serves as the catalyst resting state. Dissociation to release CuTp* allows for M-carbene formation in the rate-determining step. The amine can then attack the metal-supported carbene to generate a protic ammonium ylide, where the Cu is displaced by the CAT to generate a free enol and regenerate the Tp*Cu-CAT complex. This then facilitates enantioselective proton transfer *via* the free enol. Overall, this method is a welcome answer to the challenge of utilizing aliphatic amines in N-H insertion reactions and provides important mechanistic insight for further reaction development.

As discussed in the previous section, protic ammonium ylides can undergo subsequent [1,2]-shifts to deliver N-H insertion products. However, if the protic ammonium ylide is trapped prior to the [1,2]-H shift, it is possible to engage the anionic carbon of the ylide with an electrophile to furnish a cyclization product in an intramolecular fashion or attack another electrophile in an intermolecular manner to achieve a multi-component reaction (MCR). These reactions were shown previously (Fig. 4D) and represent a powerful way to rapidly generate complex molecules in a single step from simple precursors. The following section is divided into examples of intramolecular and intermolecular trapping, with a focus on enantioselective methods or those that represent unique advances in the recent literature.

2.7 Intermolecular electrophilic trapping of protic ammonium ylides: Multicomponent reactions

In 2011, the Hu group reported the enantioselective trapping of carbamate ammonium ylides with amines. Use of a Rh/chiral phosphoric acid (CPA) catalyst resulted in a switchable, diastereo- and enantioselective method for the generation of α,β-diamino acid derivatives (Scheme 29).[104] In initial studies, treatment of carbamate amine **214** with methyl-phenyl-diazoacetate **147** and phenyl-benzyl-imine **215** catalyzed by $Rh_2(OAc)_4$ gave the racemic MCR product **216** in 71% yield and 8:1 dr, favoring the *anti*-diastereomer (Scheme 29A, entry 1). Despite the modest yield, the reaction provided proof-of-principle and evidence for the formation of a protic ammonium ylide over a competing concerted N-H insertion process. The Hu group hypothesized that inclusion of a CPA would enable enantioselective nucleophilic addition to the imine by activation of the electrophile.

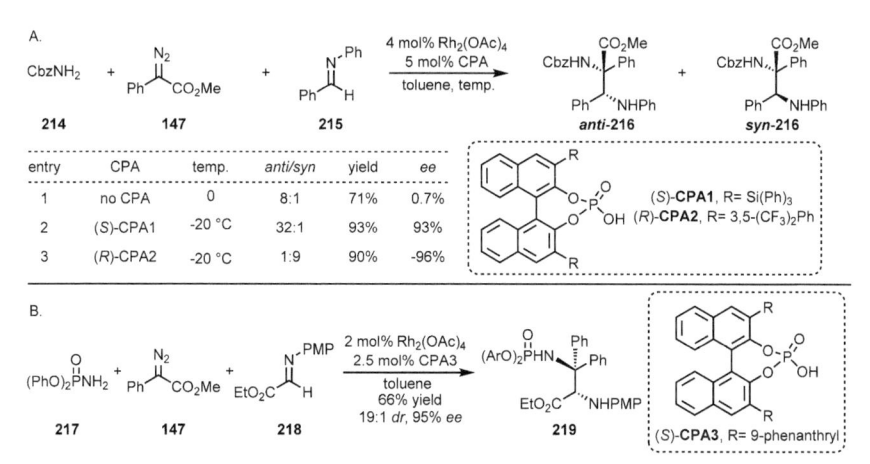

Scheme 29 (A) Optimal reaction conditions for enantioselective and diastereoswitchable syntheses of α-substituted α,β-diamines. (B) Extension to the synthesis of 2,3-diaminosuccinic acid derivatives.

A screen of catalysts, solvents, and temperatures enabled access to both diastereomers in high enantiomeric purity by selecting the proper CPA (Scheme 29A, entries 2–3). Interestingly, more electron-withdrawing CPAs afforded the *syn* diastereomer preferentially (1:9 *anti/syn*), whereas bulky catalysts gave the *anti* diastereomer (32:1 *anti/syn*) as the major product. Decreasing the catalyst loading from 10% to 2% decreased the overall *ee* (11% *ee*); this was attributed to sequestration of the chiral catalyst by the basic diamine products. Addition of an acid additive, such as tartaric acid, restored the *ee* using low CPA loadings by regenerating the acid catalyst. Cooling the reaction to −20 °C and switching to toluene further improved the *ee* to (−)-96% (when (*R*)-**CPA2** was applied) and 93% (when (*S*)-**CPA1** was applied). The reaction tolerated diverse carbamates and electronically varied imines; coupled with the ability to switch diastereoselectivity based on the choice of CPA, this chemistry represents a synthetically valuable approach to access all four diastereomers of a given α-substituted α,β-diaminoacid derivative.

Further efforts were undertaken to utilize iminoesters in this methodology.[105] Unfortunately, previously reported optimal conditions resulted low diastereoselectivity or moderate enantioselectivity (80% *ee*). Hu hypothesized replacing carbamates with phosphoramides might better matching the reactivity profile imino esters and provide a manipulable functional group handle to tune the stereoselectivity of the reaction. The use of a bulky 9-phenanthyl-derived BINOL catalyst **CPA3** enabled successful MCR with phosphoramidates, imino esters, and diazoacetates in up to 66% yield, 19:1 *dr* (in favor of *syn*), and 95% *ee* (Scheme 29B). Bulkier phosphoramidate side chains markedly improved both the stereo- and enantioselectivity. This novel MCR is complementary to Hu's initial discoveries and is the first example of the enantioselective synthesis of 2,3-succinic acid derivatives bearing a quaternary stereogenic center.

The Hu group reported an enantioselective 4-MCR for the synthesis of 1,3,4-tetrasubstituted tetrahydroisoquinolines (THIQs) using cooperative Ru(II)/CPA catalysis.[106] This MCR employed a doubly electrophilic "bifunctional" substrate, with the thought that attack by an ammonium ylide would be followed by trapping *via* an intramolecular cyclization. The bifunctional electrophile arises from reaction between aryl amine **221** and aldehyde **220**. Upon complexation with the CPA, the electrophile is primed for reaction with the ylide, which is generated from reaction of aryl diazoacetate **147** and carbamate **214**. A base-promoted intramolecular Mannich-type addition then furnishes the 1,3,4-multisubstituted THIQ **222** (Scheme 30A). Slight modifications to the solvent, CPA, and use of mandelic acid as an additive furnished the THIQ to 66% with 24:1 *dr* and 97:3 *er* (Scheme 30B, entry 2). The reaction tolerated various aryl amines and modifications to the ester groups of the bifunctional electrophile and the aryl diazoacetates.

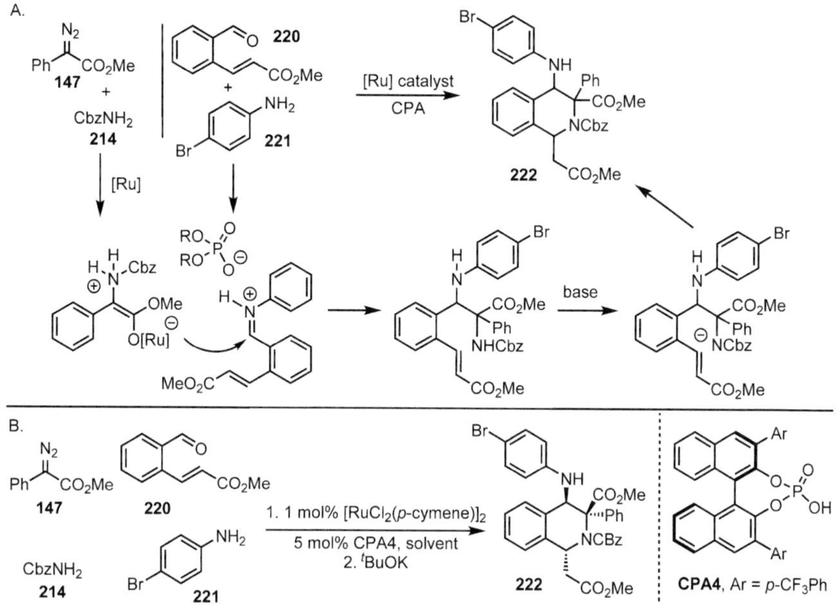

entry	CPA	additive	solvent	major/minor	yield	*er*
2	CPA4	none	DCE	85:15	51%	94:6
3	CPA4	10 mol% (+/-)-mandelic acid	DCE	96:4	66%	97:3

Scheme 30 (A) Proposed mechanism for the 4-component MCR. (B) Optimal reaction conditions.

The Hu group published examples of a divergent MCRs involving ammonium ylides in 2013.[107] $Rh_2(OAc)_4$-catalyzed reaction between arylamine **224** and 4-oxoenate **225** in the presence of diazoacetate **147** furnished two multicomponent products, an α-amino ester **227** and the pyrrolidine product **226**, which is believed to arise from an alternative ylide generation/intramolecular aldol pathway (Scheme 31A). The authors envisioned both of these MCR products might be selectively accessed *via* a simple alteration to the order of substrate addition. A mixture of **224** and **225** was pre-stirred with molecular sieves for 1 h in the presence of Al_2O_3. Addition of catalyst furnished **226** as the sole product in 88% yield and 13:1 *dr* upon slow addition of **147** (Scheme 31B). The reaction could be diverted to furnish the α-amino ester by pre-mixing **147** and aryl amine **224**,

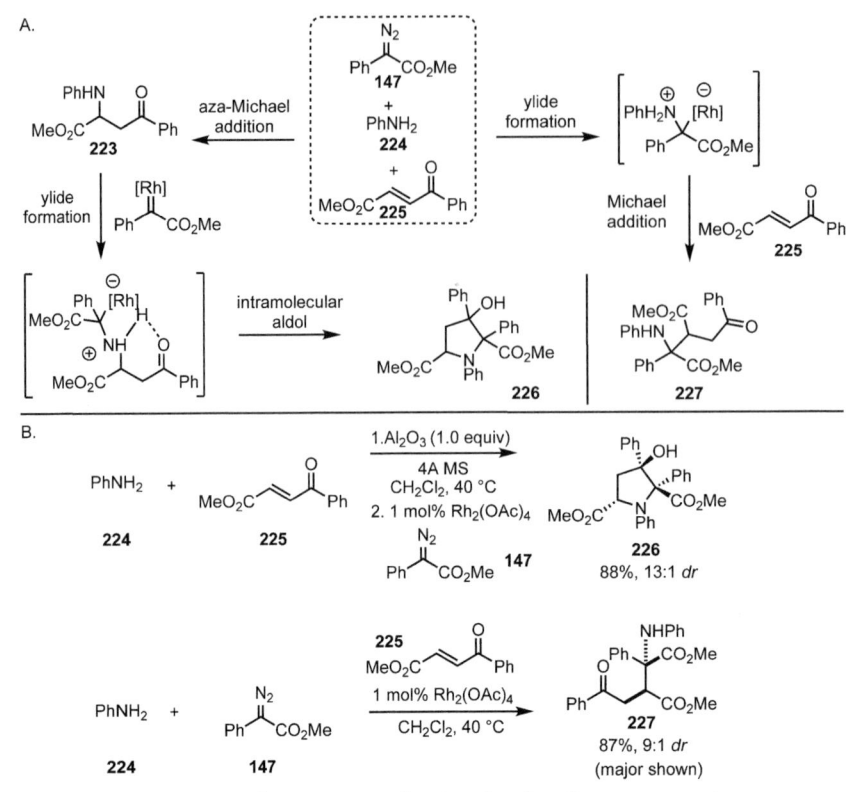

Scheme 31 (A) Proposed reaction mechanism for the divergent reaction outcomes. (B) Optimal reaction conditions for both transformations.

followed by slow addition to the catalyst and **225**. Optimal conditions yielded the α-amino ester **227** in 87% yield and 9:1 *dr*. This method represents the first divergent reaction of ammonium ylides to deliver useful pyrrolidine products in good yield and *dr*.

2.8 Intramolecular electrophilic trapping of protic ammonium ylides

Moody and coworkers reported an intramolecular trapping of a protic nitrogen ylide *via* an intramolecular Michael addition to give functionalized pyrrolidine compounds.[108] Initial studies found Rh- or Cu-catalyzed reaction of diazoester **147** and β-aminoketone **228** furnished the *N*-PMP protected pyrrolidine **230** as a single diastereomer (Scheme 32A). Cu(I) OTf was the best catalyst, giving the product in 90% yield, and was employed in studies of reaction scope. A variety of pyrrolidines were synthesized *via* this method, including spiropyrrolidines using cyclic diazoesters. Given this initial success, the Moody group investigated whether the

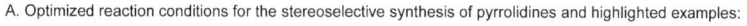

Synthetic applications in modern organic chemistry   39

Scheme 32 (A) Optimized conditions for pyrrolidine synthesis and selected examples. (B) Proposed mechanism showing the transition state leading to high *dr*.

decreased nucleophilicity of the nitrogen in ketocarbamates would be problematic. Application of Dubois' Rh$_2$(esp)$_2$ catalyst at reflux delivered the desired products as single *cis* isomers. Similar to the PMP-protected amines, carbamate derivatives were amenable to reaction with a variety of aryl diazoacetates, which were most compatible with electron-rich substrates. Successful ketocarbamates included vinyl ketones **235** and thiophenes **234**, which furnished the corresponding pyrrolidines in good yield and as single diastereomers. A proposed mechanism for pyrrolidine formation involves attack of the amine on the metal-supported carbene, which is then be trapped in an intramolecular fashion to furnish the pyrrolidine, as opposed to a competing [1,2]-H shift. A highly ordered transition state for ring closure is proposed (Scheme 32B), where proton transfer from the amine to the ketone is assisted by the ester, resulting in high *cis* selectivity.

Hexahydro-1,3,5-triazines are commonly employed as formaldimine precursors in amino-methylation and hydroaminomethylation reactions.[109,110] In 2016, the Sun group shared their work on gold-catalyzed [4+1]- and [4+3]-cycloaddition reactions, showing that hexahydro-1,3,5-triazines directly add to gold carbenes as dipolar adducts, rather than acting as formaldimine precursors (Scheme 33A).[111] Through a formal [4+1] annulation, this catalytic system afforded imidazolidines in modest yields when vinyl-substituted diazoacetates were used; seven-membered heterocycles were accessed from triazines and enol diazoacetates, with C–N bond formation occurring at the vinylogous position. In the former case,

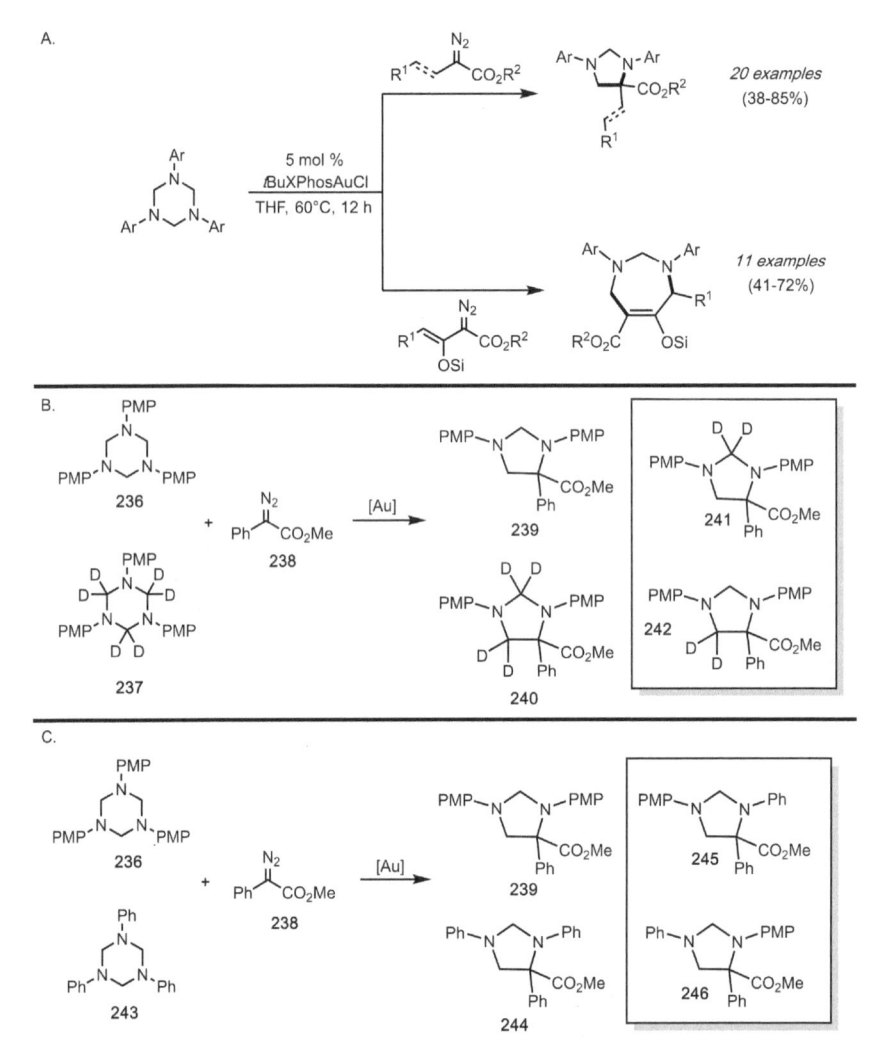

Scheme 33 (A) General scheme for Sun's gold-catalyzed formal [4+1] and [4+3] cycloadditions of diazoesters with hexahydro-1,3,4-triazines. (B and C) Deuterium-labeling and control experiments to probe the potential involvement of a formaldimine intermediate in the reaction mechanism.

nucleophilic addition occurs preferentially at the vinylogous position of vinyl gold carbenes[112–114]; although both the vinylogous and carbenic positions of metallo-vinylcarbene and metallo-enolcarbene species have electrophilic character, vinyl diazoacetates displayed a preference for carbenic reactivity in this system. In contrast, the electron-donating oxygen atom of the silyl ether group in the latter case imparts increased electrophilic character to the vinylogous position of enol diazoacetates.

In an effort to elucidate the mechanism of these transformations, deuterium labeling and control experiments were conducted. Cross-cyclization products **241** and **242** were not observed following the reaction of **236** and **237** with **238**; however, **239** and **240** were afforded in a nearly 1:1 ratio (Scheme 33B). Further probing of the mechanism gave no evidence of cross-cyclization products **245** and **246** following a control reaction between **236** and **243** with **238** (Scheme 33C). Thus, a potential pathway involving the cycloaddition of formaldimine **248** with metal carbene **250** was ruled out due to the absence of cross-cyclization products.

The proposed mechanism for [4+1] cycloaddition involves the reaction between triazine **247** with metal carbene **250** to give ammonium ylide intermediate **251** (Scheme 34A). Intramolecular electrophilic trapping of the ylide and concomitant loss of one formaldimine molecule, followed by reductive elimination, yields product **252**. On the other hand, the [4+3] cycloaddition proceeds through a proposed nucleophilic addition of **247**

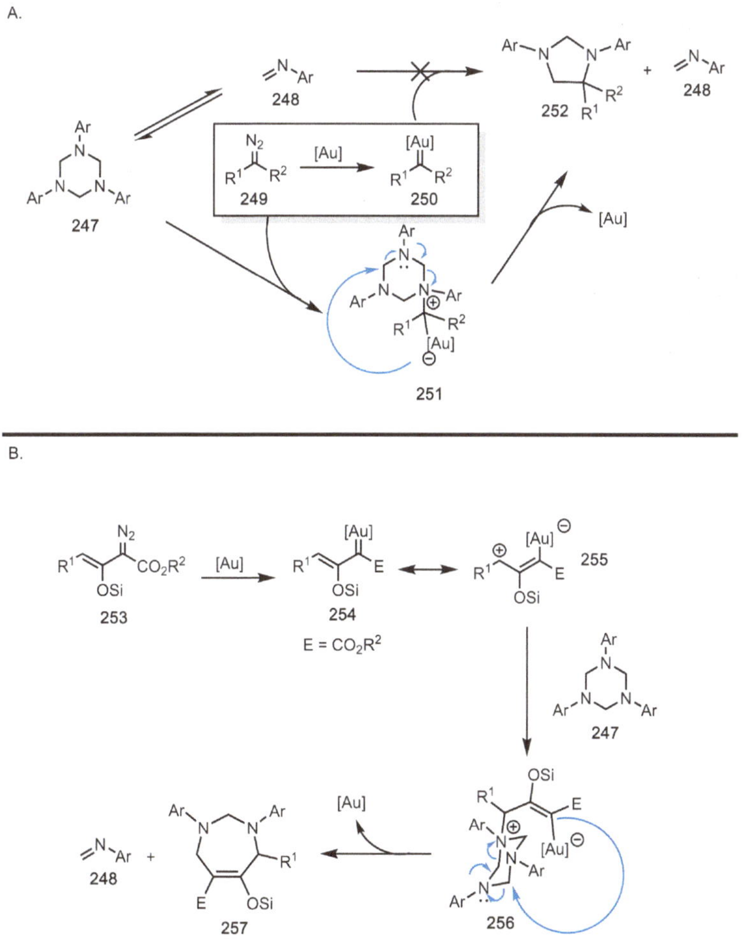

Scheme 34 (A) Proposed mechanism for the formal [4+1] cycloaddition. (B) Proposed mechanism for the formal [4+3] cycloaddition.

at the vinylogous position of the Au enolcarbene species **255** to yield intermediate **256** (Scheme 34B). An intramolecular cycloaddition and subsequent reductive elimination give the cycloaddition product **257**. DFT calculations[115] by Li and coworkers revealed that conjugation effects in the ring-opening of the 1,3,5-triazine control the regioselectivity for the [4+3]-annulation with enol diazoacetates; in the [4+1] annulation with vinyl diazoacetate, regioselectivity is postulated to be affected by coordination of a double bond to gold in the reductive elimination step.

In 2019, the Hu group reported an intramolecular Mannich trapping of a protic ammonium ylide to deliver tetrahydroquinoxalines in a formal [5+1]-annulation.[116] The group hypothesized that reaction of an ortho-amino phenyl imine with aryl diazoacetates in the presence of a Rh catalyst would deliver a protic ammonium ylide that could be trapped intramolecularly by the adjacent amine before a proton shift could occur. Treatment of ortho-amino-phenyl **258** and methyl phenyl diazoacetate **147** proceeded well to deliver the desired Mannich-type trapping product in 46% yield and >19:1 *dr*, favoring the *syn* diastereomer (Scheme 35). Cu catalysts were also compatible with this reaction but provided no benefit to the overall reaction yields. Key was elevation of the temperature to 40 °C to enable more facile generation of the initial metal carbenoid, which raised the overall yield to 75% with excellent *dr*. This reaction was amenable to many aryl diazoacetates, although *p*-NO$_2$ and *o*-Br groups were incompatible, suggesting that strongly electron-withdrawing or sterically congested aryl diazoacetates are not viable. A wide variety of amino phenyl imines were also compatible with the reaction conditions, including those bearing nitro, benzoyl, bromo, and methoxy substituents (selected examples **260–264**). This reaction represents a novel Mannich [5+1] cyclization and demonstrates a viable method for the synthesis of tetrahydroquinoxalines bearing quaternary stereogenic centers with excellent diastereoselectivity.

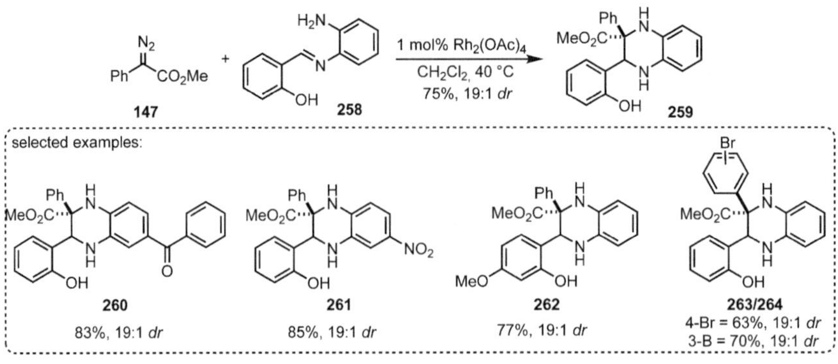

Scheme 35 Optimal reaction conditions and selected examples.

An interesting dual catalytic approach toward the synthesis of N-heterocycles via trapping Rh mediated carbenes with aminoalkynes was reported by the Sharma group in 2018.[117] Dual Rh and Au catalysis is utilized in a cascade sequence, where a protic ylide is trapped with an Au-activated alkyne that undergoes Conia-ene cyclization to furnish spirocyclic pyrrolidine derivatives. $Rh_2(esp)_2$/PPh_3AuCl/$AgSbF_6$-catalyzed reactions of an aminoalkyne and a 2-tetralone diazo compound furnished the major spiropyrrolidine product **265** in 60% yield (Scheme 36A). The scope of the reaction was moderate to yield various fused aromatic spiropyrrolidines but

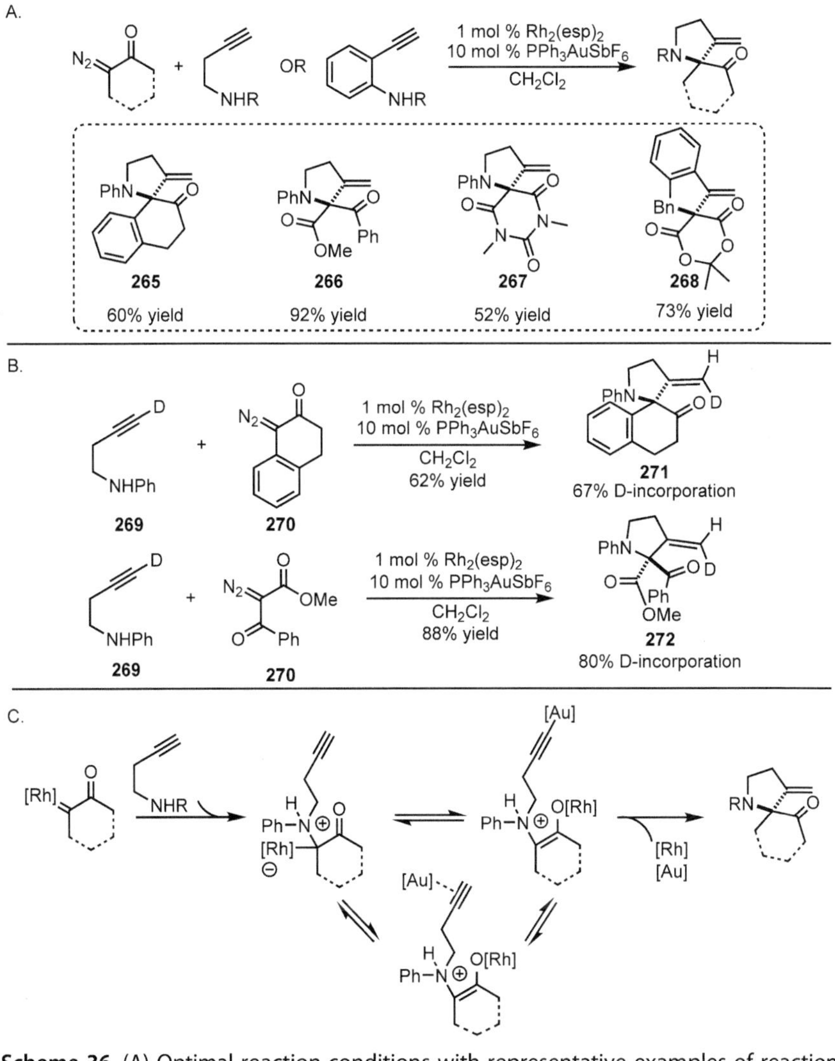

Scheme 36 (A) Optimal reaction conditions with representative examples of reaction scope. (B) Deuterium labeling experiment. (C) Proposed mechanism for the synthesis of spiropyrrolidines.

was expanded to untethered alkyne amines to furnish the corresponding spiropyrrolidine derivatives **266**. Deuterium labeling experiments revealed scrambling, a result inconsistent with other reports in the literature (Scheme 36B). The authors hypothesized that the *Au*-acetylide may be in equilibrium with an alkyne π-complex; a proposed reaction mechanism is shown in (Scheme 36C).

2.9 Concluding remarks

The reactions presented in this section represent an overview of the history and recent developments in the sphere of the ammonium ylides. A better understanding of the mechanisms of this reaction and their interactions with metal catalysts have led to developments in their application in organic synthesis in recent years. Although applications of these methods have become more common, they are still relatively underutilized especially compared to their base generated counterparts. Future work is likely to focus on improving the generality of these methods, especially looking toward different sets of catalysts, ligands, and reaction additives which will allow for milder reactions conditions.

3. Oxonium ylides

Oxonium ylides are highly reactive trivalent oxygen intermediates that are typically formed from the reaction of an ethereal oxygen and a metal-supported carbene that arises from the reaction of a diazo carbene precursor and a metal (Scheme 37).[118–120] Alcohols are also known to participate in oxonium ylide formation, although these precursors are more prone to side reactions than their ethereal counterparts. The equilibrium is believed to disfavor formation of the oxonium ylide; this hypothesis is supported by early investigations that show alkene cyclopropanation is the major reaction in ethereal solvents.[121]

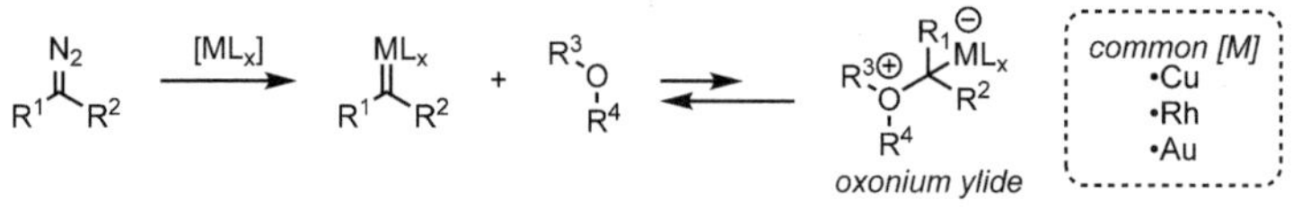

Scheme 37 Formation of oxonium ylides *via* reaction of ethers with metal-supported carbenes.

The reactions of oxonium ylides can be subdivided into a few general categories. The most common reactions involve [2,3]-sigmatropic rearrangements (Fig. 5A) and [1,2]-shifts (similar to the analogous Stevens rearrangement of ammonium ylides, Fig. 5B). The [2,3]-sigmatropic rearrangement is a symmetry-allowed pericyclic reaction in which an ethereal oxygen substituted with an allyl or propargyl group attacks a metal-supported carbene, thus transposing the allyl or propargylic group to the anionic ylide carbon. This rearrangement has been most prominently studied in its intramolecular form, although intermolecular versions are known and are becoming more common. In contrast, the [1,2]-shift is a symmetry-forbidden process widely believed to occur via a stepwise radical mechanism. Although [1,2]-shifts can compete with [2,3]-sigmatropic rearrangements, the latter tends to dominate, due to the higher activation

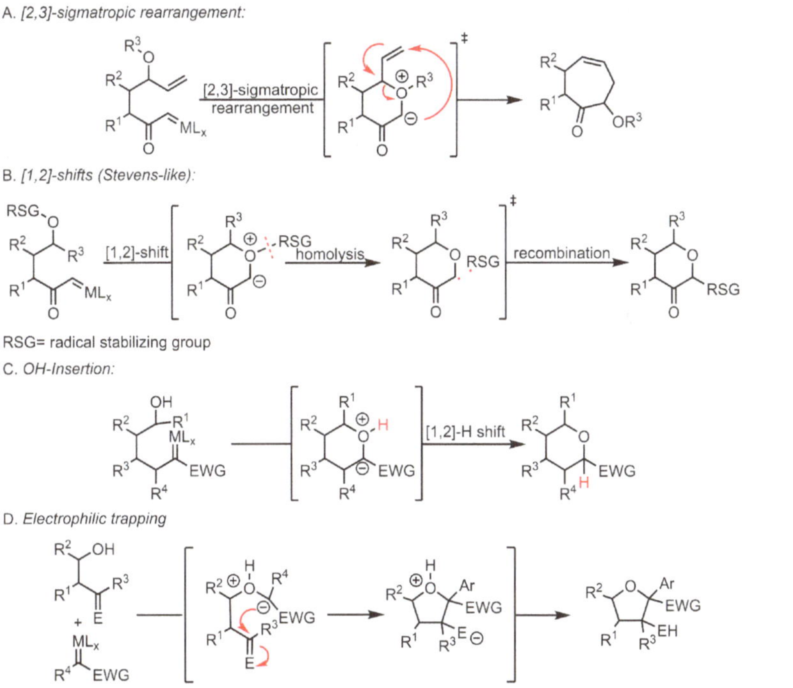

Fig. 5 Reactions of oxonium ylides. (A) General representation of [2,3]-sigmatropic rearrangement and mechanism (B) general reaction scheme and mechanism of the [1,2]-shift. (C) General representation of O-H insertion reactions. (D) Representation of intramolecular electrophilic trapping (can be intermolecular in a multicomponent reaction).

energy barrier of the initial homolytic step in the former.[118–120] In the case of a [1,2]-shift, the transposition of the migrating group from the oxygen atom to the anionic ylide carbon furnishes the product.

Another class of reactions commonly encountered in ylide chemistry are the X-H (Fig. 5C) and C-H insertions (not pictured) reactions. A C-H insertion occurs when another viable reaction path, such as rearrangement, is not present. If rearrangement is possible, the O-H insertion reaction will proceed directly from the trivalent oxonium ylide and undergoes a 1,2-H-shift to generate a new ether (Fig. 5C). Reactions involving direct insertion of the metal-supported carbene into the O–H bond without ylide formation will not be discussed in this review.

A final reaction type that will be covered in this chapter involves the trapping of oxonium ylide intermediates with various electrophiles. An alcohol first generates the intermediate oxonium ylide; the amphiphilic nature of the trivalent ylide then enables the nucleophilic carbanion of the ylide to intercept an electrophile. This oxonium ylide then undergoes a delayed H shift or protonation after electrophilic trapping to form the heterocyclic product. The electrophile can be intercepted either in an intermolecular, multicomponent fashion or trapped intramolecularly (Fig. 5D); both instances will be covered herein.

The remainder of this section is dedicated to presenting selected historic examples that set the stage for each reaction type, followed by notable examples from the past ∼20 years. References are provided for the most recent review covering each reaction type, with subsequent years surveyed for more recent examples. While not comprehensive in nature, the goal is to demonstrate key advancements in the field, with a focus on heterocycle syntheses, intermolecular variants, asymmetric reactions, and reactions with applications in other fields. In cases where material is not covered, interested readers will be directed to other oxonium ylide reviews that address issues such as the choice of metal,[120] reaction type,[119] and applications in total synthesis.[118]

3.1 [2,3]-Sigmatropic rearrangements

The [2,3]-sigmatropic rearrangement is one of the most common and well-studied reactions of oxonium ylides. Early examples of intramolecular [2,3]-sigmatropic rearrangements of these reactive intermediates are covered briefly here. Emphasis is placed on intermolecular [2,3]-sigmatropic rearrangements, including strategies developed to control chemo-, stereo-, and enantioselectivity in synthetically useful methodologies.

3.2 Intramolecular [2,3]-sigmatropic rearrangements

While the first examples of oxonium ylides were reported by Nozaki in 1965,[122a,b] interest in the field was renewed with simultaneous 1986 reports from the groups of Pirrung[123] (Scheme 38A) and Johnson[124] (Scheme 38B) on intramolecular [2,3]-sigmatropic rearrangements. Treatment of allyl-oxy diazoketones 273 and 276 with $Rh_2(OAc)_4$ promoted intramolecular [2,3]-sigmatropic rearrangements in good-to-excellent yields to furnish tetrahydrofuranones 274, 275, and 277. Johnson demonstrated a high degree of stereospecificity in the transformation of 273 to 274-75 in favor of the *trans*-diastereomer 274 (12:1). Importantly, the reaction showed excellent chemoselectivity for the [2,3]-sigmatropic rearrangement over the competing [1,2]-shift. These early examples provided a foundation for further developments in intramolecular [2,3]-sigmatropic rearrangements and have been reviewed elsewhere.[119]

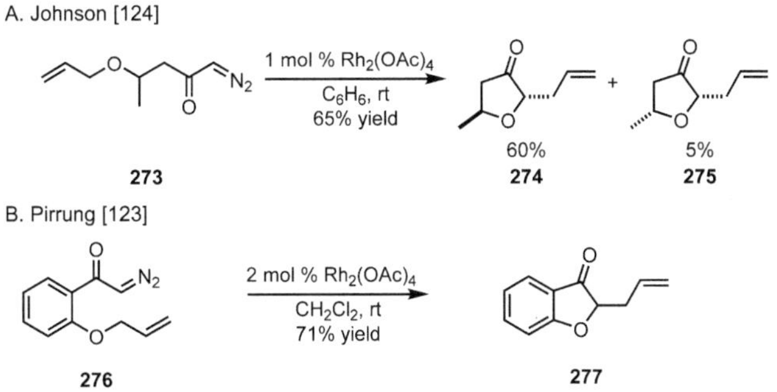

Scheme 38 (A) Johnson's demonstration of a [2,3]-sigmatropic rearrangement of allyloxydiazoketones. (B) Pirrung's [2,3]-rearrangement of cyclic allyl ethers to benzoannulated tetrahydrofuranones.

In 2018, Deska reported an interesting strategy that merged the chemoenzymatic desymmetrization of carboxylic acids with Cu-catalyzed intramolecular [2,3]-sigmatropic rearrangement to prepare enantiomerically pure tetrahydrofuranones (Scheme 39).[125] Optimal conditions for the enantioselective hydrolysis of allyl ether **278** involved treatment with lipase B from *Candida antarctica* to furnish the corresponding carboxylic acid in 99% yield, 96:4 *dr*, and 96% *ee*. The acids were then transformed to the corresponding diazoketone **279** *via* a chlorination and diazomethylation sequence. Treatment of **279** with Cu(acac)$_2$ promoted [2,3]-sigmatropic rearrangement in good yield with high *trans* selectivity and complete retention of enantiopurity to deliver *trans*-tetrahydrofuranone **280** (Scheme 39). The high preference for the *trans*-tetrahydrofuranone likely originates from the sterically less encumbered *endo* transition state. Interestingly, initial optimization indicated that altering the counteranion of the Cu acetonitrile complex further enhanced the *ee*. The yield was also sensitive to the counteranion identity, with Cu(MeCN)$_4$BF$_4$ furnishing <5% of the [2,3]-sigmatropic product. This method provides a simple way to prepare enantiopure tetrahydrofuranones in high *dr* without the need for expensive chiral transition metal catalysts.

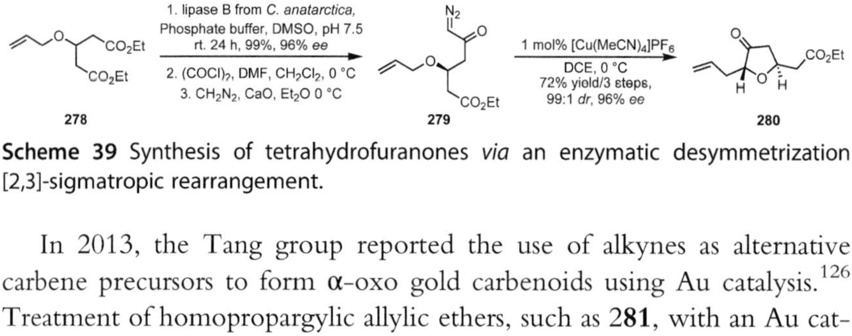

Scheme 39 Synthesis of tetrahydrofuranones *via* an enzymatic desymmetrization [2,3]-sigmatropic rearrangement.

In 2013, the Tang group reported the use of alkynes as alternative carbene precursors to form α-oxo gold carbenoids using Au catalysis.[126] Treatment of homopropargylic allylic ethers, such as **281**, with an Au catalyst in the presence of a pyridine *n*-oxide furnished the corresponding fused bicyclic dihydrofuranones **282** in good yield and stereoselectivity after [2,3]-sigmatropic rearrangement (Scheme 40A). The authors expanded this method to benzoannulated derivatives **284** and found allyl ethers gave formal [2,3]-sigmatropic rearrangement products at room temperature *via* the isolable vinyl ether intermediate **286**. Deuterium labeling studies confirmed that inversion of the allyl ether did not occur until **286** was heated to 80 °C (Scheme 40B, top). This information, coupled with identification of the intermediate vinyl ether, led the authors to propose a 1,4-allyl migration/Claisen rearrangement pathway leading to benzoannulated products. In contrast, deuterium labeling of propargylic ether **288** confirmed inversion of the allyl substituent to yield **289**. This finding supports a [2,3]-sigmatropic rearrangement mechanism for these substrates (Scheme 40B, bottom) and highlights the dependence of reaction outcome on the nature of the substrate.

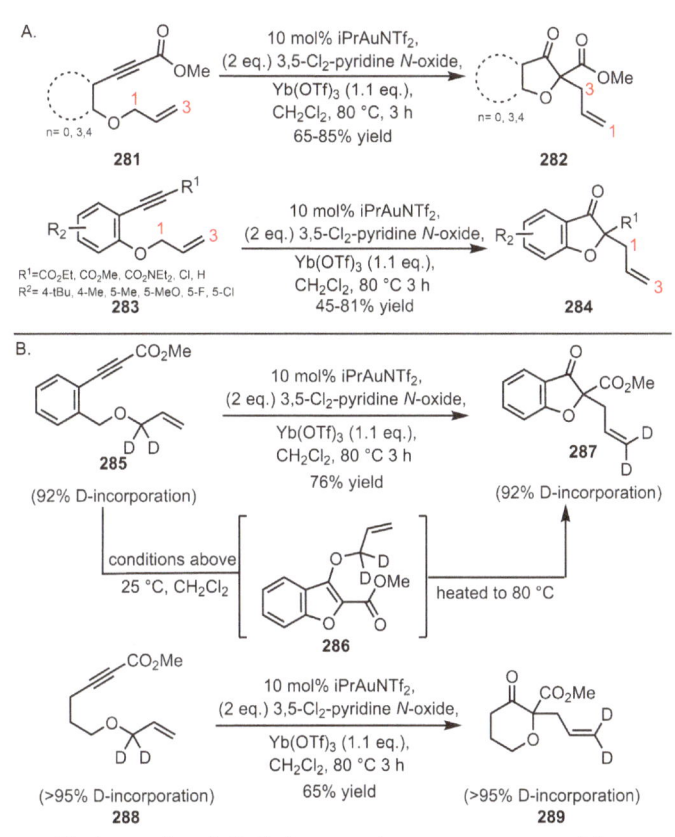

Scheme 40 (A) Au-catalyzed [2,3]-sigmatropic rearrangement of homopropargylic allylic ethers to deliver fused bicyclic dihydrofuranones. (B) Deuterium labeling experiments indicate either a [2,3]-sigmatropic rearrangement or a 1,4-allyl migration/Claisen rearrangement pathway.

Shortly after this initial report on Au catalysis, Tae and coworkers reported a similar [2,3]-sigmatropic rearrangement of allyl homopropargylic ether **290/291** to deliver 2,5-disubstituted dihydrofuranones **292/293** (Scheme 41A).[127] This reaction showed a preference for the *trans* diastereomer, likely due to minimization of steric interactions between the propargylic substituent and the allyl group in the transition state. Notably, the *dr* in the reaction depended on the identity of the propargyl substituent. Electron-donating substituents (R=PhOMe) delivered exclusively the *trans* diastereomer, while electronically neutral alkyl substituents (R= $(CH_2)_4CH_3$) furnished a more modest *dr* of 2:1. To further support the synthetic utility of this method, Tae completed a formal synthesis of (−)-kumausallene (Scheme 41B).

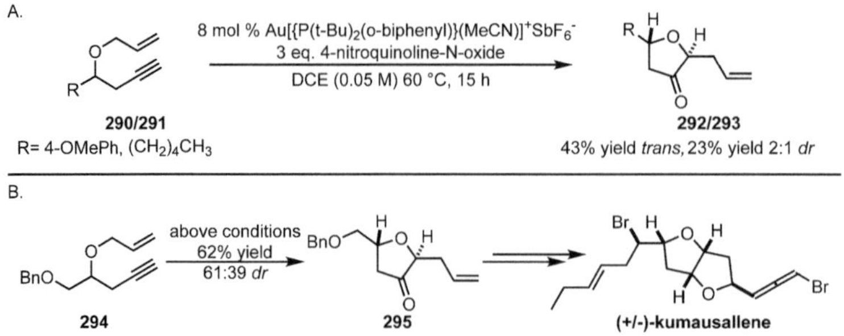

Scheme 41 (A) Selected examples of Au-catalyzed rearrangement of allyl homopropargylic ethers to deliver 2,5-disubstituted dihydrofuranones. (B) Application to the formal synthesis of (+/−)-kumausallene.

The number of available carbene precursors for oxonium ylide formation and subsequent [2,3]-sigmatropic rearrangement was expanded in 2014 with Boyer's report demonstrating that di- and trisubstituted 1,2,3-triazoles could be employed in Rh-catalyzed [2,3]-sigmatropic rearrangements of homopropargylic allyl ethers (Scheme 42A).[128] This methodology efficiently delivers dihydrofuran-3-imines in good yields and moderate-to-excellent *dr* (*trans*-selective). Products are readily functionalized to furnish dihydrofuran-3-ones such as **298** (*via* hydrolysis) or allylated to give trisubstituted tetrahydrofurans. This method proved robust for a variety of substituted homopropargylic allyl ethers, with the *dr* dictated by the steric bulk of the substituent (R=Me, 5:1 *dr*, R=^{i}Pr, 20:1 *dr*). Application to the total synthesis of (+)-petromyroxol further demonstrated the utility of this chemistry (Scheme 42B).[129]

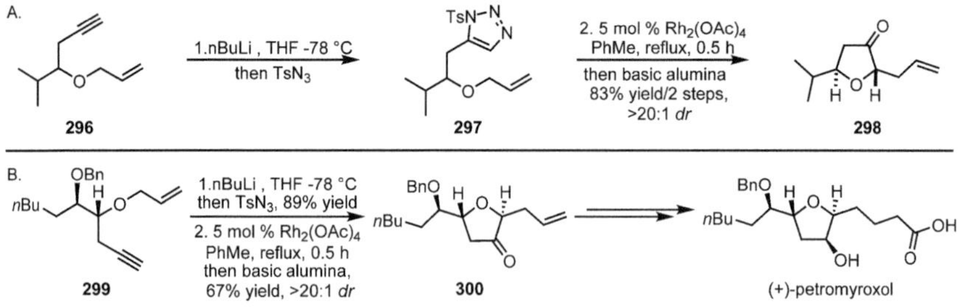

Scheme 42 (A) Use of 1,2,3-triazoles for the Rh$_2$(OAc)$_4$-catalyzed [2,3]-sigmatropic rearrangement of homopropargylic allyl ethers to deliver dihydrofuran-3-imines that are hydrolyzed dihydrofuran-3-ones. (B) Application to the total synthesis of (+)-petromyroxol.

In 2013, Clark published an elegant mechanistic ^{13}C labeling study to assess whether metal-bound or free ylides were involved in [2,3]-sigmatropic rearrangements to form benzoannulated furanones (Scheme 43).[130] Clark hypothesized that if metal-free ylides were involved in the rearrangement, there should be no catalyst dependence on the reaction mechanism. The group utilized scrambling of a ^{13}C isotopic label to observe differing product ratios resulting from formal [1,2]-shifts and [2,3]-sigmatropic rearrangements. Ultimately, the data suggested that product ratios are heavily dependent on the identity of the metal catalyst, supporting the likelihood that different mechanisms are operative (Scheme 43).

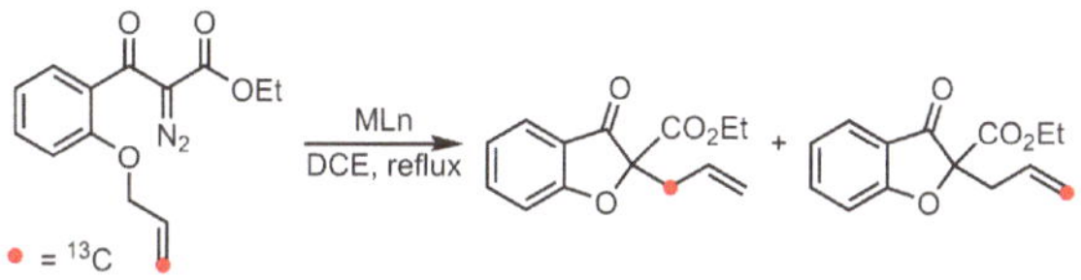

catalyst	yield	ratio 302:303
Rh$_2$(OAc)$_4$	76%	>99:1
[Cu(acac)$_2$]	43%	84:16
[Ir(cod)Cl]$_2$	37%	44:56

Scheme 43 ^{13}C labeling studies showing a mechanistic dependence on metal identity in [2,3]-sigmatropic rearrangements to furnish benzoannulated furanone derivatives.

3.3 Intermolecular [2,3]-sigmatropic rearrangements

Intermolecular [2,3]-sigmatropic rearrangements were initially overlooked, due to the potential for competing reactivity between oxonium ylide formation and alkene cyclopropanation. Much of the early work in the field of C-H insertion and cyclopropanation showed these reactions proceed well in ethereal solvents, with no evidence of products from oxonium ylide formation. This further called into question the synthetic potential of intermolecular rearrangements[121]; fortunately, recent work has revealed several useful examples of this chemistry.

The first instance of an intermolecular [2,3]-sigmatropic rearrangement of oxonium ylides was reported by Doyle in 1988, where cinnamyl methyl ether **304** is converted to **305–306** under Rh catalysis (Scheme 44).[131] Doyle found the transformation proceeded in good yield with moderate diastereoselectivity but did observe cyclopropanation as a major competing process (**307**) (Scheme 44, entry 1). Ultimately, the amount of **307** could be modulated through the identity of the carbene precursor (Scheme 44, entry 2), where the less reactive carbene precursor α-diazoacetophenone **175** suppressed cyclopropanation. Doyle also reported an enantioselective version using chiral Rh$_2$(MEOX)$_4$ catalysts to achieve excellent *ee* (Scheme 44, entry 3/4); however, the method was limited to a narrow range

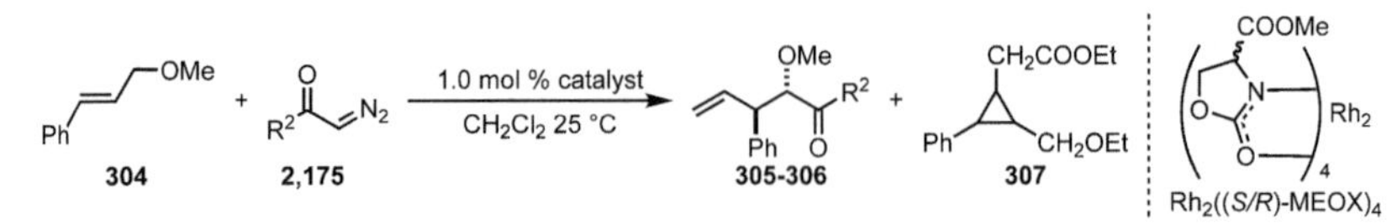

entry	R²	catalyst	yield 305-306	yield 307
1	OEt	Rh₂(OAc)₄	73% yield 5:1 dr	27% yield
2	Ph	Rh₂(OAc)₄	94% yield 10:1 dr	6% yield
3	OEt	Rh₂(4S-MEOX)₄	85% yield 6:1 dr (98% ee)	-
4	OEt	Rh₂(4R-MEOX)₄	85% yield 6:1 dr (98% ee)	-

Scheme 44 The first examples of both racemic and enantioselective Rh-catalyzed intermolecular [2,3]-sigmatropic rearrangements.

of allyl ethers.[132] This shortcoming left room for the development of more selective and general intermolecular [2,3]-sigmatropic rearrangements, many of which have been reviewed and will not be discussed further in this chapter.[118–120]

Davies expanded his early work to achieve an enantioselective intermolecular [2,3]-sigmatropic rearrangement of substituted allyl alcohols with donor/acceptor carbenes in the presence of a chiral Rh catalyst.[133] Substituted allyl alcohols, such as **308**, were explored to determine if removal of the activated C–H bond was sufficient to promote the desired reaction pathway. These efforts were successful (Scheme 45, entry 1), as exposure of the free allyl alcohols to donor/acceptor carbene **147** in the presence of a chiral Rh catalyst furnished the [2,3]-sigmatropic rearrangement product **309** in good yield and good *ee*. The Davies group extended this methodology to allyl carbene precursor **311** (Scheme 45, entry 2) to generate products such as **313**, which showed better chemoselectivity (10:1 rearrangement:O–H insertion) and *ee* (92–98% *ee*). It is important to note that less reactive donor/acceptor carbenoids were essential to achieve the desired reactivity, as more reactive carbenes gave primarily racemic O–H insertion products.

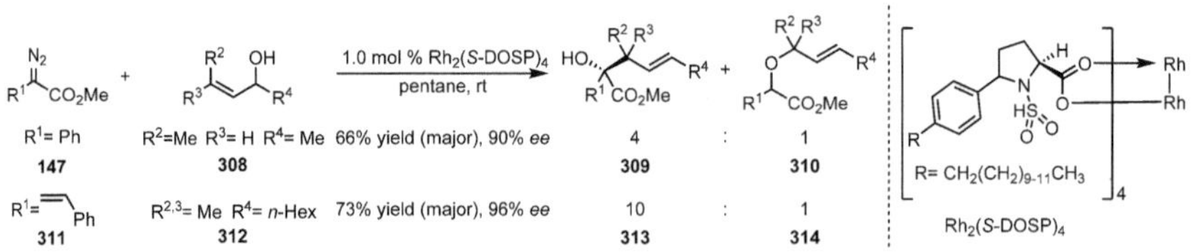

Scheme 45 Highly enantioselective intermolecular [2,3]-sigmatropic rearrangement to deliver highly substituted homoallylic alcohols.

In another study, the Davies group used an intermolecular [2,3]-sigmatropic rearrangement to construct products containing two vicinal stereocenters, such as **316/318**, with access to all four possible stereoisomers by careful consideration of substrate and catalyst.[134] The chirality at the allylic stereocenter and the alkene geometry dictated the stereochemical outcome at the allylic carbon, whereas the homoallylic stereocenter was controlled by careful selection of the chiral Rh catalyst (Scheme 46). Davies employed these highly selective methods in combination with other reactions to develop cascade reactions to furnish highly substituted cyclopentane and cyclohexane scaffolds.[135]

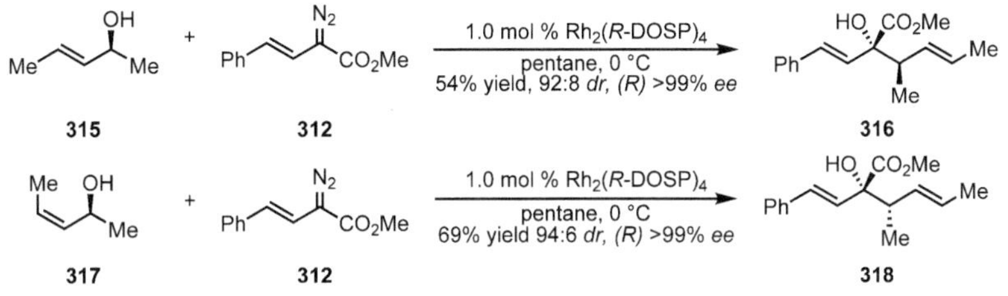

Scheme 46 Enantioselective synthesis of allylic alcohols containing vicinal stereocenters via [2,3]-sigmatropic rearrangements.

Despite the improvements to asymmetric [2,3]-sigmatropic rearrangements, Davies' system was not compatible with primary alcohols. The Prabhu group addressed this limitation in 2017 using gold carbenes generated from primary allyl alcohols to facilitate [2,3]-sigmatropic rearrangement.[136] Treatment of cinnamyl alcohol **320** and diazoacetate **319** with AuClPPh$_3$ and AgSbF$_6$ as a halide scavenger, gave the desired [2,3]-rearrangement product **321** (Scheme 47, entry 1). Further optimization revealed AgNTf$_2$ as a superior halide scavenger and PCy$_3$ as the most promising ligand for Au (Scheme 47, entry 2). The bulky PCy$_3$ ligand facilitated carbene-type reactivity via increased electron back-donation from Au to stabilize the carbene carbon and ultimately, the oxonium ylide intermediate. Optimized conditions delivered **321** in 96% yield and 24:1 dr, with only trace amounts of **322**. This method was applicable to diverse aryl diazoacetates (**323, 325, 326,**) (Scheme 47), although less sterically hindered aryl groups performed better. A variety of cinnamyl alcohols showed no significant dependence on the electronics or sterics of the substituted aromatic moiety. Secondary allylic alcohols were not tolerated, making this method complementary to Davies' work.

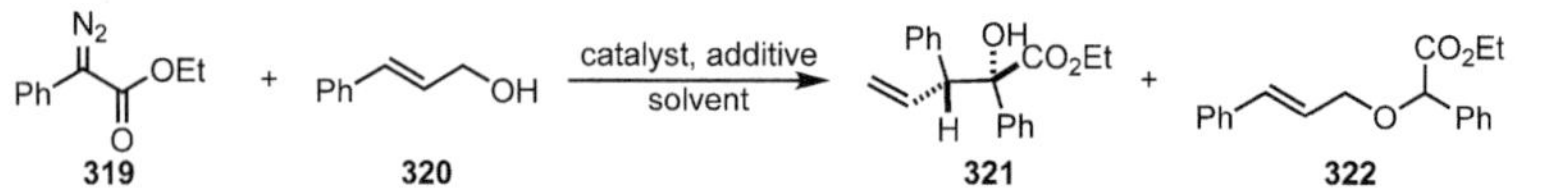

entry	conditions	yield, dr	yield
1	$PPh_3AuCl/AgSbF_6$ (1.5 mol%) DCM	45% yield 8:1 *dr*	35% yield
2	$PCy_3AuCl/AgNTf_2$ (3.0 mol%) Et_2O	96% yield 24:1 *dr*	trace

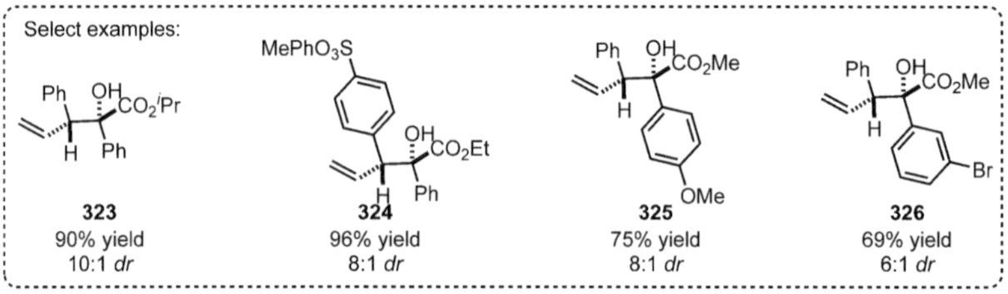

Scheme 47 Development of the intermolecular [2,3]-sigmatropic rearrangement of primary alcohols using Au catalysis and selected examples of reaction scope.

Wood studied the reaction of **194** and **327**, where $Rh_2(TFA)_4$ is hypothesized to fulfill the dual role of facilitating enol formation and the subsequent [2,3]-sigmatropic rearrangement, to furnish α-hydroxyl allene **328** (Scheme 48A).[137] The Davies group developed a similar enantioselective method to furnish highly substituted, enantiomerically pure α-hydroxyl allenes *via* intermolecular [2,3]-sigmatropic rearrangement.[138] They hypothesized the use of diazoester **311**, instead of a diazoketone such as **194**, would suppress enol formation and give a rearrangement where the chiral Rh catalyst remains intact in the intermediate oxonium ylide. Previously reported reaction conditions (described above), in combination with a chiral $Rh_2(R\text{-DOSP})_4$ catalyst and highly substituted propargyl alcohol substrates, gave the desired enantiopure allene **330** in 77% yield and 96% *ee* with

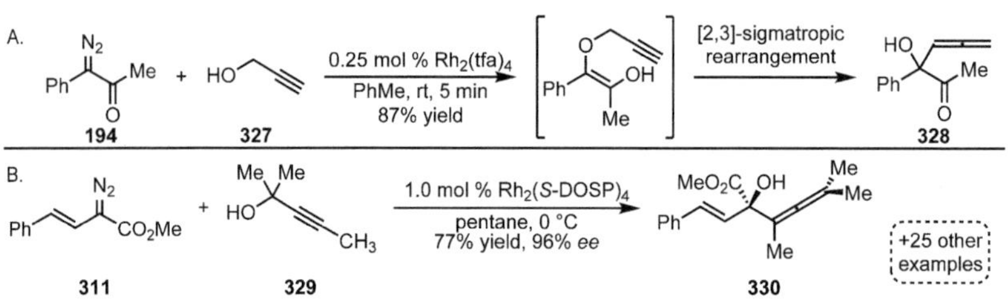

Scheme 48 (A) Wood's racemic Rh-catalyzed α-hydroxyl allene synthesis. (B) Davies' enantioselective Rh-catalyzed synthesis of α-hydroxyl allenes.

negligible amounts of O-H insertion product (Scheme 48B). A variety of electronically diverse aryl diazoacetates and differentially substituted propargyl ethers were well-tolerated; the method could also be extended to kinetic resolution of racemic chiral propargylic alcohols by running the reaction to 50% completion.

Intermolecular [2,3]-sigmatropic rearrangements have also evolved to include vinyl epoxides and oxetanes. In 2006, Quinn reported the Cu(acac)$_2$-catalyzed intermolecular rearrangement of divinyl epoxides to furnish dihydropyrans.[139] This transformation was initially explored with **331**; however, carbene transfer led to only 22% of the expected dihydropyran **333**. The poor reaction outcome was attributed to facial selectivity in the oxonium ylide formation, which could either occur from the β- or the α-face of the epoxide (Scheme 49A) to provide two diastereomeric ylide intermediates. It was postulated only one of these intermediates contains the proper relationship between the carbene and vinyl substituent for rearrangement; the incorrect conformer decomposes to yield the deoxygenation byproduct **334**. Quinn sought to circumvent unproductive ylide formation by utilizing a C2 symmetric divinyl epoxide **335**, in which ylide rearrangement from both faces of the epoxide are possible. A Cu(hfacac)$_2$ (copper hexafluoroacetylacetonate) catalyst furnished dihydropyrane **336** in 69% yield (Scheme 49B), supporting the hypothesis that deoxygenation occurs as a result of unproductive oxonium ylide formation and highlighting the importance of conformation in productive [2,3]-sigmatropic rearrangements.

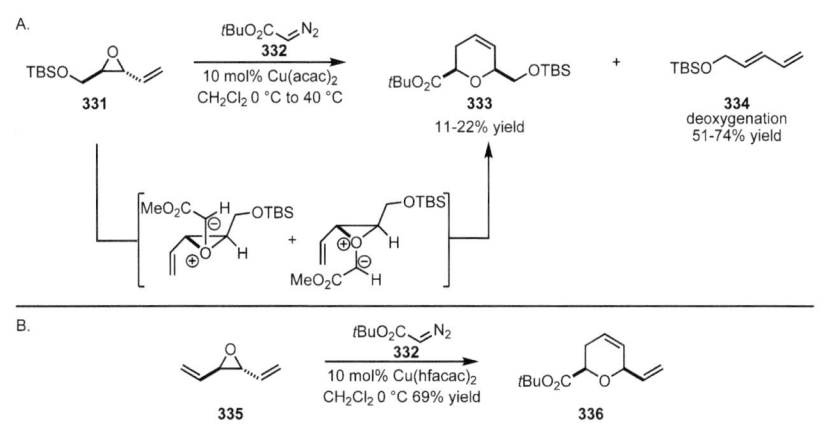

Scheme 49 (A) Initial studies to prepare dihydropyrans. (B) Application of symmetric divinyl epoxides to ensure productive oxonium ylide formation.

Njardarson reported the conversion of vinyl oxetanes and vinyltetrahydrofurans to their corresponding sigmatropic rearrangement products in the presence of diazomalonate **347** or ethyl diazoacetate **2** using a Cu(hfacac)$_2$ catalyst (Scheme 50).[140] Competitive [1,2]-shifts were highly dependent on the substrate and carbene precursor combination. When vinyl oxetane **337** was treated with **2** and **347**, **340–341** were the major products, resulting from a [1,2]-shift (Scheme 50, top). Selectivity was reversed in the reaction of vinyl tetrahydrofuran **342** with **2** to furnish primarily the [2,3]-sigmatropic rearrangement products **343–344**. Although limited in scope, both methods demonstrated that oxygen heterocycles can be employed for intermolecular formation of oxonium ylides, efforts that are sure to stimulate further work in understanding and controlling their reactivity.

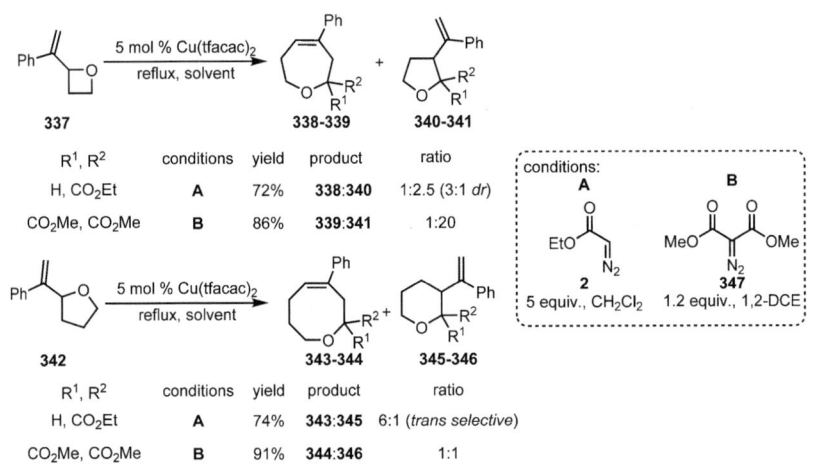

Scheme 50 Intermolecular oxonium ylide formation between oxygen heterocycles of varying sizes. Identities of carbene precursor and vinyl heterocycle have a significant impact on reaction outcome.

3.4 1,2-Shifts
3.4.1 Intramolecular [1,2]-shifts
The [1,2]-shift is a common reaction mode of oxonium ylides, where the migrating group is transferred from the oxygen of the oxonium ylide directly to the anionic carbon without inversion (Fig. 5B). The rearrangement is a symmetry-forbidden process and is proposed to occur *via* a stepwise radical path. As such, the migrating substituent must be capable of stabilizing a radical to promote the desired pathway over degradation of the oxonium ylide. The first example of a [1,2]-shift in oxonium ylides was reported by

Johnson,[124] who found that treating diazoketone-methyl ether **348** with a Rh catalyst gave cyclobutanes **349–350** (Scheme 51A). This method was expanded upon by West[141] to generate cyclic ethers such as **352**, in which the substrates were designed to undergo [1,2]-shift of the exocyclic ether substituent (Scheme 51B). Updates regarding the development of [1,2]-rearrangements have been recently covered in 2009,[119] along with examples in total synthesis covered in 2014.[118] Thus, this section focuses on a brief discussion of historical context, coupled with new mechanistic insight and applications of the [1,2]-Stevens rearrangement in synthetic chemistry.

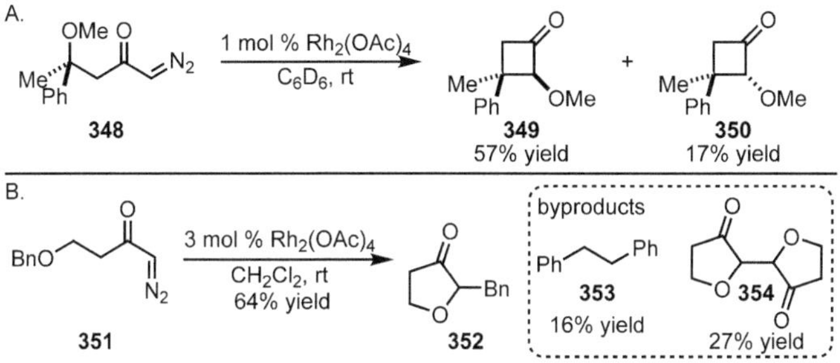

Scheme 51 (A) Johnson's initial findings on [1,2]-shift of oxonium ylides. (B) West's expansion of this chemistry to include the synthesis of cyclic ethers.

West's early work provided experimental evidence for a radical pathway (Scheme 51B). Along with the expected cyclic ether product derived from the [1,2]-shift, tetrahydrofuranone dimers and dibenzyl dimers **353–354** were identified (Scheme 51B).[141] These byproducts can only be produced if radical intermediates escape their solvent cage and recombine. In 2017, the West group reexamined the [1,2]-shift using radical clocks as ether substituents to determine whether radical ring-opening products could be observed.[142] Treatment of **355** with Cu(hfacac)$_2$ yielded [1,2]-shift product **366** with small amounts of **354**, but no cyclopropane cleavage was observed (Scheme 52A). Given the inconclusive nature of the data, the group applied a hypersensitive mechanistic probe **367**, where two divergent cyclopropane cleavage pathways can occur, depending upon the involvement of radical or cationic intermediates at the cyclopropylcarbinyl position (Scheme 52B). Interestingly, the only products obtained were diastereomeric tetrahydrooxepines **368**, which arise *via* ring closure of the enolate oxygen onto the oxonium ion. No corresponding homolytic fragmentation products were observed. This is the first experimental evidence suggesting a

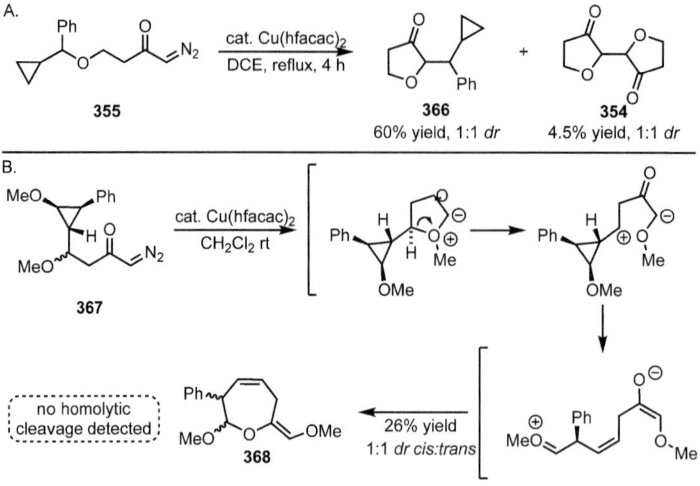

Scheme 52 (A and B) West's use of cyclopropane-derived radical clock as mechanistic probes to determine the homo- or heterolytic nature of the [1,2]-shift.

heterolytic [1,2]-shift path is possible; moreover, there may be multiple possible mechanisms for these shifts that are highly dependent on the nature of the substrate.

Further efforts to develop synthetically useful intra- and intermolecular [1,2]-shifts have been carried out. Similar to [2,3]-sigmatropic rearrangements described in the previous section, [1,2]-Stevens rearrangements have been coupled with enzymatic desymmetrization of glutarate derivatives to furnish enantiopure tetrahydrofuranones such as **372** (Scheme 53).[125] *Candida antarctica* type B furnished the corresponding enantiopure carboxylic acid **370** in 93% yield and 99% ee. Sequential diazo formation and carbene transfer using Rh$_2$(oct)$_4$ at low temperatures delivered diastereomeric tetrahydrofuranones **372** in a combined yield of 81% (only major diastereomer shown) and with

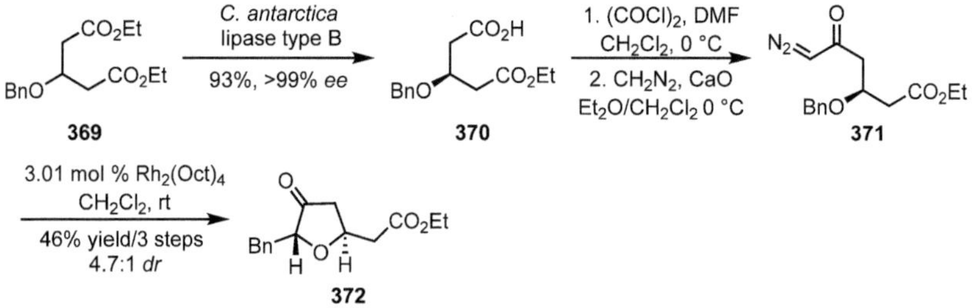

Scheme 53 Application of chemoenzymatic desymmetrization in [1,2]-Stevens rearrangements to deliver enantiopure tetrahydrofuranones.

high enantioretention. This method tolerated various aromatic substitutions, including electron-donating and withdrawing substituents, as well as piperonyl and naphthyl substitution. Despite the moderate *dr*, the method provides rapid access to enantiopure benzyl-substituted tetrahydrofuranones, attractive building blocks for the synthesis of complex molecules.[125]

3.4.2 Intermolecular [1,2]-shifts

The intermolecular [1,2]-shift was expanded by the Krasavin group in 2019[143] to achieve Rh-catalyzed spirocyclization of α-diazo-homophthalimides with cyclic ethers. This transformation was initially discovered as a byproduct of a [3 + 2] cycloaddition reaction between α-diazo-homophthalimide **373** and a nitrile in THF as a solvent (Scheme 54A). Although the desired **374** was produced, the spirocyclic byproduct **375** was hypothesized to result from oxonium ylide formation and subsequent [1,2]-Stevens rearrangement between solvent and **373**. Krasavin optimized for **375** and found that solvent levels of cyclic ethers in the presence of $Rh_2(OAc)_4$ preferentially generated

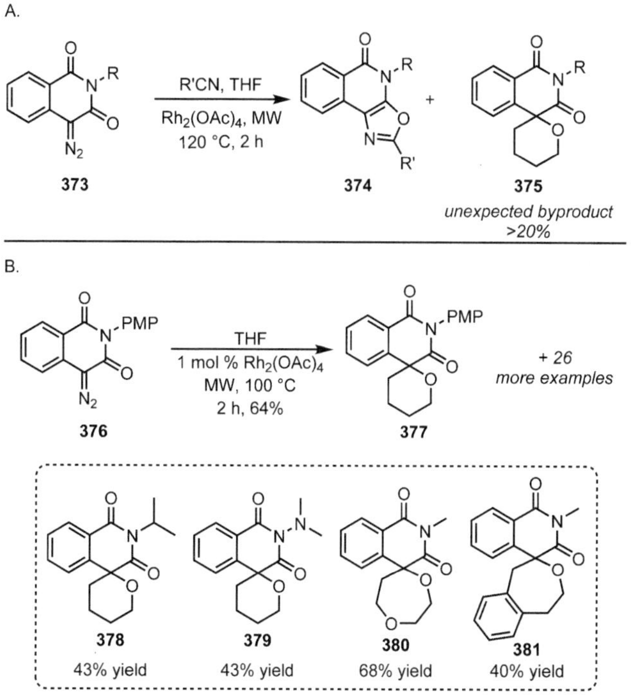

Scheme 54 (A) Spirocyclic byproduct believed to be derived from a [1,2]-Stevens rearrangement between solvent and the corresponding α-diazo-homophthalimide, (B) Selected examples of the reaction scope.

spirocyclic products in yields of 23–68% (Scheme 54B). Notably, only Rh catalysts were successful; no other ylide-forming catalysts furnished the desired product. The reaction was compatible with a variety of α-diazo-homophthalimides containing diverse substitution on the nitrogen and was not impacted by sterics or electronics (**378, 379**). While a variety of cyclic ethers were tolerated, such as 1,4-dioxanes (**380**), benzene-fused cyclic ethers presented limitations (**381**).

3.5 O-H insertion

Metal-supported carbenes are well-known to insert into heteroatom-H bonds (X-H insertion). The seminal report by Casanova and Reichstein[144] described the first example of O-H insertion in the conversion of diazo-pregenenolone derivative **382** to an α-methoxy ketone **383** in the presence of CuO and methanol (Scheme 55A). Although the intended Wolff rearrangement product was not obtained, this work opened the door for further explorations into X-H insertions using metal-supported carbenes. Shortly after this initial report, Yates described the first systematic study of carbene X-H insertions into N-H, S-H, and O-H bonds in the presence of a copper catalyst (Scheme 55B).[145] Since these initial observations, this methodology has been significantly expanded to include intramolecular and enantioselective variants, primarily catalyzed by Rh and Cu catalysts.

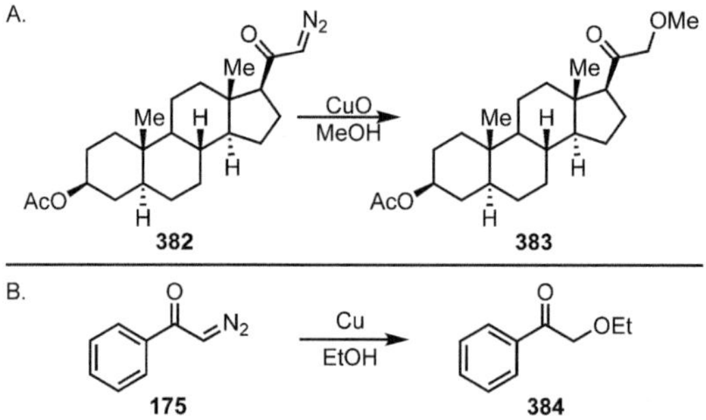

Scheme 55 (A) First example of the O-H insertion in the literature to deliver pregenenolone derivatives. (B) First systematic exploration of multiple X-H insertion reactions with only O-H insertion shown here.

Despite their popularity in early studies, Rh catalysts have been primarily used in racemic methods, as chiral Rh catalysts do not give products in high *ee*. The low enantiopurity in these reactions is thought to result from a metal-free ylide intermediate. Elegant mechanistic work conducted by Yu[146] studied the enantioselectivity of the O–H insertion of water to form α-hydroxyl esters and found that Rh dissociation to form a free ylide intermediate was 6.5 kcal/mol more stable than the metal-associated ylide intermediate (Scheme 56B). This contrasts with the same reaction catalyzed by a chiral Cu complex, where the metal-associated ylide is favored by 10 kcal/mol (Scheme 56A). The utilization of chiral Cu complexes in the reactions described below ensures that the stereocenter in question can be formed in a chiral environment when the metal is associated to the ylide. As this is not the case in Rh-catalyzed reactions, meaningful levels of enantioselectivity have yet to be achieved. The following section provides an update on O–H insertion reactions hypothesized to proceed through an oxonium ylide pathway. Emphasis is placed on enantioselective methods, as well as mechanistic studies of these reactions. Early racemic work will not be covered, as these reports have been described in other reviews.[95]

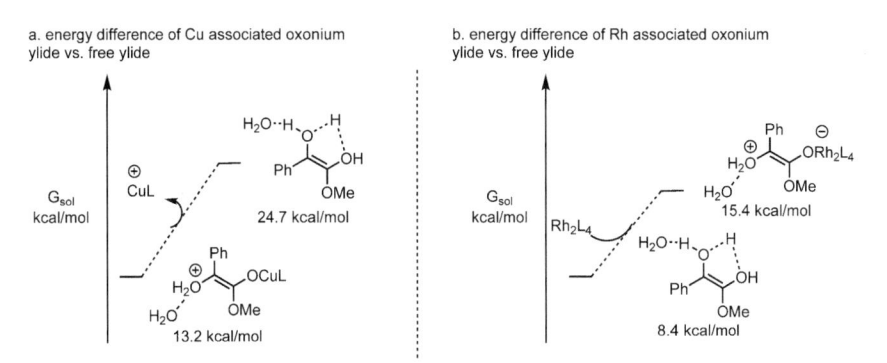

Scheme 56 (A) Energetic differences between a Cu-associated ylide *vs* free ylide. (B) Energetic differences between Rh-associated ylide *vs* a free ylide.

In 2006, the Fu group reported the first example of enantioselective O–H insertion with *ee* >8%.[147] Drawing on previously established conditions, a standard Cu/bisazaferrocene complex was applied to generate various α-hydroxy esters in excellent enantioselectivities (Scheme 57). Variations from standard conditions consistently decreased the overall yield and *ee*. Although an elevated *ee* (40%) was achieved with PyBOX ligands, all other Cu catalysts/ligand combinations, including chiral BOX, semicorrin, DUPHOS, and BINAP ligands furnished less than 9% *ee*. An essential discovery was that the addition of water (4%), in combination with the Cu/bisazaferrocene complex, enhanced the *ee* from 22% to 86%. The authors

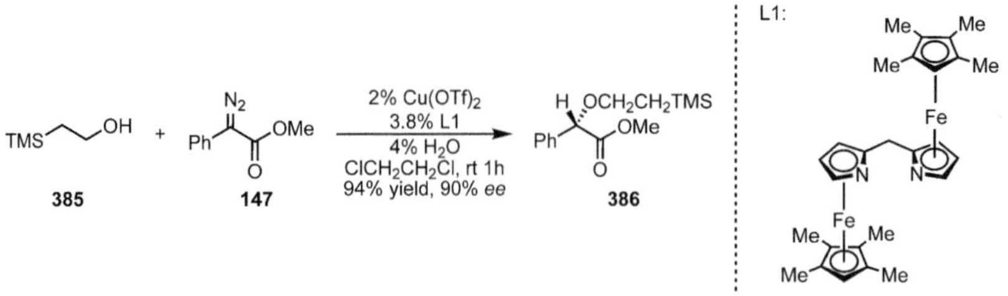

Scheme 57 Enantioselective O-H insertion of TMS-substituted alcohols to deliver α-hydroxy esters.

describe this as a serendipitous discovery, as it is not clear why water enhances the *ee*. A variety of alcohols were subjected to the reaction conditions, with **385** producing **386** in 94% yield and 90% *ee* (Scheme 57).

The Zhou group further developed a variety of inter/intramolecular O-H insertion reactions using Cu and Fe catalysts, reporting a 2007 phenol version of Fu's ester α-hydroxylation.[148] While Fu's Cu/bisazaferrocene catalyst gave only 11% *ee* with phenol, the Zhou group reported much better performance with a Cu/chiral spirobisoxazoline system also utilized in enantioselective N-H insertion reactions (Scheme 25). While other simple chiral BOX ligands gave moderate *ee*, the spirobox ligand delivered up to 99% *ee* (Scheme 58, entry 1). Similar to Fu's report, aliphatic alcohols gave lower *ee* (59–61% *ee*) (Scheme 58, entry 2). It is important to note that methyl substitution on the diazoester is essential for the reaction to proceed. When this group is altered to be larger, such as Ph, negligible *ee* is obtained (∼10%). This contrasts with the bulk of the ester substituent, which has almost no effect on the reaction outcome.

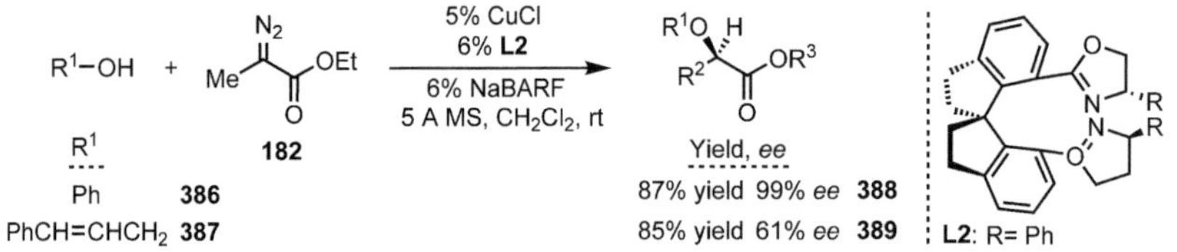

Scheme 58 Asymmetric O-H insertion of phenols to deliver enantiopure α-hydroxy esters.

The Cu/spirobox system was also applied to intramolecular O-H insertion in 2010[149] to furnish chiral 2-carboxy cyclic ethers such as **393–395**. Interestingly, the authors found that a much more sterically encumbered ligand was necessary to achieve high levels of enantioselectivity. The previously employed Ph-substituted spiroligand furnished **393** in 93% yield, but in only 53% *ee*; however, with a *t*Bu substituent, a 74% yield and 88% *ee* was achieved. The Cu source was also found to influence enantioselectivity, with CuOTf furnishing the highest *ee*. It was found that five-, six-, and seven-member rings were tolerated in this system, as well as fused rings, to give good yields and excellent enantioselectivity (82–97% *ee*) (Scheme 59, selected examples shown).

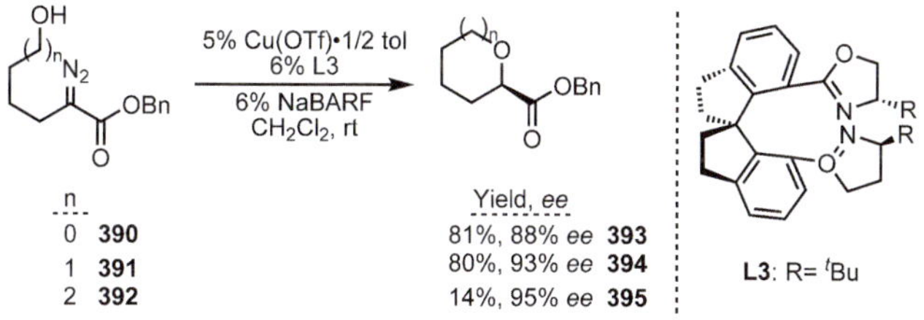

Scheme 59 Application of Cu/spirobisoxazoline system to the synthesis of enantiopure oxygen heterocycles.

The Zhou group further extended this method to include water as a substrate to furnish enantiopure α-hydroxy esters.[150a] This reaction was attempted by Landais,[150b] but only 8% *ee* was achieved using Rh catalysis, while Fu noted 15% *ee* using the Cu/bisazaferrocene system.[147] Zhou found utilizing the Cu/spirobox **L2** proved optimal to furnish **396** in 91% yield and 90% *ee* (Scheme 60A). Examination of the scope of the aromatic diazoester showed the sterics of the ester substituent had a slight effect on the reaction, with smaller ester substituents performing better. All *p*- and *m*-substituted aromatic moieties furnished α-hydroxylated products in excellent yield, with the exception of *o*-substituted aryl diazoacetates. This reaction was subsequently studied utilizing DFT calculations to

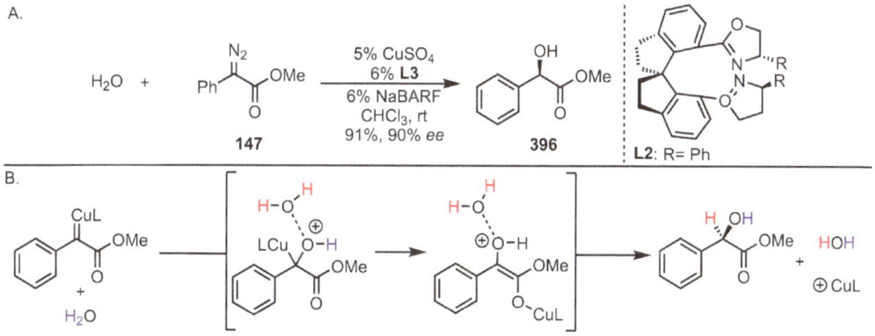

Scheme 60 (A) Application of the Cu/spirobisoxazoline system to water O-H insertion to deliver α-hydroxy esters. (B) Proposed water-assisted 1,2-H shift.

demonstrate that high enantioselectivity likely arises from a Cu-associated oxonium ylide. This intermediate was calculated to be ∼10 kcal/mol lower in energy than the free ylide. Interestingly, the calculated activation energy barrier for the [1,2]-H shift to furnish **396** was kinetically unfeasible at 10 °C, with an activation energy barrier of 32.7 kcal/mol. Based on this insight, the authors proposed that a water-catalyzed [1,2]-proton shift (only 19.8 kcal/mol) is the operative pathway (Scheme 60B).

Due to the high selectivity and specificity of these O-H insertion reactions, they have found applications in both total synthesis and chemical biology. In 2010,[151] the Yang group employed a racemic Rh mediated O-H insertion of **397** for the formation of a challenging seven-member cyclic ether **398** (Scheme 61) in the total synthesis of (±)-maoecrystal V. Cyclopropanation was outcompeted under these conditions to furnish **398** in 60% yield.

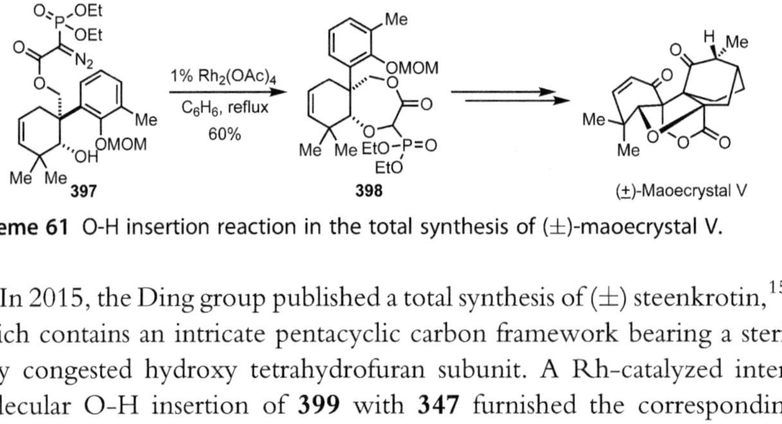

Scheme 61 O-H insertion reaction in the total synthesis of (±)-maoecrystal V.

In 2015, the Ding group published a total synthesis of (±) steenkrotin,[152] which contains an intricate pentacyclic carbon framework bearing a sterically congested hydroxy tetrahydrofuran subunit. A Rh-catalyzed intermolecular O-H insertion of **399** with **347** furnished the corresponding diester **400** in 73% yield (Scheme 62). This reaction enabled subsequent synthesis of the hydroxytetrahydrofuran moiety accessed *via* a key intramolecular carbonyl-ene cyclization to furnish **403**.

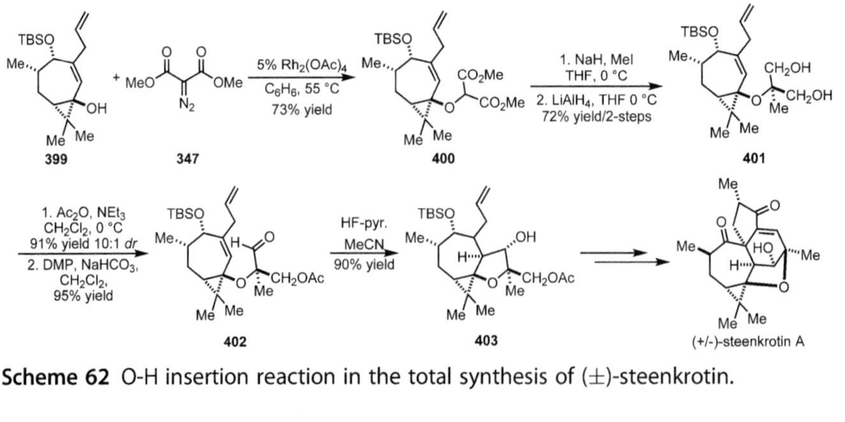

Scheme 62 O-H insertion reaction in the total synthesis of (±)-steenkrotin.

3.6 Electrophilic trapping

Protic oxonium ylides can be trapped by reaction of the carbanionic terminus of the ylide with an electrophile, followed by a delayed proton transfer (Fig. 5C). Recent advances in this area highlight the fact that the heightened reactivity of protic oxonium ylides does not prevent productive ylide formation and supports the ability of oxonium ylide reactivity to outcompete O–H insertion. Two reaction types will be discussed in this section, including intermolecular electrophile trapping, where the product arises from a novel multicomponent reaction (MCR). The progress of MCRs in oxonium ylide chemistry has been recently reviewed in 2013[36]; thus, the focus of this section will be to introduce the initial findings and highlight recent examples. The second topic involves intramolecular variants of this chemistry, where the protic oxonium ylide intercepts an internal electrophile to deliver a cyclic product.

3.7 Multicomponent reactions (intermolecular electrophile trapping)

The advent of MCRs in oxonium ylide chemistry began with a report of a novel Rh-catalyzed, three-component MCR to deliver aldol products.[153] The Hu group employed reaction conditions developed for an analogous reaction with ammonium ylides[154] to achieve a three-component reaction between **147**, **405** and **407** in 70% yield and modest dr (Scheme 63, entry 1).

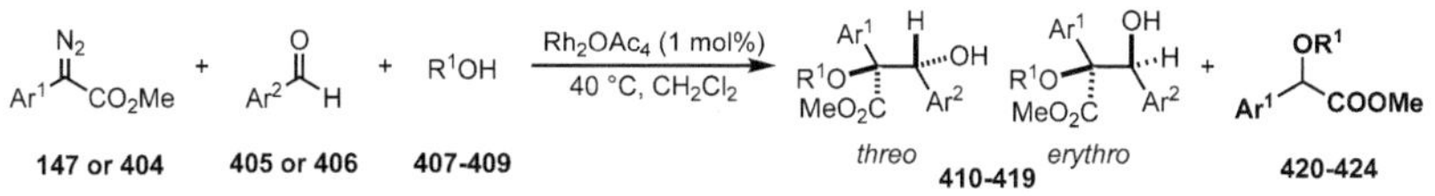

entry	Ar¹	Ar²	R¹	additives	yield (%) threo:erythro	3-MCR: OH insertion
1	Ph **147**	p-NO₂Ph **405**	Bn **407**	no additive	70% 1:1 dr **410-411**	7:1 **410-411: 420**
2	Ph **147**	p-NO₂Ph **405**	PMB **408**	no additive	83% 1.1.4 dr **412-413**	19:1 **412-413: 421**
3	Ph **147**	p-NO₂Ph **405**	Me **409**	no additive	48% 1:1 dr **414-415**	2:1 **414-415:422**
4	Ph **147**	p-MeOPh **406**	Bn **407**	no additive	0% **416-417**	–
5	Ph **147**	p-MeOPh **406**	Bn **407**	Ti(OⁱPr)₄ 1.1 equiv	56% 1:2 dr **416-417**	3:1 **416-417:423**
6	PMP **404**	p-NO₂Ph **405**	Bn **407**	no additive	87% 1:2 dr **418-419**	12:1 **418-419:424**

Scheme 63 Initial findings in the 3-MCR to yield a variety of hydroxy esters.

Control experiments confirmed formation of the MCR product does not occur *via* epoxide ring-opening or O–H insertion, supporting the hypothesis that it proceeds through a discreet ylide intermediate. The reactivity and chemoselectivity of this transformation were highly dependent on the electronics of the alcohol, carbene precursor and electrophile. Electron-rich alcohols and carbene precursors favored oxonium ylide formation and gave rise to higher yields (Scheme 27, entries 2–3, 6). The reaction was also dependent on the electronics of the electrophilic trap, with electron-deficient aldehydes such as **405** yielding best results. Electron-rich aldehydes, such as **406**, were unreactive (Scheme 63, entry 4), but proved successful in the presence of Lewis acidic titanium isopropoxide (Scheme 63 entry 5). This work was later expanded to encompass Cu-catalyzed[155] reactions of electron-rich aryl and aliphatic aldehydes without the need for an exogenous Lewis acid. In addition to aldehydes, imines were suitable electrophiles for the synthesis of β-amino-α-hydroxy esters. MCRs of oxonium ylides have been rendered diastereoselective[156–165]; however, success depends heavily on the identities of the alcohol and electrophile. These reactions have been recently covered in another review and are not discussed herein.[36]

More recently, the scope of oxonium ylide mediated MCRs was expanded to include asymmetric catalysis using a cooperative chiral Brønsted acid/Rh catalyst system.[166] Previous reports showing proton donors facilitate the addition of oxonium ylides to electrophiles[153,154] was leveraged to construct β-amino-α-hydroxy esters. Addition of a chiral phosphoric acid (CPA) formed an activated iminium species to promote an asymmetric Mannich-type reaction between an oxonium ylide and the iminium ion (Scheme 64B). **CPA5** was employed under previously reported $Rh_2(OAc)_4$ catalysis to promote reaction between **147**, **407**, and imine **425** to furnish the corresponding β-amino-α-hydroxy ester in 84% isolated yield, 81:19 *dr*, and 56% *ee* (Scheme 64A, entry 1). The stereoselectivity was improved utilizing bulky alcohols in combination with a 9-phenathryl derived BINOL phosphoric acid **CPA3** at low temperatures to give **428** in 86% yield, >99:1 *dr* and 93% *ee* (Scheme 64A, entry 3). Hu later applied this method to the synthesis of the β-amino-α-hydroxyl ester side chain of Taxol **433** (Scheme 64C).[167]

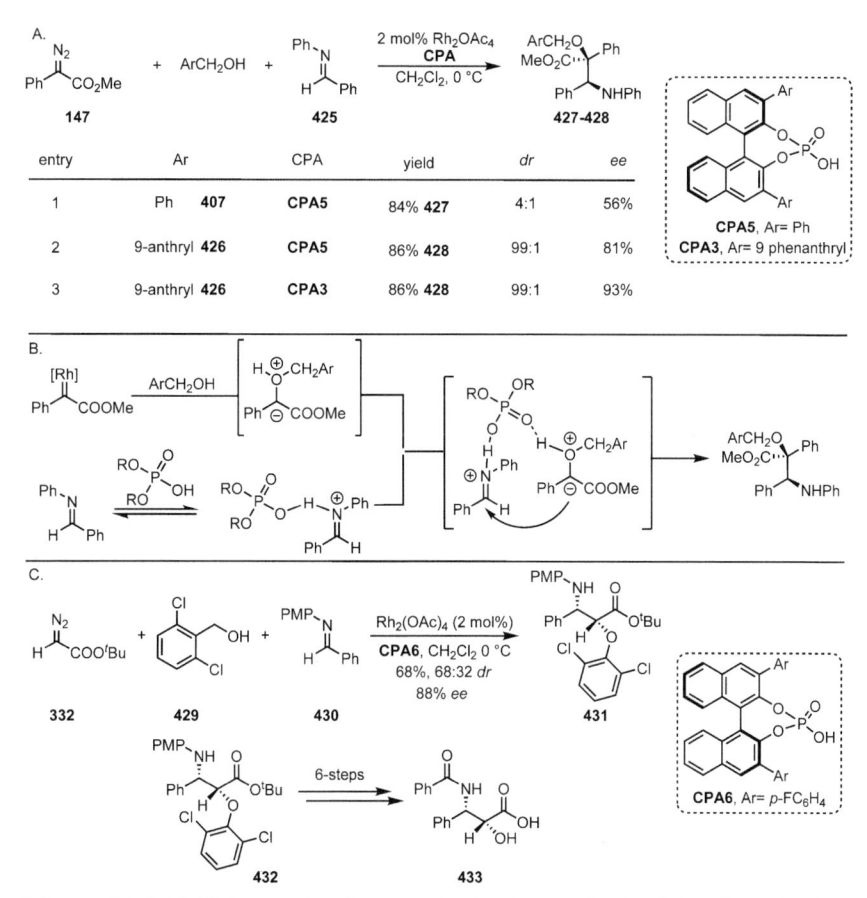

Scheme 64 (A) Initial screening for enantioselective synthesis of β-amino-α-hydroxy esters via Rh/CPA cooperative catalysis. (B) Proposed mechanism for the synthesis of β-amino-α-hydroxy esters. (C) Application of this method to the synthesis of the β-amino-α-hydroxyl ester side chain of Taxol.

This method was further expanded to construct vicinal quaternary centers using the same conditions developed in the initial asymmetric MCR report.[168] Reaction of 3-hydroxyisoindolinones, aryl diazoacetates, and alcohols in the presence of Rh/**CPA3** catalyst generates enantiopure isoindolinone derivatives through a formal SN1 type pathway in moderate dr and excellent ee (Scheme 65). Notably, this transformation was also accomplished using Pd catalysis, albeit in lower yields. The electron-rich

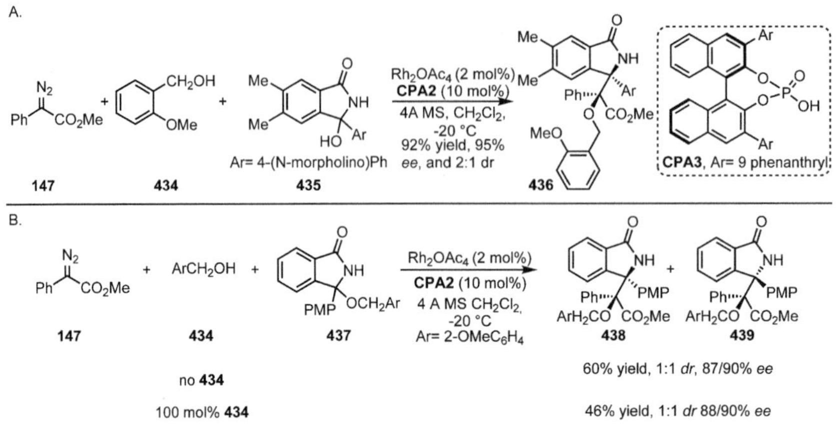

Scheme 65 (A) Optimization of an asymmetric 3-MCR to yield enantiopure isoindolinone derivatives. (B) Control experiments revealing that although the reaction can proceed in a bimolecular fashion.

alcohol **434** in combination with isoindolinone **435** furnished the best yield of 92% in 95% ee and 2:1 dr (Scheme 65A). Interestingly, while control experiments revealed the reaction proceeds in a higher overall yield, dr, and ee, the presence of exogenous alcohol is not required. The asymmetric reaction still proceeds in a bimolecular fashion to furnish the corresponding isoindolinone **438/439** in 60% yield, 1:1 dr, and 87% ee (major syn diastereomer) (Scheme 65B).

Hu's oxonium ylide MCRs have been applied to the selective modification of the C-40 -OH of rapamycin.[169] Modification of this group regulates the signaling strength of the mTORC1 and mTORC2 pathways. Previous attempts to generate selective C-40 -OH analogs were marginally successful, as the similarly reactive C28 -OH group was competitive. Huang applied Hu's chemistry to a Rh-catalyzed, selective introduction of isatins to the C-40 -OH group in a MCR using aryl diazoacetates. The reaction was tested with **147**, various isatins, and rapamycin to exclusively deliver C-40 O-H functionalization without the need for hydroxyl protecting groups (Scheme 66A). Evaluation of the scope of the MCR showed electron-donating diazoacetates were tolerated, as well as isatins bearing substitution in the 6 and 7 positions yielding products such as **441** (Scheme 66A). N-substitution of the isatin was also tolerated, enabling the synthesis of MCR products with functionalized nitrogen functionality, such as **442**. The biological activity of these analogs against A549, HeLa, and SKBR3 cell lines was assessed; several derivatives maintained their bioactivity with reduced cytotoxicity (Scheme 66B). This application demonstrates

Scheme 66 (A) Representative example of a Rh-catalyzed MCR to selectively deliver rapamycin C-40 -OH analogs. (B) Bioactivity profiles of derivatives **441** and **442**.

the utility of oxonium ylide MCR chemistry for the mild and selective late-stage functionalization of complex molecules.

3.8 Intramolecular electrophile trapping

In 2019, Wang published a unique application for Cu-catalyzed synthesis of enantioenriched 2,3-dihydrobenzofuranes.[170] The Cu catalyst serves dual roles to generate the metal-supported carbene and as a Lewis acid to promote intramolecular electrophile trapping. Wang found that Cu/BOX catalysts and bulky organic bases, such as DIPEA, promoted oxonium ylide formation and cyclization to yield 2,3-dihydrobenzofurans **444/446** from the corresponding 2-acyl or 2-iminyl phenols **443/445** in excellent yield, dr, and ee (Scheme 67A). The organic base is believed to act as a proton shuttle

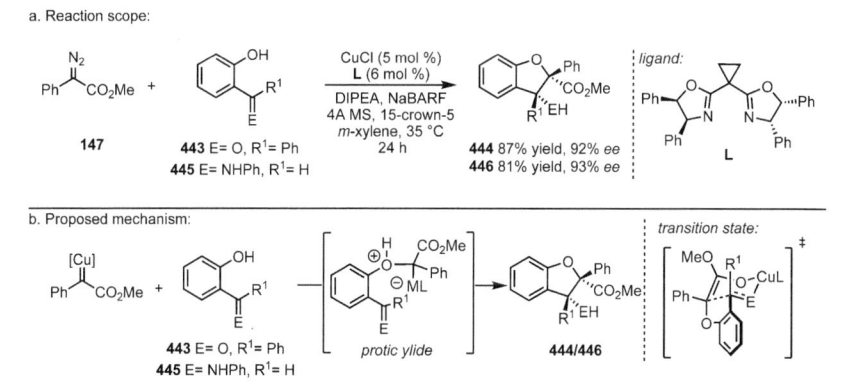

Scheme 67 (A) Representative examples of the enantioselective synthesis of 2,3-dihydrobenzofurans. (B) Proposed mechanism for the enantioselective synthesis of 2,3-dihydrobenzofuranes.

with no competitive binding to the Cu catalyst. The scope of this reaction was independent of steric and electronic modifications to both the aryl diazoacetate and phenol substrates, yielding products as a single diastereomer. The high degree of selectivity was attributed to a chair-like transition state model, where the Cu catalyst coordinates to both the oxygen atom of the enolate and the nitrogen (or oxygen) of the corresponding electrophile to adopt a conformation that minimizes steric repulsions (Scheme 67B, right). The authors ruled out a potential pathway involving intermediate O-H insertion by independent synthesis of the O-H insertion product and exposure to the reaction conditions. No 2,3-dihydrobenzofuran was obtained, further supporting the intermediacy of a protic oxonium ylide (Scheme 67B, left). Ultimately, this method serves as a unique way to synthesis 2,3-dihydrofurans bearing a quaternary center in both a stereo- and enantioselective fashion.

In 2020, the Hu group reported the enantioselective synthesis of spirochroman-3,3-oxindoles *via* asymmetric Michael addition of *o*-hydroxylmethyl chalcones and diazoindolinones under cooperative Rh and CPA catalysis (Scheme 68).[171] Rh-based catalysts were superior, as Cu catalysts

Scheme 68 (A) Selected scope in intramolecular asymmetric Michael addition of *o*-hydroxylmethyl chalcones and diazoindolinones or aryl diazoacetates under cooperative Rh and CPA catalysis. (B) A proposed mechanism for the asymmetric Michael addition.

yielded only isobenzofurans from direct intramolecular oxo-Michael addition and no incorporation of the diazoindolinone. Screening of a series of CPAs revealed a bulky triphenylsilyl-bearing **CPA1** gave up to 96% *ee*. The scope of this reaction was robust, tolerating both electronic and steric modifications to the aromatic group and the ketone of the Michael acceptor. The diazoindolinone scope was more limited; however, this was circumvented by extending the scope of the aryl diazoacetates to give isochromans (selected examples, Scheme 68A). Exposing independently synthesized O-H insertion products to the reaction conditions yielded no desired product; the same results was observed with a pre-synthesized isobenzofuran, suggesting these derivatives are not reaction intermediates. Based on this insight, a mechanism was proposed where a *o*-hydroxylmethyl chalcone reacts with a Rh-supported carbene to furnish an oxonium ylide that undergoes intramolecular Michael addition mediated by **CPA1** (Scheme 68B). Ultimately, this report demonstrates the first example of catalytic asymmetric intramolecular Michael-type trapping of an oxonium ylide enabled by the Rh/CPA dual H-bonding activation model to yield chiral spirochroman-3,3-oxindoles.

3.9 Concluding remarks

The reactions presented herein represent an overview of historical and recent developments in the chemistry of oxonium ylides and their applications in synthesis. A better understanding of the mechanistic underpinnings of reactions involving these unusual intermediates has led to great strides in the development of highly diastereo- and enantioselective methodologies. These chemistries have enabled an expanded exploration of new chemical space and late-stage functionalization of complex molecules. Future work is likely to focus on improving the generality of these methods, with an eye toward other metals, ligands, and precursors that enable highly enantioselective reactions.

4. Sulfur ylides
4.1 Introduction to sulfur ylides and their reactivity

Sulfur ylides are zwitterionic compounds in which a carbanion is flanked by a vicinal, positively charged sulfur atom. The utility of these versatile reagents has been demonstrated in a number of classical transformations. The most representative examples in the literature are rearrangement reactions ([2,3]-sigmatropic or [1,2]-Stevens), insertion into C–H and X–H bonds, and the synthesis of small heterocyclic ring systems (epoxidation, aziridination and cyclopropanation reactions) through the addition of sulfur ylides to electron-poor π systems. A number of review articles and book chapters have been published in which examples of these transformations have been highlighted.[1,2,34,35,95,172–179]

It has been reported that the preparation of sulfonium ylides is easier than the analogous oxonium or ammonium ylide formations, due to the enhanced stabilization imparted by the vacant d-orbitals of sulfur atoms.[177,180] The adjacent negative charge can be stabilized by the sulfur atom through its vacant low energy 3d-orbitals and participation in back-bonding. Sulfur-based ylides are commonly accessed through the deprotonation of sulfonium or sulfoxonium salts.[2,177,178,180] However, this approach is hampered by limitations that include functional group incompatibility with basic conditions, the potential for dealkylative side reactions with the electrophilic sulfonium, and regioselectivity issues arising from deprotonation of salts that contain more than one acidic site. In recent years, the application of sulfur ylides in organic synthesis has been further expanded by exploring the reactivity of the ylides in combination with transition-metal catalysis.[1,2,34,35,95,172–179] The reaction of sulfur nucleophiles with metal-supported carbenes, easily accessed *via* transition metal-catalyzed decomposition of diazo compounds, has proven to be a highly efficient approach for the formation of sulfur ylides. Methods to access sulfur-containing heterocycles are of particular interest, as their structure is pervasive in the scaffolds of pharmacologically active ingredients and natural products.[181]

4.2 Generation of sulfur ylides from metal carbenes

In contrast to indirect base-promoted methods for the generation of sulfur ylides, the facile addition of a sulfur nucleophile to an electrophilic metal carbene complex allows for the direct formation of these reactive intermediates (Fig. 6). In addition, the reaction proceeds under neutral conditions and tolerates a wide range of functional groups. These advantages have resulted in great attention being devoted to studies of sulfonium ylides derived from metal-supported carbenes. Different sulfur nucleophiles can be employed including sulfides or thiols, sulfoxides, and thiocarbonyls, which form sulfonium, sulfoxonium and thiocarbonyl ylides, respectively. Typically, the metal carbenes are generated from the transition metal-catalyzed decomposition of α-diazocarbonyl compounds. Though copper catalysis is still occasionally employed, rhodium (II) acetate has emerged as the preferred catalyst for diazo decomposition. However, partial catalyst poisoning may occur due to

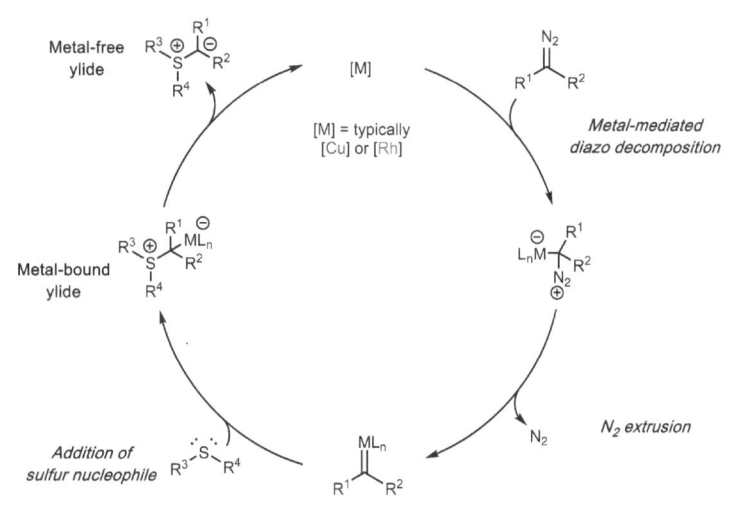

Fig. 6 General formation of a sulfonium ylide *via* reaction between a sulfur nucleophile with a metal carbenoid species.

coordination of the sulfide to the rhodium metal center, thus requiring ylide formation to be conducted at elevated reaction temperatures.[43] Other metals, including ruthenium,[4,5] iron,[15–17] palladium,[20] and silver[24] have also been explored. To highlight the versatility of sulfur ylides for applications in synthetic organic chemistry, representative examples from the previous two decades of transformations involving metal carbene-derived sulfonium ylides will be discussed in the remainder of this section.

The electronic nature of the substituents adjacent to the carbanion affect the stability of the formed ylides, as well as their reactivity and subsequent transformations.[177,180] For example, ylide **451** (Fig. 7A) can undergo either a [1,2]-shift (Stevens rearrangement) or α′,β-elimination, depending on the presence or absence of a β-hydrogen on the substituents attached to the sulfur atom. In the case of allyl (or propargyl) sulfide nucleophiles (Fig. 7B), the resulting sulfonium ylide **452** favors [2,3]-rearrangement to give **453**; in contrast, thiol nucleophiles (Fig. 7C), generate sulfonium ylides **454** that are prone to S–H insertion. Isolable sulfonium ylides can be obtained if they are derived from a carbenoid species substituted with two electron-withdrawing groups; these stable ylides typically rearrange only when exposed to harsh, forcing conditions.

4.3 [2,3]-Sigmatropic rearrangements

The initial discovery of the [2,3]-sigmatropic rearrangement reaction of allyl sulfides with diazoalkanes was reported by Kirmse and Kapps in 1968 with

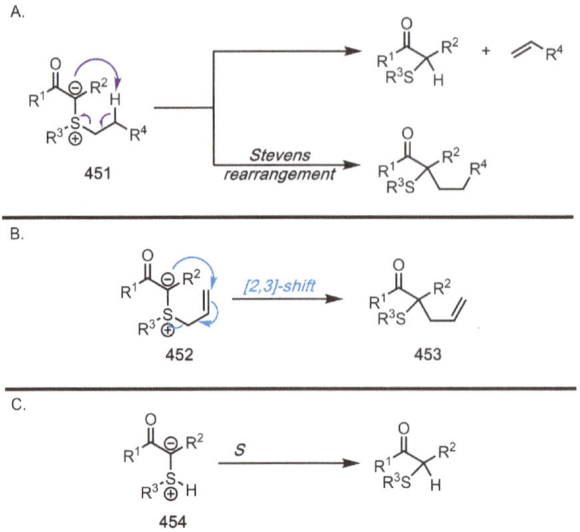

Fig. 7 Possible transformations of sulfonium ylides generated from metallocarbenes.

their seminal work on the copper-catalyzed reaction of an allylic thiol with diazomethane.[182] This transformation was largely ignored, only to reemerge following a study on ylide formation from reactions of allyl amines and sulfides with various rhodium(II) carboxylates described by Doyle and coworkers in 1981.[43,183] The transformation, now known as the Doyle–Kirmse reaction, has since been widely employed to access functionalized sulfur-containing compounds in organic synthesis.

4.4 Intermolecular [2,3]-sigmatropic rearrangements

Despite reports[184] of chirality transfer from sulfur to carbon through [2,3]-sigmatropic rearrangements that proceed with high selectivity, attempts to develop an asymmetric Doyle–Kirmse transformation have been hampered by poor enantioselectivity.[32,185–191] These limitations have been attributed to a shortage in available chiral catalysts or suitable substrates. In 2005, Wang and coworkers attempted to overcome these limitations by employing diazo compounds tethered with Oppolzer's camphor sultam chiral auxiliary.[192] The respective alcohol products derived from alkyl, alkenyl, and aryl-substituted diazo compounds were obtained in high yields with good enantioselectivities following removal of the chiral auxiliary (Scheme 69A). It was ultimately determined that asymmetric induction occurred due to the chiral auxiliary, not the ligand-bound catalyst; exposing the reaction components to either enantiomer of the chiral salen ligand **456**

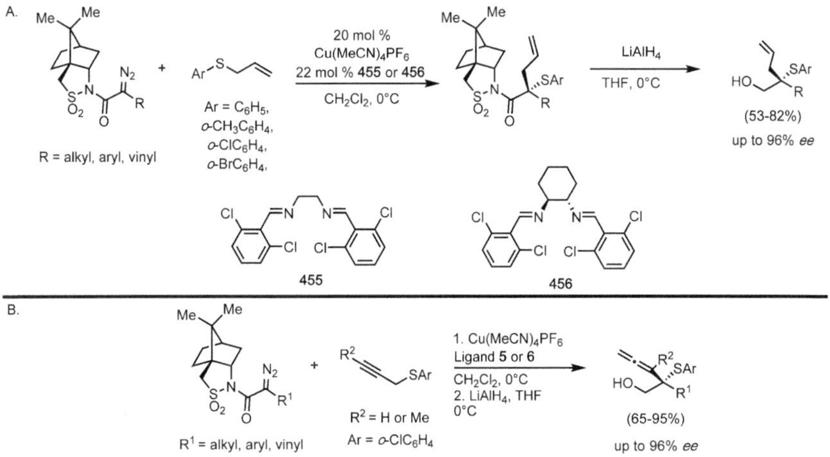

Scheme 69 Wang's asymmetric auxiliary-controlled Doyle-Kirmse reactions of (A) allylic and (B) propargyl sulfides.

gave the same enantiomer of the alcohol product in approximately 80% *ee*. The double asymmetric induction approach was further extended to generate allenyl alcohols in good yields and with high levels of enantioselectivity using a range of chiral sultam-diazo compounds and propargyl sulfides (Scheme 69B).

Later in 2017, the Wang group shared another investigation into the development of an enantioselective Doyle-Kirmse reaction.[193] In this study, high levels of enantioselectivity were achieved without the use of a chiral auxiliary in the synthesis of trifluoromethylthio-substituted compounds (Scheme 70A). Nucleophilic addition of the sulfide to the rhodium-carbene intermediate was identified as the stereodetermining step, while facial differentiation in the allyl trifluoromethylsulfides due to differences in the steric and electronic properties of the substituents was proposed as the reason for the high degree of stereocontrol. The steric and electronic properties of the chiral Rh(II)-catalyst were also considered, although it is still not clear whether a metal-bound ylide or a free ylide is involved in the rearrangement step of the Doyle-Kirmse reaction.[32,185–191] Control experiments were conducted to evaluate the hypothesis generated from previous studies that the rearrangement most likely proceeds through a free ylide. In the first study (Scheme 70B), it was observed that the diastereoselectivities of the C–C bond forming step were essentially independent of the nature of the rhodium catalyst used. Further support for the free ylide mechanism was obtained in the second study (Scheme 70C), as the product **457** was

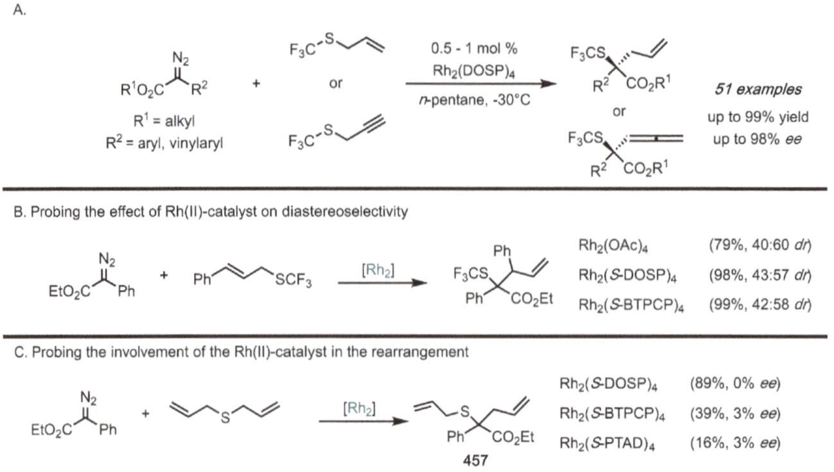

A.

Scheme 70 (A) General scheme for Wang's enantioselective Doyle–Kirmse reaction of sulfonium ylides derived from allyl or propargyl trifluoromethyl sulfides. (B and C) Control experiments evaluating the free ylide hypothesis in the [2,3]-sigmatropic rearrangement mechanism.

essentially racemic, no matter which chiral rhodium catalyst was employed; if the rearrangement did involve a metal-bound ylide, the resulting products would be expected to display considerable levels of enantioselectivity.

If the π-system involved in the [2,3]-sigmatropic rearrangement is in an aromatic ring, the transformation is termed a Sommelet-Hauser rearrangement.[194–196] Applications of catalytic thia-Sommelet-Hauser rearrangements have been underexplored in comparison to the traditional Doyle-Kirmse reaction. In 2008, Wang reported a Rh-catalyzed thia-Sommelet-Hauser reaction between aryl diazoacetates and aryl sulfides.[197] An aryl sulfide **458** and a diazo compound **459** initially generate an ylide species **460** that undergoes a subsequent proton transfer to form sulfur ylide species **461** (Scheme 71). The second ylide intermediate, which is stabilized by an additional electron withdrawing group, undergoes a [2,3]-sigmatropic rearrangement and a [1,3]-shift rearomatization to give the *ortho*-substituted product **462**. Despite the dearomatizative [2,3]-sigmatropic shift and [1,3]-shift rearomatization, the overall transformation proceeds smoothly at room temperature with a selective *ortho*-substitution and allows access to di- and trisubstituted arenes.

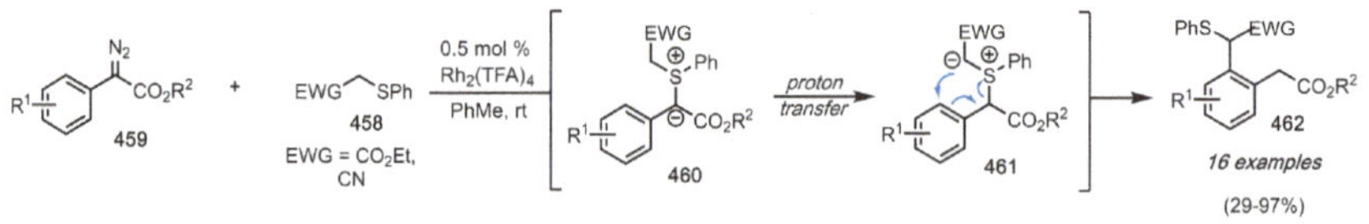

Scheme 71 Wang's Rh-catalyzed thia-Sommelet–Hauser rearrangement of aryl sulfides.

Inspired by Gassman's work[198] on the base-promoted thia-Sommelet-Hauser rearrangement of sulfonium ylides, Wang and coworkers pursued the development of a catalytic variation of this transformation.[199] It was originally proposed this might be achieved by accessing the sulfonium ylide intermediate **463** through a metal-catalyzed reaction between a sulfenamide and a carbene precursor. The resulting 3-arylthio-1,3-disubstituted oxindole products **464** were readily synthesized from the Rh$_2$(OAc)$_4$-catalyzed reaction between sulfenamides **465** and diazoacetates **466** (Scheme 72). The catalytic thia-Sommelet-Hauser rearrangement allows for the efficient introduction of substituents to the *ortho* position of aryl acetates. The products are generated in moderate-to-good yields under neutral conditions at low temperature. Attempts to accomplish an enantioselective transformation were unsuccessful, as only racemic products resulted when a chiral Rh(II)-catalyst was used. Overall, this transformation demonstrates the efficacy of a metal carbene-mediated thia-Sommelet-Hauser rearrangements for the synthesis of oxindole derivatives.

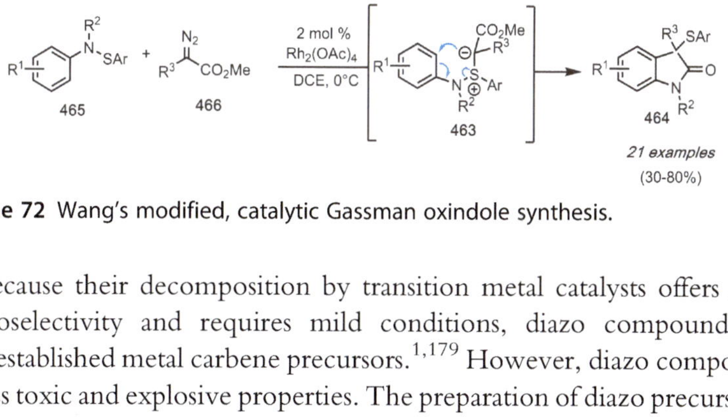

Scheme 72 Wang's modified, catalytic Gassman oxindole synthesis.

Because their decomposition by transition metal catalysts offers good chemoselectivity and requires mild conditions, diazo compounds are long-established metal carbene precursors.[1,179] However, diazo compounds possess toxic and explosive properties. The preparation of diazo precursors is limited to stabilized compounds that bear one or two electron-withdrawing groups. Alternatives to the direct use of diazo compounds as carbene precursors have been developed, including the *in-situ* generation of diazo intermediates or the formation of metal carbenes from unsaturated C–C bonds and π-acid complexes.[13,200–202] Alternative approaches to forming metal carbenes that do not require the intermediacy of diazo compounds have also been developed.[203–208]

In pursuit of identifying diazo surrogates for the formation of allyl sulfonium ylides, Davies and Albrecht identified an alternative method of generating carbenes utilizing alkynes as ylide precursors.[209] Inspiration came from the findings of Toste[210] and Zhang[211] showing that α-keto gold carbenes could be generated from reaction between a sulfoxide group and a tethered alkyne. The carbene and sulfide components required for ylide formation arise from an internal redox process. The scope of dihydrothiophenones and dihydrothiopyranones that could be accessed through a sequential oxidative annulation and [2,3]-sigmatropic rearrangement of terminal or internal allyl alkynyl sulfoxides was explored (Scheme 73A). The proposed catalytic cycle is initiated by π-acid activation of the alkyne and nucleophilic attack of the sulfoxide on the alkyne, which ultimately generates metal carbene **467** (Scheme 73B). Formation of the sulfur ylide **468** occurs following intramolecular trapping of the metal mediated carbene by the tethered sulfide. The ylide then undergoes [2,3]-sigmatropic rearrangement to furnish the cyclic sulfide product **469**. The desired products were furnished in modest yields for substrates containing an internal alkyne that were subjected to platinum(II) catalysis. Improved yields for these substrates were later achieved upon switching to a dichloro(pyridine-2-carboxylate)gold(III) catalyst **470** system.

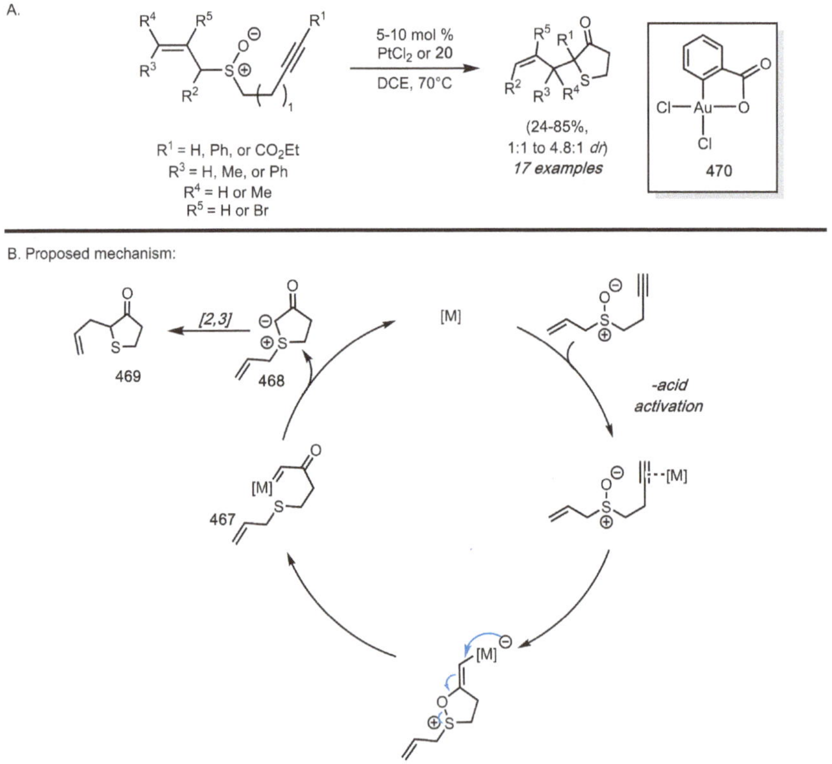

Scheme 73 (A) General scheme of Doyle and Albrecht's synthesis of sulfur heterocycles via sulfonium ylides derived from alkyne precursors. (B) Proposed mechanism of the intramolecular redox reaction, ylide formation, and 2,3-sigmatropic rearrangement.

4.5 1,2-Stevens (thia-Stevens) rearrangement

Stevens rearrangements involve 1,2-migrations of alkyl substituents from the cationic sulfur atom to the anionic site.[176] Traditionally, a Stevens rearrangement refers to a base-promoted transformation of a sulfonium or quaternary ammonium salt to a sulfide or tertiary amine, followed by the 1,2-migration of an alkyl group from the central nitrogen or sulfur atom. Though underexplored in comparison with its oxonium and ammonium ylide counterparts, the equivalent sulfur transformation, referred to as a thia-Stevens rearrangement, has become an excellent tool for the formation of C–C bonds and quaternary centers.

4.6 Intermolecular 1,2-Stevens (thia-Stevens) rearrangements

In 2005, Porter reported a Cu-catalyzed ring expansion of substituted 1,3-oxathiolanes to 3-triethylsilyl-1,4-oxathiane-3-carboxylates.[212] It was initially postulated that 1,4-oxathianes **471** could be accessed through a [1,2]-Stevens rearrangement of sulfur ylides derived from 1,3-oxathiolanes **472** treated with ethyl diazoacetate **473**. However, poor yields were obtained with ethyl diazoacetate, attributed to its tendency to form dimeric byproducts, as well as the nonselective nature of the metal carbenoid (Scheme 74A). Complete conversion and higher yields were achieved by switching to sterically bulkier silylated diazoacetates **474** and **475**

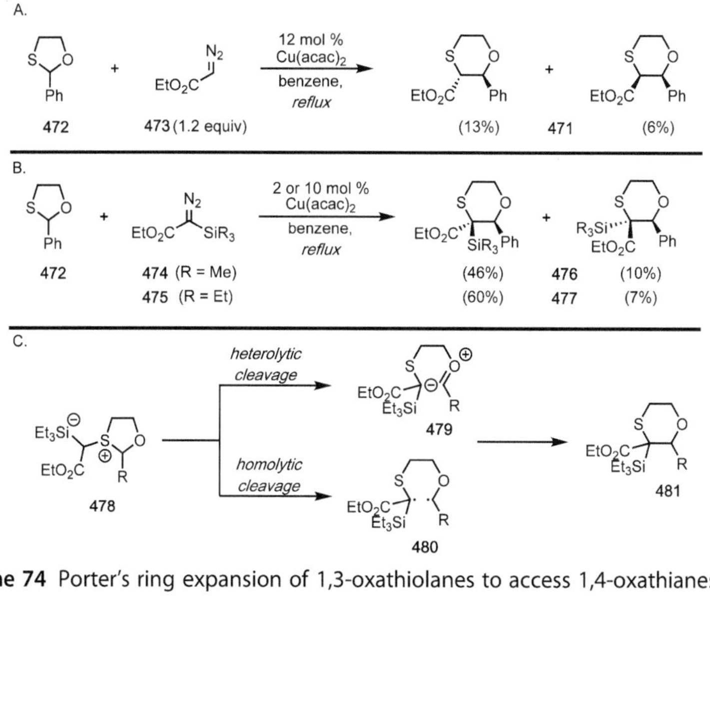

Scheme 74 Porter's ring expansion of 1,3-oxathiolanes to access 1,4-oxathianes.

(Scheme 74B). According to the proposed mechanism (Scheme 74C), the ring expansion proceeds *via* the formation of a sulfur ylide intermediate **478**, followed by the breaking of the benzylic C–S bond. The bond breakage may occur either heterolytically to give zwitterionic **479** or homolytically to give diradical intermediate **480**. The ring expansion process is completed following formation of the C–C bond, which gives the 1,4-oxathiane product **481**. The overall transformation involves insertion of CH–CO$_2$Et into the C–S bond to accomplish the one carbon ring expansion.

Zhu and coworkers demonstrated the installation of a trifluoromethyl moiety into a 1,4-oxathiane scaffold *via* a Rh$_2$(OAc)$_4$-catalyzed reaction between 1,3-oxathiolanes and trifluoromethyl diazoacetate (Scheme 75).[213] The trifluoromethyl-containing 1,4-oxathiane products were obtained in high yields and diastereoselectivity. The 1,3-oxathiolanes **483** substituted with electron-withdrawing groups gave products **487** in higher yields and lower diastereomeric ratio than those containing electron-donating groups (**484**, **488**). As spirocyclic structures are often found in the skeletons of biologically active natural and synthetic molecules,[214] attempts were made to introduce the trifluoromethyl functionality into spirocyclic 1,3-oxathiolanes **489** by using a trifluoromethyl-substituted diazoacetate **490**. Using the same reaction conditions as with the monocyclic studies, the desired spiro-1,4-oxathiolane product **491** was obtained in moderate yield.

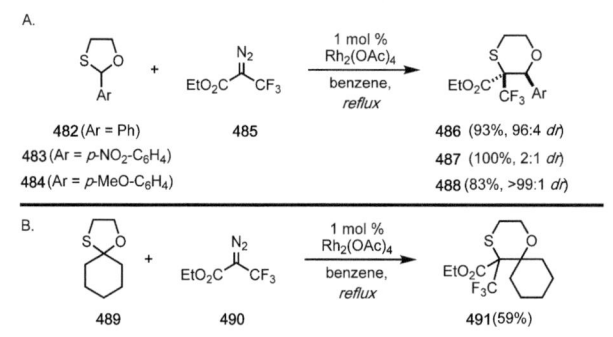

Scheme 75 (A) Selected examples of Zhu's ring expansion of 1,3-oxathiolanes to access 1,4-oxathianes and install trifluoromethyl functionality. (B) Attempted introduction of the trifluoromethyl functionality into spiro-1,3-oxathiolanes.

Kostikov and coworkers contributed their findings regarding 1,3-dithiolane ring expansions in a 2006 study.[215] The reaction between 2-phenyl-1,3-dithiolane **492** with methyl diazoacetate **493** (Scheme 76A) in the presence of catalytic Rh$_2$(OAc)$_4$ gave diastereomeric products **494** and **495** in

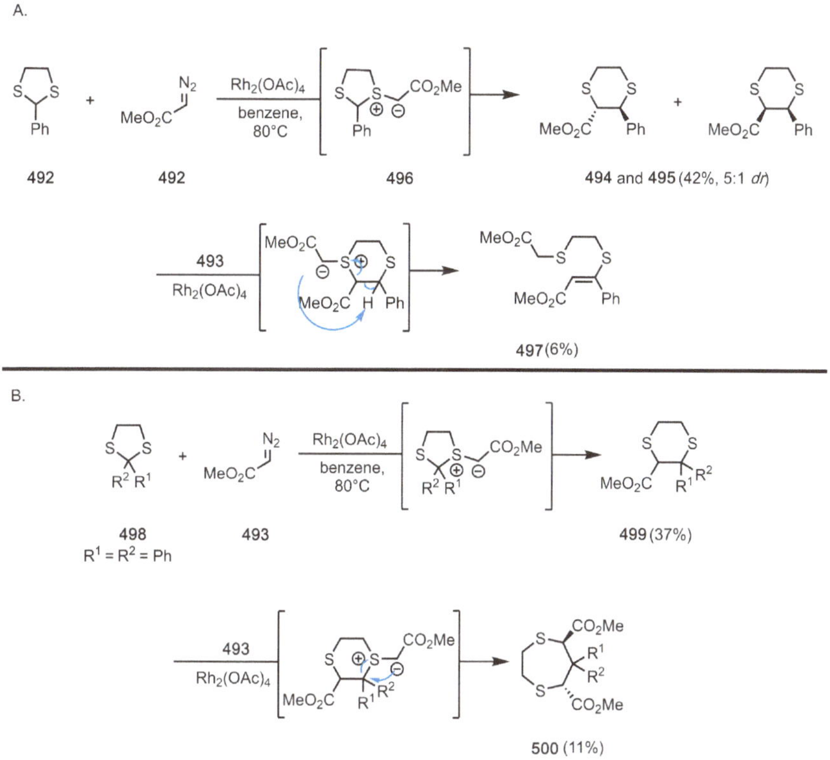

Scheme 76 (A) Kostikov's attempt at ring expansion of 1,3-dithiolane **492** and subsequent fragmentation by elimination. (B) Disubstituted dithiolane ring expansion by double insertion of methoxycarbonylcarbene.

moderate yield and selectivity (42%, 5:1 *dr*). The low yield was attributed to an additional insertion reaction between the sulfur atom of the products and the rhodium carbene, effectively giving sulfur ylide **496** which undergoes elimination to form vinyl sulfide **497**. The same double insertion phenomenon was observed with disubstituted dithiolane **498** (Scheme 76B), which initially gave 1,4-dithiane **499** in low yield. The subsequent insertion reaction between the dithiane and the rhodium carbene gave the 1,4-dithiepane **500** in a low yield.

The Diver group later exploited the double insertion phenomenon of disulfides reacting with carbene intermediates to their advantage. In 2006, they reported a double thia-Stevens rearrangement of a sulfur ylide formed from the reaction between disulfide **501** and diethyl diazomalonate **502** in the presence of catalytic $Rh_2(OAc)_4$ (Scheme 77).[216] Isolation of the ylide **503** is possible at low temperatures (in refluxing 1,2-dichloroethane or

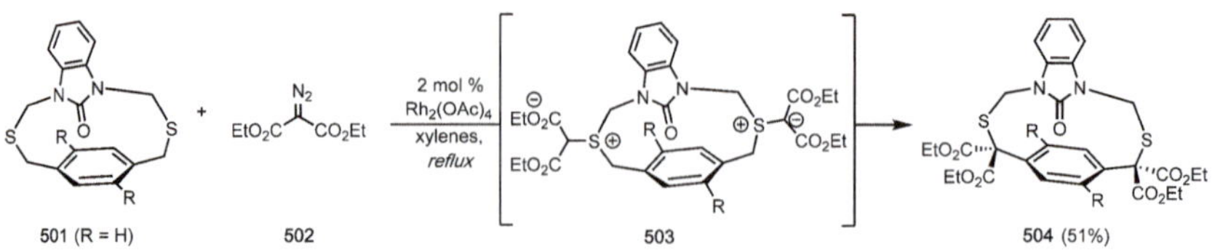

Scheme 77 Diver's macrocyclic ring expansion by a double thia-Stevens rearrangement.

benzene); however, it was observed that subjecting the isolated ylide to the reaction conditions in refluxing xylenes promoted a ready 1,2-sigmatropic shift to give the macrocyclic ring expansion product **504** in 51% yield.

The first known example of a Stevens rearrangement using spirocyclic thioketals was a 2013 report by Muthusamy and coworkers illustrating the synthesis of dispiro[1,4-dithianes/dithiepanes]bisoxindoles.[217] Sulfonium ylides **505** were formed following the rhodium(II)-catalyzed decomposition of cyclic diazoamides **506** in the presence of spiro-1,3-dithiolaneoxindole **507** or spiro-1,3-dithianeoxindole **508** (Scheme 78). A subsequent Stevens rearrangement and formation of a new carbon–carbon bond affords the desired dispiro-bisoxindole products **509** in high yields and with high diastereoselectivity.

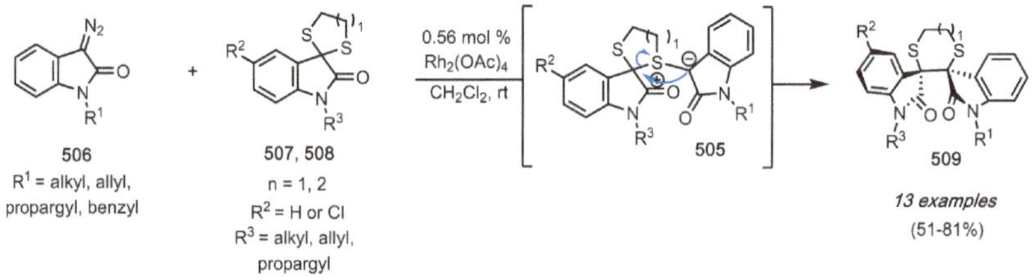

Scheme 78 Muthusamy's intermolecular spirothioketal-derived sulfonium ylide formation toward the synthesis of dispiro[1,4-dithianes/dithiepanes]bisoxindoles.

Asymmetric [1,2]-Stevens rearrangements of sulfur ylides have not been as extensively studied as enantioselective [2,3]-sigmatropic rearrangements of sulfur ylides. The first asymmetric [1,2]-Stevens transformation was reported in a study by Tang in 2009 in which the formation of 1,4-oxathianes from diazomalonates and racemic 1,3-oxathiolanes was achieved *via* a domino sulfonium ylide formation/asymmetric Stevens rearrangement sequence.[218] It was postulated that a chiral sulfur ylide could be accessed following the reaction of diazomalonate **510** with a racemic 1,3-oxathiolane **511** in the presence of a Cu(I)/chiral ligand complex; the resulting chiral sulfur ylide would then undergo a Stevens rearrangement to yield an optically active 1,4-oxathiane. Initial studies in which no ligand was present were unsuccessful in securing the desired product, even after heating the reaction to 40 °C. Formation of the desired product via [1,2]-Stevens rearrangement was later accomplished with good yield and moderate enantioselectivity by employing a CuOTf/bisoxazoline[219] complex. Reaction optimization found that a Cu complex using chiral ligand **512** gave the 1,4-oxathiane products **513** in good yields and enantioselectivities. General trends affecting the yields and enantioselectivities of the 1,4-oxathiane products were identified based on the R^2 substituent of the 1,3-oxathiolane (Scheme 79A). An electron-donating or an electron-withdrawing substituted phenyl group at R^2 gave the desired products **513b–d** in high yields. In considering enantioselectivity trends, 1,3-oxathiolanes with aryl substituents bearing electron-withdrawing groups **511b–d** gave products that displayed higher *ee* than those with electron-donating groups **511e**. Other bisoxazoline ligands were screened to improve the enantioselectivity of the transformation, including ligand **514** which gave comparable yields and *ee* to the optimized conditions. Interestingly,

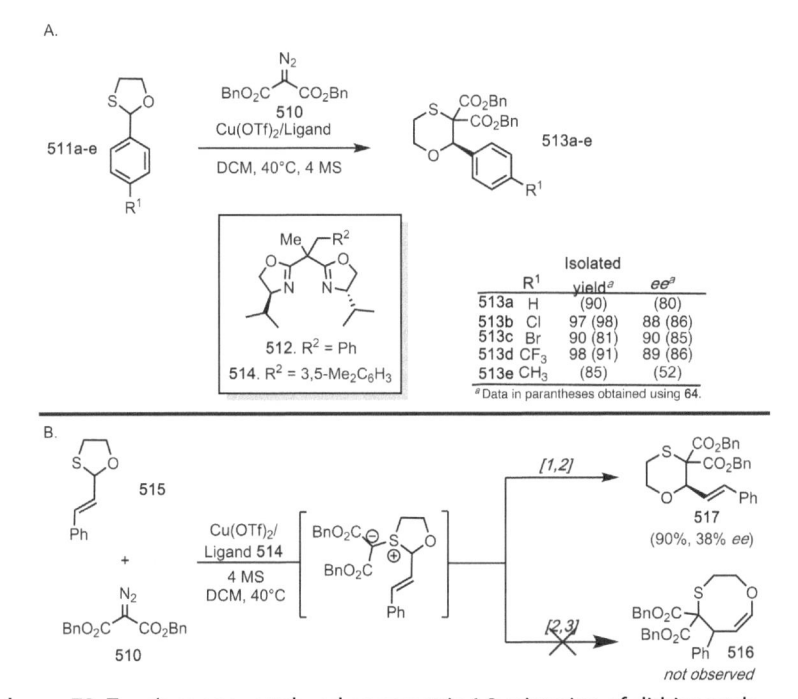

Scheme 79 Tang's copper-catalyzed asymmetric 1,2-migration of dithioacetals.

though allylic substrate **515** also had the potential to undergo an [2,3]-sigmatropic rearrangement to give **516**, it gave only the Stevens rearrangement product **517** (Scheme 79B).

4.7 Intramolecular 1,2-Stevens rearrangements

In a 2008 study toward the synthesis of tagetitoxin, the Porter group reported the application of an intramolecular Stevens rearrangement of a sulfur ylide to construct the core structure of this phytotoxin target (Scheme 80).[220] Sulfonium ylide **518** was successfully isolated as a stable compound following the Rh-catalyzed decomposition of bicyclic 1,3-oxathiolane **519**. Encouraged by previous reports of stable sulfonium ylides undergoing Stevens rearrangement at elevated temperatures, heating the reaction mixture was explored; however, initial attempts yielded only recovered ylide starting material, while prolonged heating resulted in decomposition. The desired product **520** was eventually generated in good yield *via* a photo-Stevens rearrangement of the sulfur ylide through what the authors presume is a homolytic cleavage-recombination pathway.

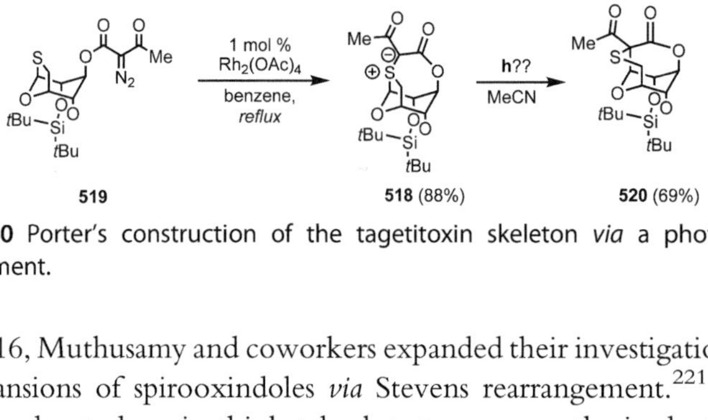

Scheme 80 Porter's construction of the tagetitoxin skeleton *via* a photo-Stevens rearrangement.

In 2016, Muthusamy and coworkers expanded their investigations of the ring expansions of spirooxindoles *via* Stevens rearrangement.[221] For this intramolecular study, spirothioketal substrates were synthesized with cyclic diazoamide tethers. Macrocyclic dispiro-1,4-dithianeoxindoles and dispiro-1,4-oxathianeoxindoles were generated in high yields and as single diastereomers following intramolecular Stevens rearrangement of the 9- to 13-membered sulfonium ylide intermediates (Scheme 81A). Studies were further extended using spirothioketal substrates with acyclic α-diazoketone tethers. Through variations in spacer chain lengths and aromatic substituents, a broad scope of macrocyclic spiro-1,4-dithianes, spiro-1,4-oxathianes, and spiro-1,4-dithiepanes was efficiently realized (Scheme 81B). Macrocyclic spiro-1,4-dithianes **523a–b** were generated from tethered 1,3-dithiolanes **524a–b** as single diastereomers; tethered 1,3-oxathiolanes **524c–d** gave a single product **523c** or a mixture of diastereomers **523d**; tethered 1,3-dithianes **524e** yielded only a mixture of diastereomers **523e**.

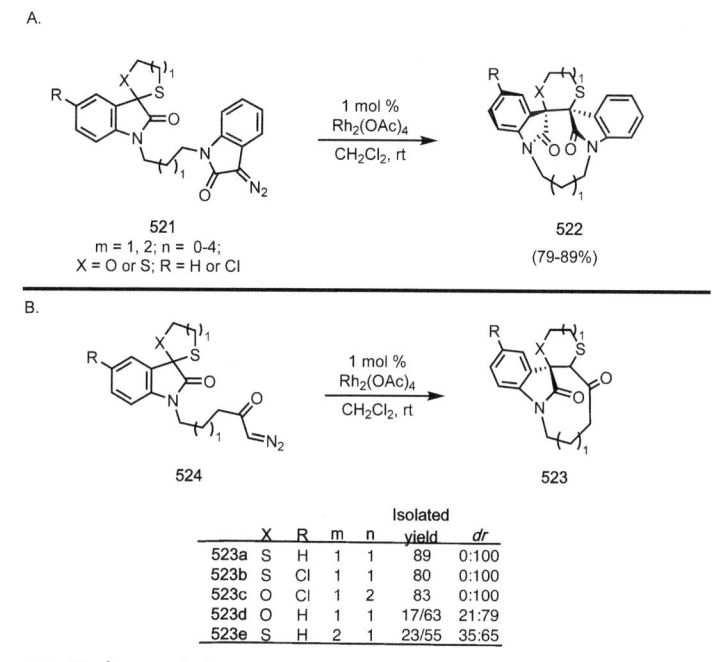

Scheme 81 Muthusamy's Stevens rearrangement of intramolecularly generated 9- to 13-membered sulfonium ylides. (A) General reaction scheme for the synthesis of macrocyclic dispiro-1,4-dithianeoxindole and dispiro-1,4-oxathianeoxindoles. (B) Selected examples from the substrate scope of macrocyclic spiro-1,4-dithianes, spiro-1,4-oxathianes, and spiro-1,4-dithiepanes derived from spirothioketals tethered with α-diazoketones.

Muthusamy and coworkers continued their studies on the Stevens rearrangement through the intramolecular generation of 11- to 21-membered macrocyclic sulfonium ylide intermediates.[222] Cyclic S, S- and O,S-acetals substrates **525** were prepared with diazoamide tethers and different chain lengths. The desired macrocyclic products **526** were obtained in high yields as single diastereomers following intramolecular macrocyclic sulfonium ylide formation and subsequent Stevens rearrangement (Scheme 82). It was noted that the length of the spacer chain influenced the product formation, as the expected macrocyclic product **526** from Stevens rearrangement was not observed with acetophenone derivative **527**; a tetracyclic macrocycle was obtained instead. Two reaction pathways were considered to account for the observed differences in product formation. Intramolecular nucleophilic addition of the thioether sulfur to the transient rhodium(II) carbenes forms sulfonium ylides **528**. For substrates with longer chain lengths (m=2, 3, 6, 9), proton abstraction *via* β-elimination and subsequent C–S bond cleavage gives the tetracyclic macrocycle **529**. However, a Stevens rearrangement of the sulfonium ylide intermediate was observed to give spiromacrocyclic product **526** when the chain length was small.

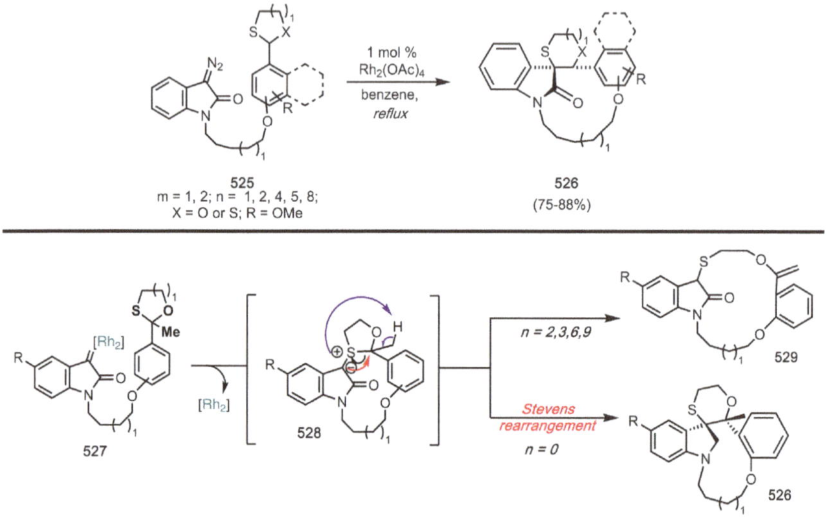

Scheme 82 (A) An extension of Muthusamy's work on the Stevens rearrangement of intramolecularly generated macrocyclic sulfonium ylides. (B) Proposed mechanism to explain the regioselective formation of macrocycles **526** and **529**.

4.8 Applications in total synthesis

In a key step of their formal synthesis of (+)-Laurencin, West and coworkers explored a thia-Stevens rearrangement of a monothioacetal-derived sulfur ylide.[223] Initial studies with model 1,3-oxathiane substrates **530a–b** gave the respective sulfur bridged cyclic ether products **531a–b** following exposure to catalytic Cu(hfacac)$_2$ in toluene at 100 °C (Scheme 83A). However, subjecting diazo ketoester substrate **532** to the same conditions gave a monocyclic olefin **533** as the major product in 43% yield as a result of α′,β-elimination of the ylide intermediate; the desired bicyclic oxocane **534** was obtained in only 16% yield (Scheme 83B). Diazoketoester **532** was converted into the desired oxocane **534** in 60% yield after exploring alternative conditions (CH$_2$Cl$_2$ at reflux, Scheme 83C). Remarkably,

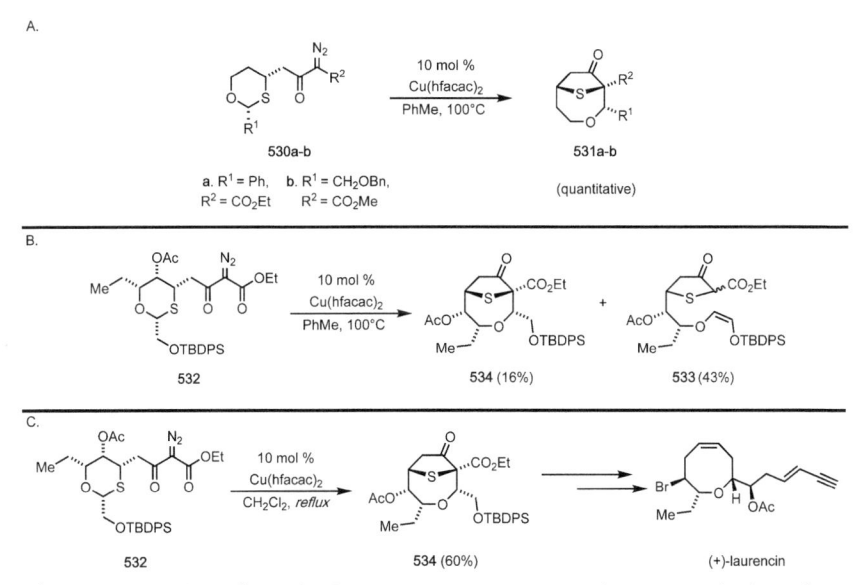

Scheme 83 West's study on the Stevens rearrangement as a key step in the formal synthesis of (+)-Laurencin. (A) Initial model study for evaluation of a [1,2]-shift to access sulfur bridged oxacycles. (B) Application of the initial model conditions to the key sulfonium ylide rearrangement. (C) Application of optimized conditions to the key sulfonium ylide rearrangement.

spectroscopic analysis revealed the stereochemical information at the anomeric center was retained following the 1,2-migration.

The preparation of thiolanes *via* carbene transfer between thietanes and metal carbenes has not been extensively explored in comparison to the analogous transformation to generate tetrahydrofuran derivatives from oxetanes and carbenes.[177] Thiolanes are commonly generated by trapping thiocarbonyl ylides with dipolarophiles, such as electron-deficient alkenes and carbonyl compounds. In 2017, Zakarian and coworkers reported an asymmetric total synthesis of (+)-6-hydroxythiobinupharidine and (−)-6-hydroxythionuphlutine.[224] In a divergence from previous approaches to the synthesis of *Nuphar* alkaloids, the Zakarian group utilized a Cu-catalyzed Stevens rearrangement of a sulfonium ylide as the key step[225] to access the bis-spirocyclic thiolane core. Using optimized conditions, spirocyclic thietane ring **535** and diazo ester **536** were exposed to microwave irradiation at 100 °C in the presence of Cu(hfacac)$_2$ to give key thiolane intermediates **537** and **538** (Scheme 84). Only two of the four expected diastereomeric products were identified as being present both by NMR analysis of the crude reaction mixture and after product isolation by column chromatography.

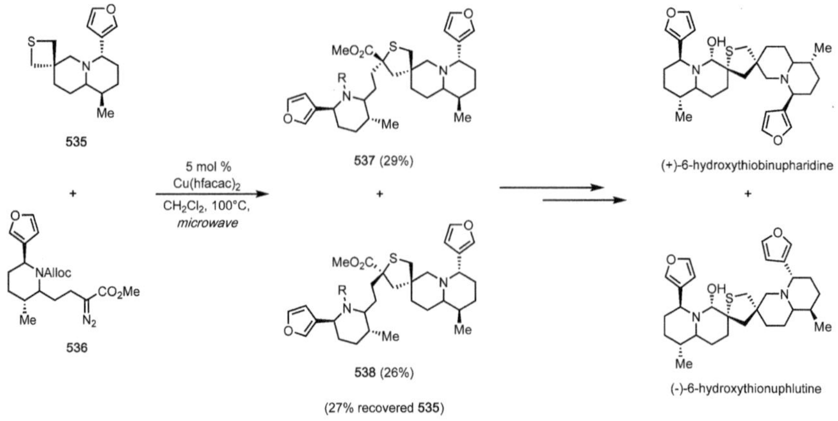

Scheme 84 Zakarian's application of a copper-catalyzed Stevens rearrangement of a sulfonium ylide toward the synthesis of the bis-spirocyclic thiolane core of (+)-6-hydroxythiobinupharidine and (−)-6-hydroxythionuphlutine.

4.9 Concluding remarks

In this section, recent accounts from the past two decades involving the formation of sulfonium ylides from metallocarbenes and sulfur nucleophiles and their subsequent transformations have been described, including [2,3]-rearrangements and [1,2]-Stevens rearrangements. However, it is clear from these discussions that this area of synthetic organic chemistry, specifically the use of transition metal catalysts in combination with sulfur ylides, still remains fairly underexplored. For example, few accounts have been reported involving the transformations of metal-carbene-derived sulfonium ylides directed toward the synthesis of complex natural products. We hope the scarcity of work in this potentially fruitful area encourages exciting investigations in the coming years.

Acknowledgment
We thank the NIH R01 GM132300 and the ACS-PRF No. 53146-ND1 for financial support.

References
1. Ford A, Miel H, Ring A, Slattery CN, Maguire AR, McKervey MA. Modern organic synthesis with α-diazocarbonyl compounds. *Chem Rev.* 2015;115:9981–10080.
2. Padwa A, Hornbuckle SF. Ylide formation from the reaction of carbenes and carbenoids with heteroatom lone pairs. *Chem Rev.* 1991;91:263–309.
3. Wittig G, Geissler G. Zur Reaktionsweise Des Pentaphenyl-Phosphors Und Einiger Derivate. *Justus Liebigs Ann Chem.* 1953;580:44–57.

4. Simonneaux G, Galardon E, Paul-Roth C, Gulea M, Masson S. Ruthenium–porphyrin-catalyzed carbenoid addition to allylic compounds: application to [2,3]-sigmatropic rearrangements of ylides. *J Organomet Chem.* 2001;617–618:360–363.
5. Zhou C-Y, Yu W-Y, PWH C, Che C-M. Ruthenium porphyrin catalyzed tandem sulfonium/ammonium ylide formation and [2,3]-sigmatropic rearrangement. A concise synthesis of (±)-platynecine. *J Org Chem.* 2004;69:7072–7082.
6. Ho C-M, Zhang J-L, Zhou C-Y, et al. A water-soluble ruthenium glycosylated porphyrin catalyst for carbenoid transfer reactions in aqueous media with applications in bioconjugation reactions. *J Am Chem Soc.* 2010;132:1886–1894.
7. Koduri ND, Scott H, Hileman B, et al. Ruthenium catalyzed synthesis of enaminones. *Org Lett.* 2012;14:440–443.
8. Koduri ND, Wang Z, Cannell G, et al. Enaminones via ruthenium-catalyzed coupling of thioamides and α-diazocarbonyl compounds. *J Org Chem.* 2014;79:7405–7414.
9. Egger L, Guénée L, Bürgi T, Lacour J. Regioselective and enantiospecific synthesis of dioxepines by (cyclopentadienyl)ruthenium-catalyzed condensations of diazocarbonyls and oxetanes. *Adv Synth Catal.* 2017;359:2918–2923.
10. Wang M-Z, Xu H-W, Liu Y, Wong M-K, Che C-M. Stereoselective synthesis of multifunctionalized 1,2,4-triazoli-dines by a ruthenium porphyrin-catalyzed three-component coupling reaction. *Adv Synth Catal.* 2006;348:2391–2396.
11. Xu H-W, Li G-Y, Wong M-K, Che C-M. Asymmetric synthesis of multifunctionalized pyrrolines by a ruthenium porphyrin-catalyzed three-component coupling reaction. *Org Lett.* 2005;7:5349–5352.
12. Avis I, Gross Z. Iron(III) corroles and porphyrins as superior catalysts for the reactions of diazoacetates with nitrogen- or sulfur-containing nucleophilic substrates: synthetic uses and mechanistic insights. *Chem Eur J.* 2008;14:3995–4005.
13. Hock KJ, Mertens L, Hommelsheim R, Spitzner R, Koenigs RM. Enabling iron catalyzed Doyle–Kirmse rearrangement reactions with in situ generated diazo compounds. *Chem Commun.* 2017;53:6577–6580.
14. Holzwarth MS, Alt I, Plietker B. Catalytic activation of diazo compounds using electron-rich, defined iron complexes for carbene-transfer reactions. *Angew Chem Int Ed.* 2012;51:5351–5354.
15. Carter DS, Van Vranken DL. Iron-catalyzed Doyle–Kirmse reaction of allyl sulfides with (trimethylsilyl)diazomethane. *Org Lett.* 2000;2:1303–1305.
16. Prabharasuth R, Van Vranken DL. Iron-catalyzed reaction of propargyl sulfides and trimethylsilyldiazomethane. *J Org Chem.* 2001;66:5256–5258.
17. Zhu S-F, Zhou Q-L. Iron-catalyzed transformations of diazo compounds. *Natl Sci Rev.* 2014;1:580–603.
18. Zhu C, Chen P, Zhu R, Lin Z, Wu W, Jiang H. C=N bond formation via palladium-catalyzed carbene insertion into N=N bonds: inhibiting the general 1,2-migration process of ylide intermediates. *Chem Commun.* 2017;53:2697–2700.
19. Jiang H, Chen F, Zhu C, et al. Two C–O bond formations on a carbenic carbon: palladium-catalyzed coupling of N-tosylhydrazones and benzo-1,2-quinones to construct benzodioxoles. *Org Lett.* 2018;20:3166–3169.
20. Greenman KL, Carter DS, Van Vranken DL. Palladium-catalyzed insertion reactions of trimethylsilyldiazomethane. *Tetrahedron.* 2001;57:5219–5225.
21. Urbano J, Belderraín TR, Nicasio MC, Trofimenko S, Díaz-Requejo MM, Pérez PJ. Functionalization of primary carbon–hydrogen bonds of alkanes by carbene insertion with a silver-based catalyst. *Organometallics.* 2005;24:1528–1532.
22. Krishnamoorthy P, Browning RG, Singh S, Sivappa R, Lovely CJ, Rias HVR. Silver-catalyzed [2,3]-rearrangement of halonium ylides derived from allyl and propargyl halides and alkyl diazoacetates. *Chem Commun.* 2007;731–733.

23. Caballero A, Despagnet-Ayoub E, Díaz-Requejo MM, et al. Silver-catalyzed C-C bond formation between methane and ethyl diazoacetate in supercritical CO2. *Science.* 2011;332:835–838.
24. Davies PW, Albrecht SJC, Assanelli G. Silver-catalysed Doyle–Kirmse reaction of allyl and propargyl sulfides. *Org Biomol Chem.* 2009;7:1276–1279.
25. Lankelma M, Olivares AM, de Bruin B. [Co(TPP)]-catalyzed formation of substituted piperidines. *Chem Eur J.* 2019;25:5658–5663.
26. Dzik WI, Xu X, Zhang XP, Reek JNH, de Bruin B. 'Carbene radicals' in Co^{II}(por)-catalyzed olefin cyclopropanation. *J Am Chem Soc.* 2010;132:10891–10902.
27. Das BG, Chirila A, Tromp M, Reek JNH, de Bruin B. Co^{III}–carbene radical approach to substituted 1*H*-indenes. *J Am Chem Soc.* 2016;138:8968–8975.
28. Lu H, Dzik WI, Xu X, Wojtas L, de Bruin B, Zhang XP. Experimental evidence for cobalt(III)-carbene radicals: key intermediates in cobalt(II)-based metalloradical cyclopropanation. *J Am Chem Soc.* 2011;133:8518–8521.
29. Chen Y, Zhang XP. Asymmetric cyclopropanation of styrenes catalyzed by metal complexes of D_2-symmetrical chiral porphyrin: superiority of cobalt over iron. *J Org Chem.* 2007;72:5931–5934.
30. Huang L, Chen Y, Gao G-Y, Zhang XP. Diastereoselective and enantioselective cyclopropanation of alkenes catalyzed by cobalt porphyrins. *J Org Chem.* 2003;68: 8179–8184.
31. Skaggs AJ, Lin EY, Jamison TF. Cobalt cluster-containing carbonyl ylides for catalytic, three-component assembly of oxygen heterocycles. *Org Lett.* 2002;4:2277–2280.
32. Fukuda T, Irie R, Katsuki T. Catalytic and asymmetric [2,3]sigmatropic rearrangement: Co(III)-salen catalyzed S-ylide formation from allyl aryl sulfides and their rearrangement. *Tetrahedron.* 1999;55:649–664.
33. Fukuda T, Katsuki T. Co(III)-salen catalyzed carbenoid reaction: stereoselective [2,3] sigmatropic rearrangement of S-ylides derived from allyl aryl sulfides. *Tetrahedron Lett.* 1997;38:3435–3438.
34. Sweeney J. Sigmatropic rearrangements of 'onium' ylids. *Chem Soc Rev.* 2009;38: 1027–1038.
35. Sheng Z, Zhang Z, Chu C, Zhang Y, Wang J. Transition metal-catalyzed [2,3]-sigmatropic rearrangements of ylides: an update of the most recent advances. *Tetrahedron.* 2016;73:4011–4022.
36. Guo X, Hu W. Novel multicomponent reactions via trapping of protic onium ylides with electrophiles. *Acc Chem Res.* 2013;46:2427–2440.
37. Bur S, Albert P. Ammonium ylides as building blocks for alkaloid synthesis. In: *Modern Tools for the Synthesis of Complex Bioactive Molecules.* John Wiley & Sons, Inc.; 2012:433–484.
38. Fu Y, Wang H-J, Chong S-S, Guo Q-X, Liu L. An extensive ylide thermodynamic stability scale predicted by first-principle calculations. *J Org Chem.* 2009;74:810–819.
39. Aggarwal VK, Harvey JN, Robiette R. On the importance of leaving group ability in reactions of ammonium, oxonium, phosphonium, and sulfonium ylides. *Angew Chem Int Ed.* 2005;44:5468–5471.
40. Coldham I, Hufton R. Intramolecular dipolar cycloaddition reactions of azomethine ylides. *Chem Rev.* 2005;105:2765–2810.
41. Moderhack D. N-ylides of 1,2,3-triazoles and tetrazoles—an overview. *Heterocycles.* 2014;89:2053–2089.
42. Sowmiah S, Esperança JMSS, Rebelo LPN, Afonso CAM. Pyridinium salts: from synthesis to reactivity and applications. *Org Chem Front.* 2018;5:453–493.
43. Doyle MP, Tamblyn WH, Bagheri V. Highly effective catalytic methods for ylide generation from diazo compounds. mechanism of the rhodium- and copper-catalyzed reactions with allylic compounds. *J Org Chem.* 1981;46:5094–5102.

44. Clark JS, Hodgson PB, Goldsmith MD, Street LJ. Rearrangement of ammonium ylides produced by intramolecular reaction of catalytically generated metal carbenoids. Part 1. Synthesis of cyclic amines. *J Chem Soc Perkin Trans 1.* 2001; 3312–3324.
45. Clark JS, Hodgson PB, Goldsmith MD, Blake AJ, Cooke PA, Street LJ. Rearrangement of ammonium ylides produced by intramolecular reaction of catalytically generated metal carbenoids. Part 2. Stereoselective synthesis of bicyclic amines. *J Chem Soc Perkin Trans 1.* 2001;3325–3337.
46. Clark JS, Middleton MD. Synthesis of novel α-substituted and α,α-disubstituted amino acids by rearrangement of ammonium ylides generated from metal carbenoids. *Org Lett.* 2002;4:765–768.
47. Heath P, Roberts E, Sweeney JB, Wessel HP, Workman JA. Copper(II)-catalyzed [2,3]-sigmatropic rearrangement of n-methyltetrahydropyridinium ylids. *J Org Chem.* 2003;68:4083–4086.
48. Roberts E, Sançon JP, Sweeney JB, Workman JA. First efficient and general copper-catalyzed [2,3]-rearrangement of tetrahydropyridinium ylids. *Org Lett.* 2003;5(25):4775–4777.
49. Roberts E, Sançon JP, Sweeney JB. A new class of ammonium ylid for [2,3]-sigmatropic rearrangement reactions: ene-endo-spiro ylids. *Org Lett.* 2005;7:2075–2078.
50. Xu H-D, Jia Z-H, Xu K, Zhou H, Shen M-H. One-pot protocol to functionalized benzopyrrolizidine catalyzed successively by $Rh_2(OAc)_4$ and $Cu(OTf)_2$: a transition metal–Lewis acid catalysis relay. *Org Lett.* 2015;17:66–69.
51. Honda K, Shibuya H, Yasui H, Hoshino Y, Inoue S. Copper-catalyzed intermolecular generation of ammonium ylides with subsequent [2,3]sigmatropic rearrangement. Efficient synthesis of bifunctional homoallylamines. *Bull Chem Soc Jpn.* 2008;81: 142–147.
52. O'Hagan D. Pyrrole, pyrrolidine, pyridine, piperidine and tropane alkaloids. *Nat Prod Rep.* 2000;17:435–446.
53. Davis FA, Wu Y, Xu H, Zhang J. Asymmetric synthesis of cis-5-substituted pyrrolidine 2-phosphonates using metal carbenoid nh insertion and δ-amino β-ketophosphonates. *Org Lett.* 2004;6:4523–4525.
54. Deng Q-H, Xu W-H, Yuen AW-H, Xu Z-J, Che C-M. Ruthenium-catalyzed one-pot carbenoid N−H insertion reactions and diastereoselective synthesis of prolines. *Org Lett.* 2008;10:1529–1532.
55. Dong C, Mo F, Wang J. Highly diastereoselective addition of the lithium enolate of α-diazoacetoacetate to N-sulfinyl imines: enantioselective synthesis of 2-oxo and 3-oxo pyrrolidines. *J Org Chem.* 2008;73:1971–1974.
56. Bott TM, Vanecko JA, West FG. One-carbon ring expansion of azetidines via ammonium ylide [1,2]-shifts: a simple route to substituted pyrrolidines. *J Org Chem.* 2009;74:2832–2836.
57. Li G-Y, Chen J, Yu W-Y, Hong W, Che C-M. Stereoselective synthesis of functionalized pyrrolidines by ruthenium porphyrin-catalyzed decomposition of α-diazo esters and cascade azomethine ylide formation/1,3-dipolar cycloaddition reactions. *Org Lett.* 2003;5:2153–2156.
58. Zhu Y, Zhai C, Yue Y, Yang L, Hu W. One-pot three-component tandem reaction of diazo compounds with anilines and unsaturated ketoesters: a novel synthesis of 2,3-dihydropyrrole derivatives. *Chem Commun.* 2009;1362–1364.
59. Zhang X, Ji J, Zhu Y, Jing C, Li M, Hu W. A highly diastereoselective three-component tandem 1,4-conjugated addition–cyclization reaction to multisubstituted pyrrolidines. *Org Biomol Chem.* 2012;10:2133–2138.
60. Pinho VD, Burtoloso ACB. Preparation of α,β-unsaturated diazoketones employing a Horner−Wadsworth−Emmons reagent. *J Org Chem.* 2011;76:289–292.

61. Medvedev JJ, Galkina OS, Klinkova AA, et al. Domino [4+1]-annulation of α,β-unsaturated δ-amino esters with Rh(ii)–carbenoids—a new approach towards multi-functionalized N-aryl pyrrolidines. *Org Biomol Chem*. 2015;13:2640–2651.
62. Boralsky LA, Marston D, Grigg RD, Hershberger JC, Schomaker JM. Allene functionalization *via* bicyclic methyleneaziridines. *Org Lett*. 2011;13:1924–1927.
63. Rigoli JW, Boralsky LA, Hershberger JC, et al. 1,4-Diazaspiro[2.2]pentanes as a flexible platform for the synthesis of diamine-bearing stereotriads. *J Org Chem*. 2012;77:2446–2455.
64. Adams CS, Boralsky LA, Guzei IA, Schomaker JM. Modular functionalization of allenes to aminated stereotriads. *J Am Chem Soc*. 2012;134:10807–10810.
65. Adams CS, Grigg RD, Schomaker JM. Complete stereodivergence in the synthesis of 2-amino-1,3-diols from allenes. *Chem Sci*. 2014;5:3046–3056.
66. Burke EG, Schomaker JM. Oxidative allene amination for the synthesis of azetidin-3-ones. *Angew Chem Int Ed*. 2015;54:12097–12101.
67. Gerstner NC, Adams CS, Tretbar M, Schomaker JM. Stereocontrolled syntheses of seven-membered carbocycles by tandem allene aziridination/[4+3] reaction. *Angew Chem Int Ed*. 2016;128:13434–13437.
68. Burke EG, Gold B, Hoang TT, Raines RT, Schomaker JM. Fine-tuning strain and electronic activation of strain-promoted 1,3-dipolar cycloadditions with endocyclic sulfamates in SNO-OCTs. *J Am Chem Soc*. 2017;139:8029–8037.
69. Corbin JR, Ketelboeter DR, Fernández I, Schomaker JM. Biomimetic imino-nazarov cyclizations *via* eneallene aziridination. *J Am Chem Soc*. 2020;142:5568–5573.
70. Schmid SC, Guzei IA, Schomaker JM. A stereoselective [3+1] ring expansion of the synthesis of highly substituted methylene azetidines. *Angew Chem Int Ed*. 2017;56:12229–12233.
71. West FG, Naidu BN. New route to substituted piperidines via the Stevens [1,2]-shift of ammonium ylides. *J Am Chem Soc*. 1993;115:1177–1178.
72. West FG, Glaeske KW, Naidu BN. One-step synthesis of tertiary α-amino ketones and α-amino esters from amines and diazocarbonyl compounds. *Synthesis*. 2002;1993:977–980.
73. Padwa A, Beall LS, Eidell CK, Worsencroft KJ. An approach toward isoindolobenzazepines using the ammonium ylide/stevens [1,2]-rearrangement sequence. *J Org Chem*. 2001;66:2414–2421.
74. Vanecko JA, West FG. A novel, stereoselective silyl-directed stevens [1,2]-shift of ammonium ylides. *Org Lett*. 2002;4:2813–2816.
75. Vanecko JA, West FG. Ring expansion of azetidinium ylides: rapid access to the pyrrolizidine alkaloids turneforcidine and platynecine. *Org Lett*. 2005;7:2949–2952.
76. Mucedda M, Muroni D, Saba A, Manassero C. Concise diastereospecific pyrrolo[1,2-a][1,4]benzodiazepinone synthesis. *Tetrahedron*. 2007;63:12232–12238.
77. Takaya J, Udagawa S, Kusama H, Iwasawa N. Synthesis of N-fused tricyclic indoles by a tandem [1,2] stevens-type rearrangement/1,2-alkyl migration of metal-containing ammonium ylides. *Angew Chem Int Ed*. 2008;47:4906–4909.
78. Rosset IG, Dias RMP, Pinho VD, Burtoloso ACB. Three-step synthesis of (±)-preussin from decanal. *J Org Chem*. 2014;79:6748–6753.
79. Cordell GA, Quinn-Beattie ML, Farnsworth NR. The potential of alkaloids in drug discovery. *Phytother Res*. 2001;15:183–205.
80. Harada S, Kono M, Nozaki T, Menjo Y, Nemoto T, Hamada Y. General approach to nitrogen-bridged bicyclic frameworks by Rh-catalyzed formal carbenoid insertion into an amide C–N bond. *J Org Chem*. 2015;80:10317–10333.
81. Terada Y, Kitajima M, Taguchi F, Takayama H, Horie S, Watanabe T. Identification of indole alkaloid structural units important for stimulus-selective TRPM8 inhibition: SAR study of naturally occurring iboga derivatives. *J Nat Prod*. 2014;77:1831–1838.

82. Gorman M, Neuss N, Svoboda GH. Vinca alkaloids. IV. Structural features of leurosine and vincaleukoblastine, representatives of a new type of indole-indoline alkaloids. *J Am Chem Soc*. 1959;81:4745–4746.
83. Neuss N, Gorman M. The structure of catharanthine, a novel variant of the iboga alkaloids. *Tetrahedron Lett*. 1961;2:206–210.
84. Kono M, Harada S, Nozaki T, et al. Asymmetric formal synthesis of (+)-catharanthine via desymmetrization of isoquinuclidine. *Org Lett*. 2019;21:3750–3754.
85. Davies HML, Alford JS. Reactions of metallocarbenes derived from N-sulfonyl-1,2,3-triazoles. *Chem Soc Rev*. 2014;43:5151–5162.
86. Jia M, Ma S. New approaches to the synthesis of metal carbenes. *Angew Chem Int Ed*. 2016;55:9134–9166.
87. Gulevich AV, Gevorgyan V. Versatile reactivity of rhodium–iminocarbenes derived from N-sulfonyl triazoles. *Angew Chem Int Ed*. 2013;52:1371–1373.
88. Chattopadhyay B, Gevorgyan V. Transition-metal-catalyzed denitrogenative transannulation: converting triazoles into other heterocyclic systems. *Angew Chem Int Ed*. 2012;51:862–872.
89. Bosmani A, Guarnieri-Ibáñez A, Goudedranche S, Besnard C, Lacour J. Polycyclic indoline-benzodiazepines through electrophilic additions of α-imino carbenes to Tröger bases. *Angew Chem Int Ed*. 2018;54:7151–7155.
90. Schmid SC, Guzei IA, Fernández I, Schomaker JM. Ring expansion of bicyclic methyleneaziridines via concerted, near-barrierless [2,3]-stevens rearrangement. *ACS Catal*. 2018;8:7907–7914.
91. Eshon J, Nicastri KA, Schmid SC, et al. Intermolecular [3 + 3] ring-expansion of aziridines to dehydropiperidines through the intermediacy of aziridinium ylides. *Nat Commun*. 2020;11:1–8.
92. Dequina HJ, Eshon J, Raskopf WT, Fernández I, Schomaker JM. Rh-catalyzed aziridine ring expansions to dehydropiperazines. *Org Lett*. 2020;22:3637–3641.
93. Yates P. The copper-catalyzed decomposition of diazoketones1. *J Am Chem Soc*. 1952;74:5376–5381.
94. Salzmann TN, Ratcliffe RW, Christensen BG, Bouffard FA. A stereocontrolled synthesis of (+)-thienamycin. *J Am Chem Soc*. 1980;102:6161–6163.
95. Gillingham D, Fei N. Catalytic X–H insertion reactions based on carbenoids. *Chem Soc Rev*. 2013;42:4918–4931.
96. García CF, McKervey MA, Ye T. Asymmetric catalysis of intramolecular N–H insertion reactions of α-diazocarbonyls. *Chem Commun*. 1996;1465–1466.
97. Buck RT, Moody CJ, Pepper AG. N-H Insertion reactions of rhodium carbenoids. Part 4. New chiral dirhodium(II) carboxylate catalysts. *ARKIVOC*. 2002;16–33.
98. (a) Liu B, Zhu S-F, Zhang W, Chen C, Zhou Q-L. Highly enantioselective insertion of carbenoids into N—H bonds catalyzed by copper complexes of chiral spiro bisoxazolines. *J Am Chem Soc*. 2007;129:5834–5835. (b) Zhu S-F, Xu B, Wang G-P, Zhou Q-L. Well-defined binuclear chiral spiro copper catalysts for enantioselective N–H insertion. *J Am Chem Soc*. 2012;134:436–442.
99. Zhu S-F, Zhou Q-L. Transition-metal-catalyzed enantioselective heteroatom–hydrogen bond insertion reactions. *Acc Chem Res*. 2012;45:1365–1377.
100. Lee EC, Fu GC. Copper-catalyzed asymmetric N—H insertion reactions: couplings of diazo compounds with carbamates to generate α-amino acids. *J Am Chem Soc*. 2007;129:12066–12067.
101. Xu B, Zhu S-F, Zuo X-D, Zhang Z-C, Zhou Q-L. Enantioselective N-H insertion reaction of α-aryl α-diazoketones: an efficient route to chiral α-aminoketones. *Angew Chem Int Ed*. 2014;53:3913–3916.
102. Guo J-X, Zhou T, Xu B, Zhu S-F, Zhou Q-L. Enantioselective synthesis of α-alkenyl α-amino acids via N–H insertion reactions. *Chem Sci*. 2016;7:1104–1108.

103. Li M-L, Yu J-H, Li Y-H, Zhu S-F, Zhou Q-L. Highly enantioselective carbene insertion into N–H bonds of aliphatic amines. *Science*. 2019;366:990.
104. Jiang J, Xu H-D, Xi J-B, et al. Diastereoselectively switchable enantioselective trapping of carbamate ammonium ylides with imines. *J Am Chem Soc*. 2011;133:8428–8431.
105. Jiang J, Ma X, Liu S, et al. Enantioselective trapping of phosphoramidate ammonium ylides with imino esters for synthesis of 2,3-diaminosuccinic acid derivatives. *Chem Commun*. 2013;49:4238–4240.
106. Jiang J, Ma X, Ji C, et al. Ruthenium(II)/chiral brønsted acid co-catalyzed enantioselective four-component reaction/cascade aza-michael addition for efficient construction of 1,3,4-tetrasubstituted tetrahydroisoquinolines. *Chem Eur J*. 2014;20: 1505–1509.
107. Jing C, Xing D, Qian Y, Shi T, Zhao Y, Hu W. Diversity-oriented three-component reactions of diazo compounds with anilines and 4-oxo-enoates. *Angew Chem Int Ed*. 2013;52:9289–9292.
108. Nicolle SM, Lewis W, Hayes CJ, Moody CJ. Stereoselective synthesis of functionalized pyrrolidines by the diverted N−H insertion reaction of metallocarbenes with β-aminoketone derivatives. *Angew Chem Int Ed*. 2016;55:3749–3753.
109. Oda S, Sam B, Krische MJ. Hydroaminomethylation beyond carbonylation: allene–imine reductive coupling by ruthenium-catalyzed transfer hydrogenation. *Angew Chem Int Ed*. 2015;54:8525–8528.
110. Oda S, Franke J, Krishce MJ. Diene hydroaminomethylation via ruthenium-catalyzed C–C bond forming transfer hydrogenation: beyond carbonylation. *Chem Sci*. 2016; 7:136–141.
111. Zhu C, Xu G, Sun J. Gold-catalyzed formal [4 + 1]/[4 + 3] cycloadditions of diazo esters with triazines. *Angew Chem Int Ed*. 2016;55:11867–11871.
112. Briones JF, Davies HML. Enantioselective gold(I)-catalyzed vinylogous [3 + 2] cycloaddition between vinyldiazoacetates and enol ethers. *J Am Chem Soc*. 2013;135: 13314–13317.
113. Lonzi G, López LA. Regioselective synthesis of functionalized pyrroles via gold(I)-catalyzed [3 + 2] cycloaddition of stabilized vinyl diazo derivatives and nitriles. *Adv Synth Catal*. 2013;355:1948–1954.
114. López E, Lonzi G, López LA. Gold-catalyzed C–H bond functionalization of metallocenes: synthesis of densely functionalized ferrocene derivatives. *Organometallics*. 2014;33:5924–5927.
115. Zhou L-Y, Guo X-M, Zhang Z-C, Li J. Gold(I)-catalyzed [4 + 1]/[4 + 3] annulations of diazo esters with hexahydro-1,3,5-triazines: theoretical study of mechanism and regioselectivity. *J Organomet Chem*. 2019;897:70–79.
116. Guan X-Y, Tang M, Liu Z-Q, Hu W. A highly diastereoselective [5 + 1] annulation to 2,2,3-trisubstituted tetrahydroquinoxalines via intramolecular Mannich-type trapping of ammonium ylides. *Chem Commun*. 2019;55:9809–9812.
117. Hunter AC, Almutwalli B, Bain AI, Sharma I. Trapping rhodium carbenoids with aminoalkynes for the synthesis of diverse N-heterocycles. *Tetrahedron*. 2018;74: 5451–5457.
118. Murphy GK, West FG. Oxonium ylide rearrangements in synthesis. In: *Molecular Rearrangements in Organic Synthesis*. John Wiley & Sons, Inc.; 2015:497–538.
119. Wang J. 11.05—Synthetic reactions of MC and MN bonds: ylide formation, rearrangement, and 1,3-dipolar cycloaddition. In: DMP M, Crabtree RH, eds. *Comprehensive Organometallic Chemistry III*. Elsevier; 2007:151–178.
120. Doyle MP. 5.2—Transition metal carbene complexes: diazodecomposition, ylide, and insertion. In: Abel EW, FGA S, Wilkinson G, eds. *Comprehensive Organometallic Chemistry II*. Elsevier; 1995:421–468.

121. Doyle MP, Van Leusen D, Tamblyn WH. Efficient alternative catalysts and methods for the synthesis of cyclopropanes from olefins and diazo compounds. *Synthesis.* 1981;2002:787–789.
122. (a) Nozaki H, Takaya H, Noyori R. Reaction of carbethoxycarbene with 2-phenyloxirane and 2-phenyloxetane. *Tetrahedron.* 1966;22:3393–3401. (b) Nozaki H, Takaya H, Noyori R. The reaction of ethyl diazoacetate with styrene oxide. *Tetrahedron Lett.* 1965;6:2563–2567.
123. Pirrung MC, Werner JA. Intramolecular generation and [2,3]-sigmatropic rearrangement of oxonium ylides. *J Am Chem Soc.* 1986;108:6060–6062.
124. Roskamp EJ, Johnson CR. Generation and rearrangements of oxonium ylides. *J Am Chem Soc.* 1986;108:6062–6063.
125. Skrobo B, Schlorer NE, Neudorfl J-M, Deska J. Kirmse–Doyle- and Stevens-type rearrangements of glutarate-derived oxonium ylides. *Chem Eur J.* 2018;24:3209–3217.
126. Fu J, Shang H, Wang Z, et al. Gold-catalyzed rearrangement of allylic oxonium ylides: efficient synthesis of highly functionalized dihydrofuran-3-ones. *Angew Chem Int Ed.* 2013;52:4198–4202.
127. Han M, Bae J, Choi J, Tae J. Synthesis of 2,5-disubstituted dihydrofuran-3(2H)-ones via [2,3]-sigmatropic rearrangement of oxonium ylides generated from α-oxo gold carbenes. *Synlett.* 2013;24:2077–2080.
128. Boyer A. Rhodium(II)-catalyzed stereocontrolled synthesis of dihydrofuran-3-imines from 1-tosyl-1,2,3-triazoles. *Org Lett.* 2014;16:1660–1663.
129. Boyer A. Enantioselective synthesis of (+)-petromyroxol, enabled by rhodium-catalyzed denitrogenation and rearrangement of a 1-sulfonyl-1,2,3-triazole. *J Org Chem.* 2015;80:4771–4775.
130. Hansen E, Clark SJ. Intramolecular reactions of metal carbenoids with allylic ethers: is a free ylide involved in every case? *Chem Eur J.* 2014;20:5454–5459.
131. Doyle MP, Bagheri V, Harn NK. Facile catalytic methods for intermolecular generation of allylic oxonium ylides and their stereoselective [2,3]-sigmatropic rearrangement. *Tetrahedron Lett.* 1988;29:5119–5122.
132. Doyle MP, Forbes DC, Vasbinder MM, Peterson CS. Enantiocontrol in the generation and diastereoselective reactions of catalytically generated oxonium and iodonium ylides. metal-stabilized ylides as reaction intermediates. *J Am Chem Soc.* 1998;120: 7653–7654.
133. Li Z, Davies HML. Enantioselective C – C bond formation by rhodium-catalyzed tandem ylide formation/[2,3]-sigmatropic rearrangement between donor/acceptor carbenoids and allylic alcohols. *J Am Chem Soc.* 2010;132:396–401.
134. Li Z, Parr BT, Davies HML. Highly Stereoselective C–C bond formation by rhodium-catalyzed tandem ylide formation/[2,3]-sigmatropic rearrangement between donor/acceptor carbenoids and chiral allylic alcohols. *J Am Chem Soc.* 2012;134: 10942–10946.
135. Parr BT, Davies HML. Stereoselective synthesis of highly substituted cyclohexanes by a rhodium-carbene initiated domino sequence. *Org Lett.* 2015;17:794–797.
136. Rao S, Prabhu KR. Gold-catalyzed [2,3]-sigmatropic rearrangement: reaction of aryl allyl alcohols with diazo compounds. *Org Lett.* 2017;19:846–849.
137. Moniz GA, Wood JL. Catalyst-based control of [2,3]- and [3,3]-rearrangement in α-diazoketone-derived propargyloxy enols. *J Am Chem Soc.* 2001;123:5095–5097.
138. Li Z, Boyarskikh V, Hansen JH, Autschbach J, Musaev DG, Davies HML. Scope and mechanistic analysis of the enantioselective synthesis of allenes by rhodium-catalyzed tandem ylide formation/[2,3]-sigmatropic rearrangement between donor/acceptor carbenoids and propargylic alcohols. *J Am Chem Soc.* 2012;134: 15497–15504.

139. Quinn KJ, Biddick NA, DeChristopher BA. Ring expansion of trans-divinyl ethylene oxide by oxonium ylide [2,3] sigmatropic rearrangement. *Tetrahedron Lett.* 2006;47:7281–7283.
140. Mack DJ, Batory LA, Njardarson JT. Intermolecular oxonium ylide mediated synthesis of medium-sized oxacycles. *Org Lett.* 2012;14:378–381.
141. Eberlein TH, West FG, Tester RW. The Stevens-[1,2]-shift of oxonium ylides: a route to substituted tetrahydrofuranones. *J Org Chem.* 1992;57:3479–3482.
142. Hosseini SN, Johnston JR, West FG. Evidence for heterolytic cleavage of a cyclic oxonium ylide: implications for the mechanism of the Stevens [1,2]-shift. *Chem Commun.* 2017;53:12654–12656.
143. Guranova NI, Darin D, Kantin G, Novikov AS, Bakulina O, Krasavin M. Rh(II)-catalyzed spirocyclization of α-diazo homophthalimides with cyclic ethers. *J Org Chem.* 2019;84:4534–4542.
144. Casanova R, Reichstein T. Methoxyketone aus diazoketonen. Steroide, 5. Mitteilung. *Helv Chim Acta.* 1950;33:417–422.
145. Yates P. The copper-catalyzed decomposition of diazoketones. *J Am Chem Soc.* 1952;74:5376–5381.
146. Liang Y, Zhou H, Yu Z-X. Why is copper(I) complex more competent than dirhodium(II) complex in catalytic asymmetric O−H insertion reactions? A computational study of the metal carbenoid O−H insertion into water. *J Am Chem Soc.* 2009;131:17783–17785.
147. Maier TC, Fu GC. Catalytic enantioselective O−H insertion reactions. *J Am Chem Soc.* 2006;128:4594–4595.
148. Chen C, Zhu S-F, Liu B, Wang L-X, Zhou Q-L. Highly enantioselective insertion of carbenoids into O−H bonds of phenols: an efficient approach to chiral α-aryloxycarboxylic esters. *J Am Chem Soc.* 2007;129:12616–12617.
149. Zhu S-F, Song X-G, Li Y, Cai Y, Zhou Q-L. Enantioselective copper-catalyzed intramolecular O−H insertion: an efficient approach to chiral 2-carboxy cyclic ethers. *J Am Chem Soc.* 2010;132:16374–16376.
150. (a) Shou-Fei Z, Chen C, Cai Y, Zhou Q-L. Catalytic asymmetric reaction with water: enantioselective synthesis of α-hydroxyesters by a Copper–Carbenoid O-H insertion reaction. *Angew Chem Int Ed.* 2008;47:932–934. (b) Bulugahapitiya P, Landais Y, Parra-Rapado L, Planchenault D, Weber V. A stereospecific access to allylic systems using rhodium(II)−vinyl carbenoid insertion into Si−H, O−H, and N−H bonds. *J Org Chem.* 1997;62:1630–1641.
151. Gong J, Lin G, Sun W, Li C-C, Yang Z. Total synthesis of (±) maoecrystal V. *J Am Chem Soc.* 2010;132:16745–16746.
152. Pan S, Xuan J, Gao B, Zhu A, Ding H. Total synthesis of diterpenoid steenkrotin a. *Angew Chem Int Ed.* 2015;54:6905–6908.
153. Lu C-D, Liu H, Chen Z-Y, Hu W-H, Mi A-Q. Three-component reaction of aryl diazoacetates, alcohols, and aldehydes (or imines): evidence of alcoholic oxonium ylide intermediates. *Org Lett.* 2005;7:83–86.
154. Wang Y, Zhu Y, Chen Z, Mi A, Hu W, Doyle MP. A novel three-component reaction catalyzed by dirhodium(II) acetate: decomposition of phenyldiazoacetate with arylamine and imine for highly diastereoselective synthesis of 1,2-diamines. *Org Lett.* 2003;5:3923–3926.
155. Yue Y, Guo X, Chen Z, Yang L, Hu W. Copper(I) hexafluorophosphate: a dual functional catalyst for three-component reactions of methyl phenyldiazoacetate with alcohols and aldehydes or α-ketoesters. *Tetrahedron Lett.* 2008;49:6862–6865.
156. Huang H, Guo X, Hu W. Efficient trapping of oxonium ylides with imines: a highly diastereoselective three-component reaction for the synthesis of β-amino-α-hydroxyesters with quaternary stereocenters. *Angew Chem Int Ed.* 2007;46: 1337–1339.

157. Han X, Jiang L, Tang M, Hu W. Diastereoselective three-component reactions of aryldiazoacetates with alcohols/water and alkynals: application to substituted enelactones. *Org Biomol Chem*. 2011;9:3839–3843.
158. Han X, Gan M, Qiu H, et al. Trapping of oxonium ylides with Michael acceptors: highly diastereoselective three-component reactions of diazo compounds with alcohols and benzylidene meldrum's acids/4-oxo-enoates. *Synlett*. 2011;2011:1717–1722.
159. Zhu Y, Zhai C, Yang L, Hu W. Copper(ii)-catalyzed highly diastereoselective three-component reactions of aryl diazoacetates with alcohols and chalcones: an easy access to furan derivatives. *Chem Commun*. 2010;46:2865–2867.
160. Alcaide B, Almendros P, Aragoncillo C, Callejo R, Ruiz MP, Torres MR. Rhodium-catalyzed synthesis of 3-hydroxy-β-lactams via oxonium ylide generation: three-component reaction between azetidine-2,3-diones, ethyl diazoacetate, and alcohols. *J Org Chem*. 2009;74:8421–8424.
161. Lu C-D, Liu H, Chen Z-Y, Hu W-H, Mi A-Q. The rhodium catalyzed three-component reaction of diazoacetates, titanium(iv) alkoxides and aldehydes. *Chem Commun*. 2005;2624–2626.
162. Guo Z, Shi T, Jiang J, Yang L, Hu W. Component match in rhodium catalyzed three-component reactions of ethyl diazoacetate, H$_2$O and aryl imines: a highly diastereoselective one-step synthesis of β-aryl isoserine derivatives. *Org Biomol Chem*. 2009;7:5028–5033.
163. Ji J, Zhang X, Zhu Y, et al. Diastereoselectivity switch in cooperatively catalyzed three-component reactions of an aryldiazoacetate, an alcohol, and a β,γ-unsaturated α-keto ester. *J Org Chem*. 2011;76:5821–5824.
164. Guo X, Huang H, Yang L, Hu W. Trapping of oxonium ylide with isatins: efficient and stereoselective construction of adjacent quaternary carbon centers. *Org Lett*. 2007;9:4721–4723.
165. Guo Z, Cai M, Jiang J, Yang L, Hu W. Rh$_2$(OAc)$_4$-AgOTf cooperative catalysis in cyclization/three-component reactions for concise synthesis of 1,2-dihydroisoquinolines. *Org Lett*. 2010;12:652–655.
166. Hu W, Xu X, Zhou J, et al. Cooperative catalysis with chiral brønsted acid-Rh$_2$(OAc)$_4$: highly enantioselective three-component reactions of diazo compounds with alcohols and imines. *J Am Chem Soc*. 2008;130:7782–7783.
167. Qian Y, Xu X, Jiang L, Prajapati D, Hu W. A strategy to synthesize taxol side chain and (−)-epi cytoxazone via chiral brønsted acid-Rh$_2$(OAc)$_4$ co-catalyzed enantioselective three-component reactions. *J Org Chem*. 2010;75:7483–7486.
168. Kang Z, Zhang D, Shou J, Hu W. Enantioselective trapping of oxonium ylides by 3-hydroxyisoindolinones via a formal SN1 pathway for construction of contiguous quaternary stereocenters. *Org Lett*. 2018;20:983–986.
169. Qiu L, Su M, Wen Z, Zhu X, Duan Y, Huang Y. Semisynthesis of 3-hydroxyoxindole rapamycin analogues through site- and stereoselective trapping of oxonium ylides in RhII-catalyzed three-component reactions. *Eur J Org Chem*. 2019;2019:2914–2918.
170. Liang X-S, Li R-D, Wang X-C. Copper-catalyzed asymmetric annulation reactions of carbenes with 2-iminyl- or 2-acyl-substituted phenols: convenient access to enantioenriched 2,3-dihydrobenzofurans. *Angew Chem Int Ed*. 2019;58:13885–13889.
171. Gopi Krishna Reddy A, Niharika P, Zhou S, et al. Brønsted acid catalyzed enantioselective assembly of spirochroman-3,3-oxindoles. *Org Lett*. 2020;22:2925–2930.
172. Zhang Y, Wang J. Catalytic [2,3]-sigmatropic rearrangement of sulfur ylide derived from metal carbene. *Coord Chem Rev*. 2010;254:941–953.
173. Jones AC, May JA, Sarpong R, Stoltz BM. Toward a symphony of reactivity: cascades involving catalysis and sigmatropic rearrangements. *Angew Chem Int Ed*. 2014;53:2556–2591.

174. Li A-H, Dai L-X, Aggarwal VK. Asymmetric ylide reactions: epoxidation, cyclopropanation, aziridination, olefination, and rearrangement. *Chem Rev.* 1997;97:2341–2372.
175. Vedejs E. Sulfur-mediated ring expansions in total synthesis. *Acc Chem Res.* 1984;17:358–364.
176. Markó IE. The Stevens and related rearrangements. In: Trost BM, Fleming I, eds. *Comprehensive Organic Synthesis.* Oxford, UK: Pergamon Press; 1991:913–974.
177. Clark JS, ed. *Nitrogen, Oxygen and Sulfur Ylide Chemistry. A Practical Approach in Chemistry.* Oxford, UK: Oxford University Press; 2002.
178. Trost BM, Melvin LS. Synthesis and structures of sulfur ylides. In: Trost BM, Melvin LS, eds. *Sulfur Ylides: Emerging Synthetic Intermediates.* New York, NY: Academic Press; 1975:13–36.
179. Zhang Z, Wang J. Recent studies on the reactions of α-diazocarbonyl compounds. *Tetrahedron.* 2008;64:6577–6605.
180. Doyle MP, McKervey MA, Ye T. *Modern Catalytic Methods for Organic Synthesis With Diazo Compounds.* New York, NY: Wiley-Interscience; 1998.
181. Pathania S, Narang RK, Rawal RK. Role of sulphur-heterocycles in medicinal chemistry: an update. *Eur J Med Chem.* 2019;180:486–508.
182. Kirmse W, Kapps M. Reaktionen des Diazomethans mit Diallylsulfid und Allyläthern unter Kupfersalz-Katalyse. *Chem Ber.* 1968;101:994–1003.
183. Doyle MP, Griffin JH, Chinn MS, van Leusen D. Rearrangements of ylides generated from reactions of diazo compounds with allyl acetals and thioketals by catalytic methods. Heteroatom acceleration of the [2,3]-sigmatropic rearrangement. *J Org Chem.* 1984;49:1917–1925.
184. Trost BM, Hammen RF. New synthetic methods. Transfer of chirality from sulfur to carbon. *J Am Chem Soc.* 1973;95:962–964.
185. Nishibayashi Y, Ohe K, Uemura S. The first example of enantioselective carbenoid addition to organochalcogen atoms: application to [2,3]sigmatropic rearrangement of allylic chalcogen ylides. *J Chem Soc Chem Commun.* 1995;1245–1246.
186. Zhang X, Qu Z, Ma Z, Shi W, Jin X, Wang J. Catalytic asymmetric [2,3]-sigmatropic rearrangement of sulfur ylides generated from copper(I) carbenoids and allyl sulfides. *J Org Chem.* 2002;67:5621–5625.
187. McMillen DW, Varga N, Reed BA, King C. Asymmetric copper-catalyzed [2,3]-sigmatropic rearrangements of alkyl- and aryl-substituted allyl sulfides. *J Org Chem.* 2000;65:2532–2536.
188. Kitagaki S, Yanamoto Y, Okubo H, Nakajima M, Hashimoto S. Enantiocontrol in tandem allylic sulfonium ylide generation and [2,3] sigmatropic rearrangement catalyzed by chiral dirhodium(II) complexes. *Heterocycles.* 2001;54:623–628.
189. Zhang X, Ma M, Wang J. Catalytic asymmetric [2,3] sigmatropic rearrangement of sulfur ylides generated from carbenoids and propargyl sulfides. *Tetrahedron Asymmetry.* 2003;14:891–895.
190. Zhang X, Ma M, Wang J. Catalytic asymmetric [2, 3]-sigmatropic rearrangement of sulfur ylides generated from carbenoids and allenic 2-methylphenyl sulfide. *Chin J Chem.* 2003;2:878–882.
191. Hock KJ, Koenigs RM. Enantioselective [2,3]-sigmatropic rearrangements: metal-bound or free ylides as reaction intermediates? *Angew Chem Int Ed.* 2017;56: 13566–13568.
192. Ma M, Peng L, Li C, Zhang X, Wang J. Highly stereoselective [2,3]-sigmatropic rearrangement of sulfur ylide generated through Cu(I) carbene and sulfides. *J Am Chem Soc.* 2005;127:15016–15017.
193. Zhang Z, Sheng Z, Yu W, et al. Catalytic asymmetric trifluoromethylthiolation via enantioselective [2,3]-sigmatropic rearrangement of sulfonium ylides. *Nat Chem.* 2017;9:970–976.

194. Sommelet M. On a particular mode of intramolecular rearrangement. *Compt Rend.* 1937;205:56–58.
195. Kantor SW, Hauser CR. Rearrangements of benzyltrimethylammonium ion and related quaternary ammonium ions by sodium amide involving migration into the ring. *J Am Chem Soc.* 1951;73:4122–4131.
196. Hauser CR, Kantor SW, Brasen WR. Rearrangement of benzyl sulfides to mercaptans and of sulfonium ions to sulfides involving the aromatic ring by alkali amides. *J Am Chem Soc.* 1953;75:2660–2663.
197. Liao M, Peng L, Wang J. Rh(II)-catalyzed Sommelet – Hauser rearrangement. *Org Lett.* 2008;10:693–696.
198. Gassman PG, van Bergen TJ. General method for the synthesis of indoles. *J Am Chem Soc.* 1974;96:5508–5512.
199. Li Y, Shi Y, Huang Z, et al. Catalytic thia-Sommelet – Hauser rearrangement: application to the synthesis of oxindoles. *Org Lett.* 2011;13:1210–1213.
200. Li Y, Huang Z, Wu X, et al. Rh(II)-catalyzed [2,3]-sigmatropic rearrangement of sulfur ylides derived from N-tosylhydrazones and sulfides. *Tetrahedron.* 2012;68:5234–5240.
201. Miura T, Tanaka T, Yada A, Murakami M. Doyle–Kirmse reaction using triazoles leading to one-pot multifunctionalization of terminal alkynes. *Chem Lett.* 2013;42:1308–1310.
202. Yadagiri D, Anbarasan P. Rhodium-catalyzed denitrogenative [2,3] sigmatropic rearrangement: an efficient entry to sulfur-containing quaternary centers. *Chem Eur J.* 2013;19:15115–15119.
203. Kato Y, Miki K, Nishino F, Ohe K, Uemura S. Doyle – Kirmse reaction of allylic sulfides with diazoalkane-free (2-furyl)carbenoid transfer. *Org Lett.* 2003;5:2619–2621.
204. Davies PW, Albrecht SJC. Alkynes as masked ylides: gold-catalysed intermolecular reactions of propargylic carboxylates with sulfides. *Chem Commun.* 2008;2:238–240.
205. Santos MD, Davies PW. A gold-catalysed fully intermolecular oxidation and sulfur-ylide formation sequence on ynamides. *Chem Commun.* 2014;50:6001–6004.
206. Li J, Ji K, Zheng R, Nelson J, Zhang L. Expanding the horizon of intermolecular trapping of in situ generated α-oxo gold carbenes: efficient oxidative union of allylic sulfides and terminal alkynes via C–C bond formation. *Chem Commun.* 2014;50:4130–4133.
207. Zhang H, Wang B, Yi H, Zhang Y, Wang J. Rh(II)-catalyzed [2,3]-sigmatropic rearrangement of sulfur ylides derived from cyclopropenes and sulfides. *Org Lett.* 2015;17:3322–3325.
208. Murphy GK, West FG. [1,2]- or [2,3]-Rearrangement of onium ylides of allyl and benzyl ethers and sulfides via in situ-generated iodonium ylides. *Org Lett.* 2006;8:4359–4361.
209. Davies PW, Albrecht SJC. Gold- or platinum-catalyzed synthesis of sulfur heterocycles: access to sulfur ylides without using sacrificial functionality. *Angew Chem Int Ed.* 2009;48:8372–8375.
210. Shapiro ND, Toste FD. Rearrangement of alkynyl sulfoxides catalyzed by gold(I) complexes. *J Am Chem Soc.* 2007;129:4160–4161.
211. Li G, Zhang L. Gold-Catalyzed intramolecular redox reaction of sulfinyl alkynes: efficient generation of α-oxo gold carbenoids and application in insertion into R – CO bonds. *Angew Chem Int Ed.* 2007;46:5156–5159.
212. Ioannou M, Porter MJ, Saez F. Conversion of 1,3-oxathiolanes to 1,4-oxathianes using a silylated diazoester. *Tetrahedron.* 2005;61:43–50.
213. Zhu S, Xing C, Zhu S. Stereoselective preparation of trifluoromethyl containing 1,4-oxathiolane derivatives through ring expansion reaction of 1,3-oxathiolanes. *Tetrahedron.* 2006;62:829–832.
214. Hiesinger K, Dar'in D, Proschak E, Krasavin M. Spirocyclic scaffolds in medicinal chemistry. *J Med Chem.* 2020;64:150–183.

215. Stepakov AV, Molchanov AP, Magull J, et al. The methoxycarbonylcarbene insertion into 1,3-dithiolane and 1,3-oxathiolane rings. *Tetrahedron*. 2006;62:3610–3618.
216. Ellis-Holder KK, Peppers BP, Kovalevsky AY, Diver ST. Macrocycle ring expansion by double Stevens rearrangement. *Org Lett*. 2006;8:2511–2514.
217. Muthusamy S, Selvaraj K. Highly diastereoselective synthesis of dispiro[1,4-dithiane/dithiepane]bisoxindoles via Stevens rearrangement. *Tetrahedron Lett*. 2013;54:6886–6888.
218. Qu J-P, Xu Z-H, Zhou J, et al. Ligand-accelerated asymmetric [1,2]-Stevens rearrangement of sulfur ylides via decomposition of diazomalonates catalyzed by chiral bisoxazoline/copper complex. *Adv Synth Catal*. 2009;351:308–312.
219. Liao S, Sun XL, Tang Y. Side arm strategy for catalyst design: modifying bisoxazolines for remote control of enantioselection and related. *Acc Chem Res*. 2014;47:2260–2272.
220. Mortimer AJP, Aliev AE, Tocher DA, Porter MJ. Synthesis of the tagetitoxin core via photo-Stevens rearrangement. *Org Lett*. 2008;10:5477–5480.
221. Muthusamy S, Selvaraj K, Suresh E. Diastereoselective synthesis of macrocyclic spiro and dispiro-1,4-dithianes, -1,4-oxathianes, and -1,4-dithiepanes through intramolecular sulfonium ylides. *Asian J Org Chem*. 2016;5:162–172.
222. Muthusamy S, Selvaraj K, Suresh E. Demonstration of 11–21-membered intramolecular sulfonium ylides: regio- and diastereoselective synthesis of spiro-oxindole-incorporated macrocycles. *Eur J Org Chem*. 2016;10:1849–1859.
223. Lin R, Cao L, West FG. Medium-sized cyclic ethers via stevens [1,2]-shift of mixed monothioacetal-derived sulfonium ylides: application to formal synthesis of (+)-laurencin. *Org Lett*. 2017;19:553–555.
224. Lacharity JJ, Fournier J, Lu P, Mailyan AK, Herrmann AT, Zakarian A. Total synthesis of unsymmetrically oxidized nuphar thioalkaloids via copper-catalyzed thiolane assembly. *J Am Chem Soc*. 2017;139:13272–13275.
225. Lu P, Herrmann AT, Zakarian A. Toward the synthesis of nuphar sesquiterpene thioalkaloids: stereodivergent rhodium-catalyzed synthesis of the thiolane subunit. *J Org Chem*. 2015;80:7581–7589.

CHAPTER TWO

π-Alkene/alkyne and carbene complexes of gold(I) stabilized by chelating ligands

Miquel Navarro[†] and Didier Bourissou[*]

CNRS/Université Paul Sabatier, Laboratoire Hétérochimie Fondamentale et Appliquée (LHFA, UMR 5069), Toulouse, France
*Corresponding author: e-mail address: dbouriss@chimie.ups-tlse.fr

Contents

1. Introduction 101
2. Gold(I) π-complexes 104
   2.1 Gold(I) π-complexes with N N-chelating ligands 106
   2.2 Gold(I) π-complexes with P P and P N-chelating ligands 115
   2.3 Comparison of gold(I) ethylene complexes with chelating and hemilabile ligands 121
3. Gold(I) carbene complexes 123
4. Concluding remarks 134
Acknowledgments 135
References 135

1. Introduction

The use of gold in homogenous catalysis was scarcely investigated during the last century, since gold complexes were long considered to be chemically inert and thus catalytically useless species. This situation changed dramatically when the carbophilic properties of gold complexes were discovered, and within only two decades, they became extremely powerful and versatile catalysts for the activation and functionalization of π-CC bonds.[1] Accordingly, a number of useful catalytic transformations in organic

---

[†] Current address: Instituto de Investigaciones Químicas (IIQ), Departamento de Química Inorgánica y Centro de Innovación en Química Avanzada (ORFEO-CINQA), Consejo Superior de Investigaciones Científicas (CSIC) and University of Sevilla, Avenida Américo Vespucio 49, 41092, Sevilla, Spain.

synthesis based on π- or σ,π-coordination of alkynes, allenes and alkenes have been developed.[2] These advances in gold-catalyzed transformations have stimulated coordination and organometallic studies in order to gain more understanding of the structure and reactivity of gold complexes, in particular gold(I) species. Here, monodentate ligands, especially phosphines and N-Heterocyclic carbenes (NHCs), have been extensively used as ancillary ligands due to the inherent preference for gold(I) complexes to be two-coordinate and adopt linear geometry.[3] In contrast, gold(I) complexes featuring bidentate ligands are comparatively very rare, but have captured increasing attention because of the unique properties chelating ligands impart to gold in terms of bonding and chemical behavior. From a structural point of view, bidentate ligands force gold(I) to escape its usual two-coordinate linear form and they are preorganized to form tri or even tetra-coordinate complexes.[4] In addition, chelating ligands enforce bending at gold and thereby, noticeably modify its electronic properties. Most significant is the impact on the energy and symmetry of the frontier orbitals. As illustrated in Fig. 1 for model phosphine and diphosphine cationic gold(I) complexes, the HOMO is very much raised in energy and changes from $5d_{z^2}$ to $5d_{xz}$ in symmetry, in line with that encountered in group 10 ML_2 complexes.[5] In the meantime, the LUMO is raised in energy, but remains low. Overall, the energy gap between the vacant and occupied orbitals at gold involved in donation and backdonation (HOMO-3/LUMO for the monophosphine complex, HOMO/LUMO for the diphosphine complex) is considerably decreased.

The unique properties of bent dicoordinate L_2Au^+ fragments was shown to radically impact oxidative addition to gold. This elementary step is hardly feasible with monocoordinate Au(I) complexes,[6–9] and the reluctance of gold to cycle between the +I and +III oxidation states has limited its application in such catalytic transformations. However, when the metal fragment is activated and preorganized thanks to the use of a chelating ligand, oxidative addition to gold turns easy and general,[10–14] opening new avenues in Au(I)/Au(III) catalysis.[12,15–21]

In this review are discussed the preparation, characterization, structure and reactivity of two types of gold(I) complexes featuring chelating ligands: first, π-complexes with alkenes and alkynes side-on coordinated, and then carbene complexes. A small library of chelating ligands of L^L′, L^E(−)^L and L^X types have been exploited in gold(I) chemistry to date (Fig. 2). With this detailed survey, we hope to inspire the gold community and newcomers to further use and develop chelating ligands in gold(I) chemistry. There is clearly still a lot to discover.

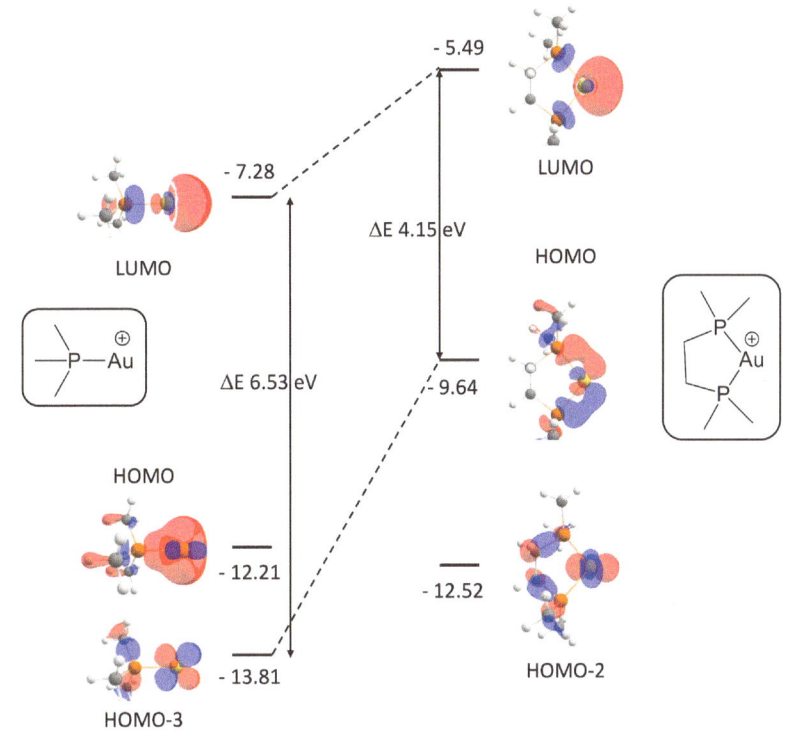

Fig. 1 Frontier orbitals of model gold(I) complexes bearing monodentate and bidentate phosphine ligands.

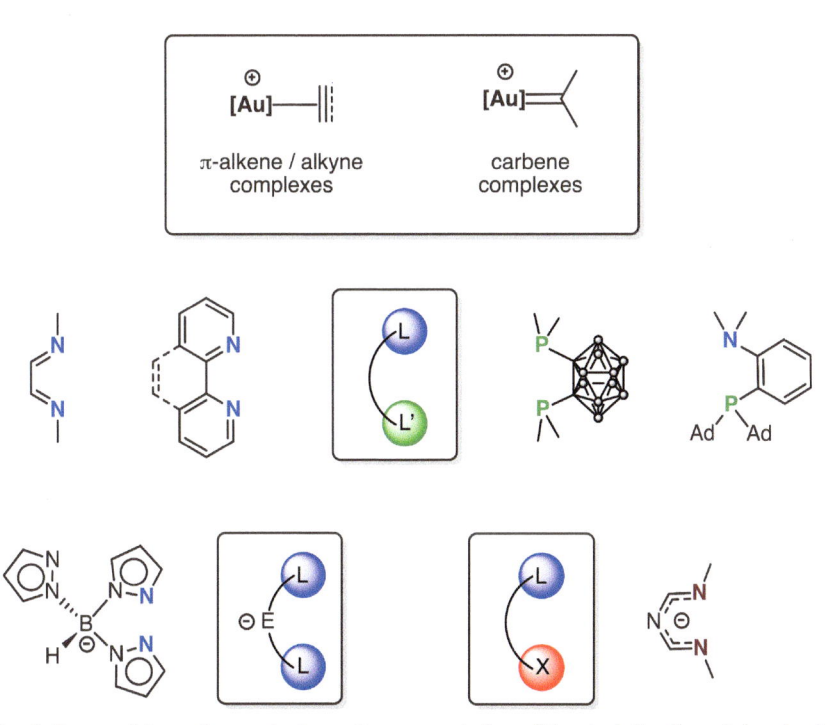

Fig. 2 Scope of the review and schematic representation of the chelating ligands involved.

2. Gold(I) π-complexes

As mentioned above, the discovery of the ability of gold complexes to activate π-bonds toward nucleophilic addition has represented a great breakthrough in homogeneous catalysis and in organic synthesis.[2,22–26] A large number of useful transformations such as the addition of oxygen-, nitrogen-, and carbon-based nucleophiles to CC multiple bonds, rearrangements and cycloaddition reactions have been developed.[27–32] All these catalytic transformations usually involve the side-on coordination and activation of a π-CC bond to gold. Consequently, the isolation of gold π-complexes has received considerable attention in the last two decades.[33–37] These compounds serve as models for the transient species and their study provide valuable insights into the factors governing the catalytic transformations. Since the determination of the crystal structure of the Zeise's salt **1** K[Cl$_3$Pt(C$_2$H$_4$)],[38] major efforts were made in order to mimic the platinum behavior to coordinate olefins. After several attempts and non-conclusive studies to isolate and characterize stable gold(I) complexes with a coordinated olefin, the group of Strähle in 1987 finally isolated and structurally characterized the first Au(I) π-complex **2**, with *cis*-cyclooctene coordinated to AuCl (Fig. 3).[39] The C=C distance is significantly longer than in the free olefin ligand (1.38(2) Å *versus* 1.332(2) Å), and the ν(C=C) stretching vibration appears at 1525 cm^{-1}, compared to 1648 cm^{-1} for the free olefin. A few years later, Fackler reported the tetranuclear complex **3** of the formula Au$_4$(dppe)$_2$[S$_2$C$_2$(CN)$_2$]Cl$_2$.[40] Here, one of the four gold(I) centers is attached to the central C=C bond of a *cis*-bis(diphenylphosphino)-ethene (ddpe) and chelated by a 1,2-dicyanoethene-1,2-dithiolate ligand, thus presenting a formal negative charge. The C=C bond sits in the SAuS

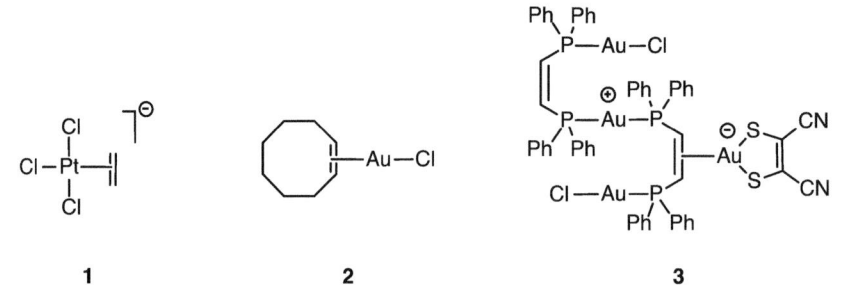

Fig. 3 Zeise's salt **1**, first isolated and structurally characterized Au(I) olefin complexes **2** and **3**.

coordination plane, enabling stabilizing interaction between the olefin π^* and the metal d_π orbitals. The C=C bond distance in complex 3 is elongated in comparison with the free olefin (1.38(6) Å versus 1.30(6) Å) indicating an important gold → π^* backdonation. In the following years, some π-alkyne gold(I) compounds were also reported.[41–44]

The isolation and characterization of the first gold(I) π-complexes have stimulated studies on olefin coordination to gold in different oxidation states. Theoretical investigations have shown that the bonding of ethylene and acetylene to gold(0) atoms is very weak,[45,46] in contrast with initial experimental results.[47–51] The binding energies are close to those of dispersion forces and gold(0) has low affinity for C=C and C≡C bonds. On the other hand, gold(III) π-complexes featuring alkenes, alkynes and arenes are known, but rare.[37] The weakness of the π-coordination to gold(III), due in part to the weak Au → π^* backdonation, makes the study of such complexes challenging. This field is still in its infancy and much remains to be done. Most of the gold(III) π-complexes reported to date derive from cyclometallated ligands. They have been isolated or spectroscopically characterized in situ at low temperature (<0 °C) under inert conditions.[52–57]

Cationic dicoordinate gold(I) π-complexes of the type [LAu(ene)]$^+$ represent the majority of the known gold π-complexes. They are generally stable both in the solid state and in solution. In most cases, N-heterocyclic carbenes (NHCs)[58–60] or phosphines[58,61–64] have been employed as ancillary ligands. The involved π-systems include alkenes, alkynes, conjugated dienes, allenes, enamines and enol ethers (Fig. 4). Tricoordinate species have also been reported with different chelating ligands. They present unique features, but remain very scarce. In this section, tricoordinate gold(I) π-complexes bearing chelating or hemilabile ligands are reviewed in detail. Special attention is given to the specific electronic and structural features of the gold(I)–π-system interaction. The impact of this unusual

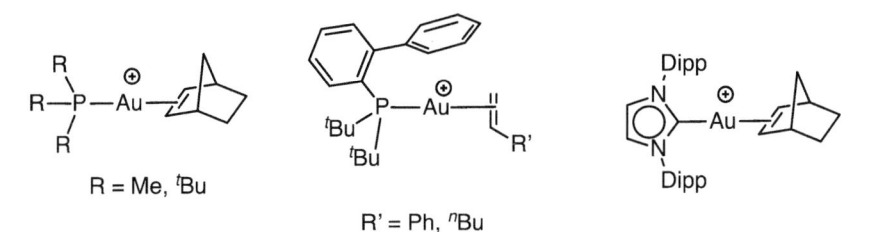

Fig. 4 Representative examples of selected NHC and phosphine π-alkene gold(I) complexes.

coordination on the alkene/alkyne properties is discussed and compared with that encountered classically in cationic dicoordinate gold(I) π-complexes.

2.1 Gold(I) π-complexes with N N-chelating ligands

In the mid 2000s, Cinellu et al. described for the first time the synthesis and structural characterization of cationic tricoordinate gold(I) π-complexes using different substituted bipyridine ligands.[65–67] Dicationic dinuclear gold(III) μ,μ-dioxo compounds were reduced by olefins and diolefins (Scheme 1) to give mononuclear gold(I) π-complexes of the type [Au(bipy)(ene)]$^+$ (23 examples) or dinuclear gold(I) complexes with a bridging diolefin (7 examples) in low to moderate yields. The [Au(bipy)(ene)]$^+$ complexes were found to be surprisingly stable thermally, with melting points above 100 °C. The olefins are bound sufficiently strongly to the gold(I) center not to dissociate either in solution or in solid state under vacuum. The olefin is not even replaced by common coordinating solvents such as acetonitrile or by CO at atmospheric pressure. However, addition of an excess of the corresponding olefin caused broadening of the ^{1}H NMR signals of the coordinated alkene, indicating fast intermolecular exchange between the coordinated and free olefin at room temperature. Variable temperature ^{1}H NMR spectroscopy showed that the dynamic process is frozen at low temperature (193 K), where the different olefinic proton signals are well-separated. In the majority of the complexes, the signals of the olefin protons are sharp and showed a significant high-field shift with respect to those of the free alkenes ($\Delta\delta^1$H in the range of 0.8–2.2 ppm). Likewise, the ^{13}C NMR signals for the olefinic carbons are upfield shifted upon coordination ($\Delta\delta^{13}$C in the range of 47.5–61.7 ppm). These spectroscopic values are in marked contrast with the minor changes observed upon coordination of olefins to gold(I) centers ligated to phosphines or NHCs. The solid-state structures of the [Au(bipy)(η^2-styrene)]$^+$ complexes **4** and **5** display trigonal-planar environments around the gold atom (Fig. 5). The olefin is coplanar with the pyridine backbone, which chelates the gold center with bite angles around 75°. The gold-carbon distances were found to be slightly shorter than in other gold(I)-alkene complexes. Most remarkable are the C—C bond lengths of the coordinated olefin, which were elongated (>0.03 Å) in comparison with the free olefin. All these data indicate that the chelating bipyridine ligand induces some metallacyclic character as the result of significant Au→alkene backdonation. DFT calculations on the model

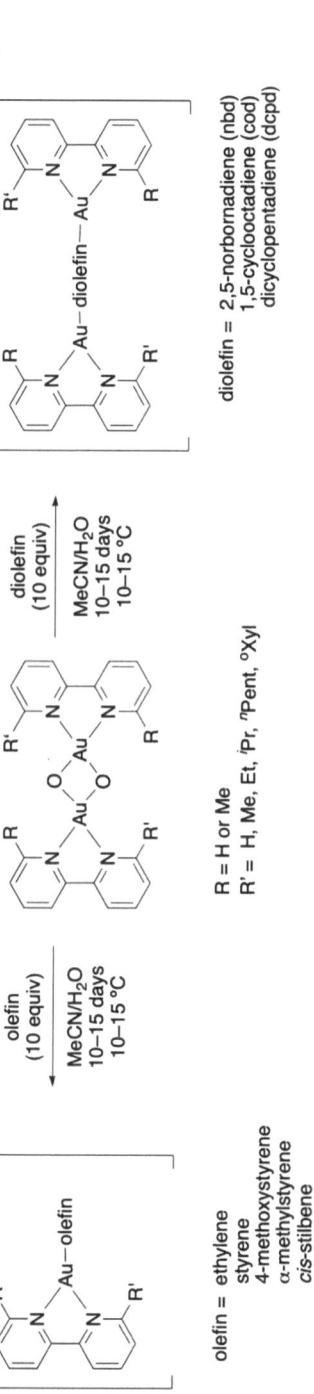

Scheme 1 Synthesis of the tricoordinate [Au(bipy)(ene)]+ π-complexes and dinuclear gold(I) complexes with bridging diolefins [(bipy)Au(diolefin)Au(bipy)]$^{2+}$.

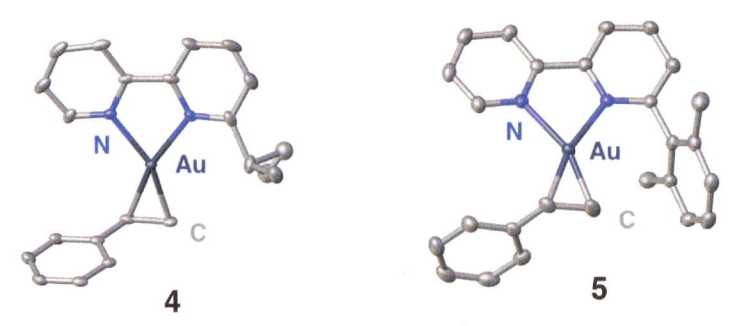

Fig. 5 Molecular structure of the [Au(bipy)(styrene)]⁺ complexes 4 and 5.

cation [Au(bipy)(η^2-CH$_2$=CH$_2$)]$^+$ were used to thoroughly analyze the Au(I)-alkene bonding situation. Natural Bond Orbital (NBO) populations, Wiberg bond indexes (WBI) and Bond Dissociation Energies (BDE) all indicate significant Au→alkene π backdonation compared to σ donation, in line with the spectroscopic and structural data.

Gold-ethylene adducts are particularly rare.[68] See Section 2.3 for a detailed comparison of the spectroscopic and structural data of the few known complexes discussed hereafter. In a pioneering work, Dias et al. described in 2007 two singular neutral gold(I) ethylene complexes with monoanionic tris(pyrazolyl)borates as supporting ligands (Scheme 2).[69–71] Reaction of the tris(pyrazolyl)borate sodium salt with gold(I) chloride under an ethylene atmosphere afforded complexes 6a,b as colorless stable solids in good yields (>70%). Remarkably, the ethylene molecule remained tightly bounded to gold(I) even when reduced pressure was applied. Also, in the presence of excess of ethylene, no alkene exchange was observed on the NMR time scale. As for the bipyridine gold(I) π-complexes, the ¹H and ¹³C NMR signals for the coordinated ethylene are markedly shifted upfield in comparison with free ethylene (3.81 and 63.7 ppm *versus* 5.43 and 116.8 ppm, respectively). However, there are no noticeable differences in chemical shifts between this neutral gold(I) ethylene complex and the related cationic gold(I) ethylene complexes, suggesting that the nature of the supporting ligand does not considerably affect the gold(I)-ethylene ligation. The single-crystal X-ray diffraction analysis showed a gold atom with trigonal-planar coordination geometry. The ethylene molecule sits in the coordination plane of the metal. It is bonded in a typical η^2-fashion with Au—C bond lengths of 2.096(6) and 2.108(6) Å. The C=C bond (1.380(10) Å) is longer than that of free ethylene (1.313 Å).[72,73] These bond distances are similar to those found in the cationic gold(I) ethylene complexes

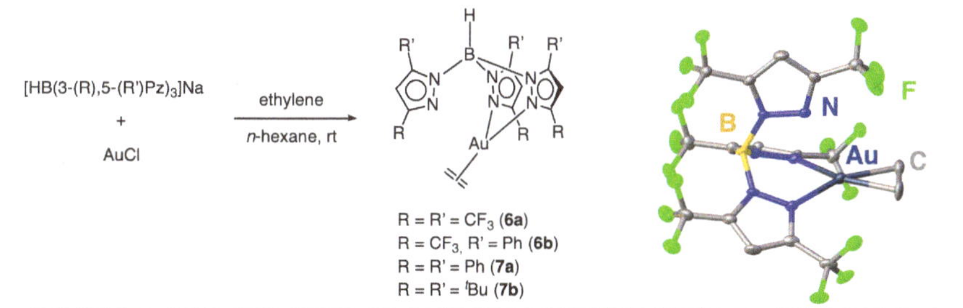

Scheme 2 Synthesis and structure of the tris(pyrazolyl)borate gold(I) ethylene complexes **6** and **7**.

bearing bipyridine ligands (see below). The tris(pyrazolyl)borate ligand coordinates to gold in a κ^2-fashion with Au—N distances of 2.2221(5) and 2.224(5) Å, while the unbounded nitrogen of the third pyrazolyl moiety is 2.710 Å away (which remains within the sum of van der Waals radii of Au and N).[74] However, complex **6a** showed fluxional behavior in solution even at −80 °C, with no differentiated signals for the coordinated and free pyrazolyl moieties in ^{1}H and ^{19}F NMR spectroscopy suggesting fast exchange between the pendant and coordinated sidearms. Analogous complexes supported by electron-rich scorpionates (R, R' = Ph, tBu) **7a,b** were recently prepared.[75] NMR, crystallographic and computational data are indicative of enhanced Au → ethylene backdonation.

In a subsequent study, the same research group capitalized on the chelating nature of triazapentadienyl ligands to stabilize a gold(I)-ethylene complex.[76] Treatment of the triazapentadienyl lithium salt with gold(I) chloride in the presence of ethylene led to the formation of the neutral gold(I)-ethylene complex **8** as a yellow solid (Scheme 3A). The ^{1}H NMR signal corresponding to the ethylene protons appeared as a singlet at 2.71 ppm. The upfield shift with respect to free ethylene is larger than that observed for related gold(I)-ethylene complexes, which is most likely due to the ring current effect caused by the aryl group rings of the ligand flanking the ethylene moiety. The ^{13}C NMR signal of the coordinated ethylene appears at 59.1 ppm, in line with the other reported cationic and neutral gold(I)-ethylene complexes. The X-ray structure of complex **8** displays a trigonal-planar coordination geometry, with the triazapentadienyl ligand binding the metal center in a κ^2 fashion and with the ethylene molecule sitting in the plane of the N⏜N–Au(I) fragment. In this case, the C=C bond length of the coordinated ethylene is also longer (1.405(4) Å) than that of free ethylene. Of note, this Au(I)-ethylene complex **8** was proved to

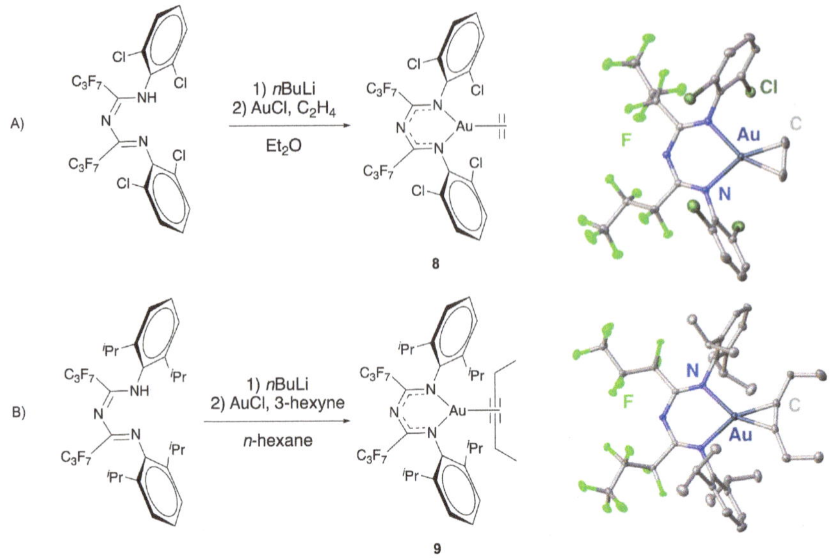

Scheme 3 Synthesis and structure of the triazapentadienyl gold(I) ethylene and 3-hexyne complexes **8** and **9**.

efficiently mediate carbene-transfer reactions from ethyl diazoacetate to saturated and unsaturated hydrocarbons (C—H insertion, cyclopropanation reactions).

A similar ligand scaffold was used to prepare the first tricoordinated gold(I) alkyne complex following the same synthetic procedure.[77] Treatment of AuCl with [N{(C$_3$F$_7$)C(Dipp)N}$_2$]Li (prepared directly from [N{(C$_3$F$_7$)C(Dipp)N}$_2$]H and *n*BuLi) in the presence of 3-hexyne in hexane afforded the (N^N)Au(EtC≡CEt) complex **9** as a yellow solid in 85% yield (Scheme 3B). This Au(I)-alkyne complex was remarkably stable and could be handled in air without decomposition. The ^{1}H NMR spectrum exhibits signals corresponding to the ethyl moieties of the alkyne at the 1.1–0.75 ppm region, which is an upfield shift compared to the corresponding signals of free 3-hexyne (2.13 and 1.09 ppm). This shift is likely due to the ring currents of the flanking *N*-aryl groups, rather than a direct electronic implication of the Au–alkyne bond. The ^{13}C NMR signal of the acetylenic carbons appeared at 91.3 ppm, which is relatively downfield shifted in comparison with free 3-hexyne (80.9 ppm), and very similar to dicoordinate gold(I)-alkyne complexes.[78–81] The X-ray structure of complex **9** presents a tricoordinate gold(I) center. The triazapentadienyl ligand coordinates the gold center in a κ^2 fashion with a N–Au–N bite angle of

85.31(13)°. The coordinated 3-hexyne sits in the plane of the [(N^N)Au]$^+$ fragment. In this case, the C≡C bond length is marginally longer (1.233(7) Å) than that computed for free 3-hexyne (1.215 Å). According to DFT calculations, the enthalpy of formation of the model π-complex [N{(CF$_3$)C(C$_6$H$_5$)N}$_2$]M(EtC≡CEt) is 36.6 kcal/mol. NBO analysis showed that σ-donation from the alkyne to the metal center dominates over Au → alkyne π-backdonation (109.8 versus 48.1 kcal/mol). In addition, the coordination of the alkyne was found to be stronger for gold than copper and silver, as well as the degree of backdonation versus donation.

A few cationic Au(I)-ethylene complexes have also been described. In 2016, Daugulis et al. reported a stable complex featuring an hindered α-diimine ligand.[82] Reaction of tris(ethylene)-gold hexafluoroantimonate **10**[83] and the diimine ligand under ethylene atmosphere afforded the tricoordinate gold(I)-ethylene complex **11** in moderate yields (42%) (Scheme 4). Complex **11** decomposed in solution within an hour at room temperature and had to be handled under inert atmosphere. Both ^{1}H and ^{13}C NMR signals corresponding to the ethylene moiety are upfield shifted in comparison with the free olefin (3.31 and 3.28 versus 5.43 ppm and 65.4 versus 116.8 ppm, respectively). Olefin exchange in presence of excess ethylene was found to occur, albeit slowly on the NMR time scale even at 50 °C. To determine the exchange rate, complex **11-d$_4$** was prepared in good yield (81%) by purging a solution of complex **11** with C$_2$D$_4$. Treatment with an excess of C$_2$H$_4$ was then monitored by ^{1}H NMR spectroscopy giving $\Delta G^{\ddagger}_{298}$ = 12.9 ± 0.1 kcal/mol, $\Delta H^{\ddagger}$ = 10.0 ± 1.4 kcal/mol and $\Delta S^{\ddagger}$ = −22.5 ± 4.8 eu, consistent with an associative exchange mechanism. The complex displayed a tricoordinate Au(I) center with the metal, the diimine ligand and the ethylene molecule lying in the same plane. The C=C bond of

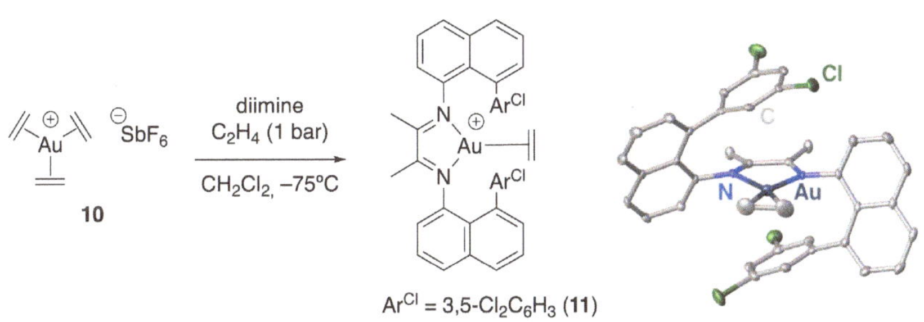

Scheme 4 Synthesis and structure of the [Au(diimine)(ethylene)][SbF$_6$] complex **11**.

the η²-coordinated ethylene molecule (1.455(13) Å) is longer than those observed in structurally characterized dicoordinate Au(I)-ethylene complexes (1.35–1.39 Å).

In 2018, the group of Russell used the synthetic methodology developed by Daugulis et al.[82] to prepare cationic [Au(bipy)(ene)]⁺ complexes.[84] The reaction of 2,2′-bipyridines with the unstable gold(I) tris-ethylene complex [Au(C₂H₄)₃][NTf₂] afforded [Au(bipy)(C₂H₄)][NTf₂] complexes **12** and **13** in good yields (∼60%) (Scheme 5). These complexes showed impressive stability. They can be handled in air and used for subsequent reactions in non-anhydrous solvents (for oxidative addition of C(sp²)–I and C(sp)–I bonds in particular).[84,85] The molecular structure of the complex **13** revealed a distorted trigonal planar gold center with the bipyridine and ethylene ligands in the same plane and symmetrically κ²/η² coordinated. Analogous to the [Au(bipy)(styrene)]⁺ complexes previously described by Cinellu et al.[65,66] the Au—C bond distances are contracted and the C=C bond is significantly elongated compared to the two-coordinate [Au(L)(ene)]⁺ complexes (L = PR₃ or NHC). The ¹H and ¹³C NMR signals for the coordinated ethylene are both remarkably upfield shifted in comparison with free ethylene (i.e., 3.90 and 63.8 versus 5.43 and 116.8 ppm respectively). Again, this is the result of enhanced π-backdonation induced by the chelating bipyridine ligand.

Different studies by Shi, Liu and Waser et al. on gold-catalyzed alkynylations have shown that Au(I) complexes can catalyze alkynylation reactions.[86–91] The alkynylation product likely results from reductive elimination of an alkynyl Au(III) complex, which is derived from oxidative addition of an alkynyl iodonium salt to gold(I). In order to gain more insight into this catalytic

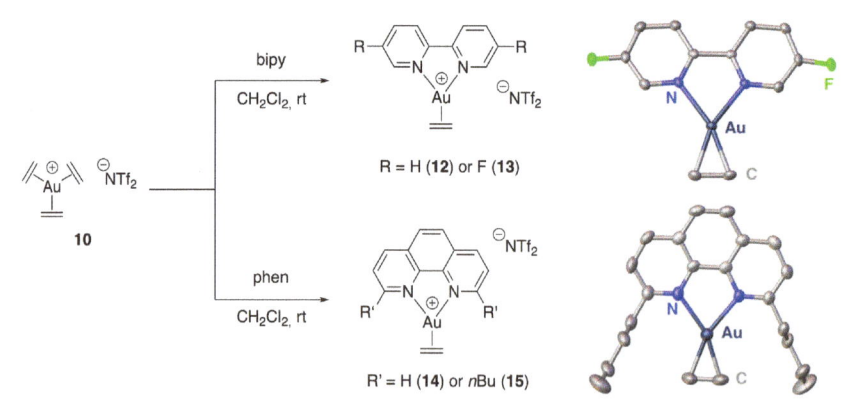

Scheme 5 Syntheses and structures of the gold(I)-ethylene bipyridine and phenanthroline complexes reported by the Russell's and Hashmi's groups.

transformation, the group of Hashmi synthetized in 2019 [(Phen)Au(C$_2$H$_4$)][NTf]$_2$ complexes **14** and **15**[92] (Scheme 5) following the same synthetic procedure than that used by Daugulis[82] and Russell.[84] Complex **15** proved to be a competent pre-catalyst for the alkynylation of cyclopropenes. As for the bipyridine analogs, ^{1}H and ^{13}C NMR spectroscopy showed a marked upfield shift of the ethylene signals. In addition, the C=C bond distance (1.411(10) Å) is larger than in dicoordinate gold(I)-ethylene complexes. Note that the phenanthroline ligand is coordinated in a symmetric κ^2 fashion, in marked contrast with that found in related (Phen)Au(phosphine)$^+$ complexes.[93]

In a recent study, Pintus, et al. have also utilized phenanthrolines as N^N-chelating ligands to prepare a series of stable tricoordinate Au(I) π-complexes with norbornene as alkene (six examples).[94] In a first attempt, [Au(phen)(nb)]$^+$-type complexes (nb = norbornene) were synthetized from the dimeric Au(III)-oxo species following the synthetic procedure described by Cinellu et al. The desired products were obtained in low yields and after long reaction times (>10 days) (Scheme 6, route A). As observed in the (di) substituted 2,2′-bipyridine derivatives,[65,66] a mixture of the gold(I) alkene [Au(phen)(nb)][PF$_6$] complexes and the aura-oxetane [Au(κ^2-O,C-2-oxynorbornyl)(NCP)][PF$_6$] was obtained. Although the molar ratio of the two complexes could be optimized, separation of the mixture required tedious work-up. A cleaner and more effective synthetic route was thus developed. The [Au(phen)(nb)][PF$_6$] complexes **16** were obtained in higher yields (up to 80%) and very short reaction times (<40 min) by reacting the phenanthroline ligand with [Au(*tht*)Cl] in dichloromethane, followed by the addition of AgPF$_6$ and norbornene (Scheme 6, route B).

The [Au(phen)(nb)][PF$_6$] complexes **16** are stable both in solution and in the solid state showing no sign of olefin dissociation or degradation. In the ^{1}H NMR spectrum, the olefin protons are significantly high-field shifted with respect to those of free norbornene, in accordance with significant π-backdonation enforced by the chelating nature of the ligand, with $\Delta\delta^1$H up to 1.70 ppm, while the other aliphatic protons remain almost unchanged or resonate at lower field. Likewise, the resonance of the olefin C=C carbon atoms are shifted upfield ($\Delta\delta^{13}$C up to 53.3 ppm). According to XRD analyses, all the complexes display tricoordinate Au(I) centers with the C=C bond of the norbornene ligand lying in the plane of the Au(N^N) fragment (Fig. 6). The C=C distances are elongated with respect to the free alkene, in agreement with π-backdonation from the metal center. DFT calculations confirmed the bonding features, and further elucidated

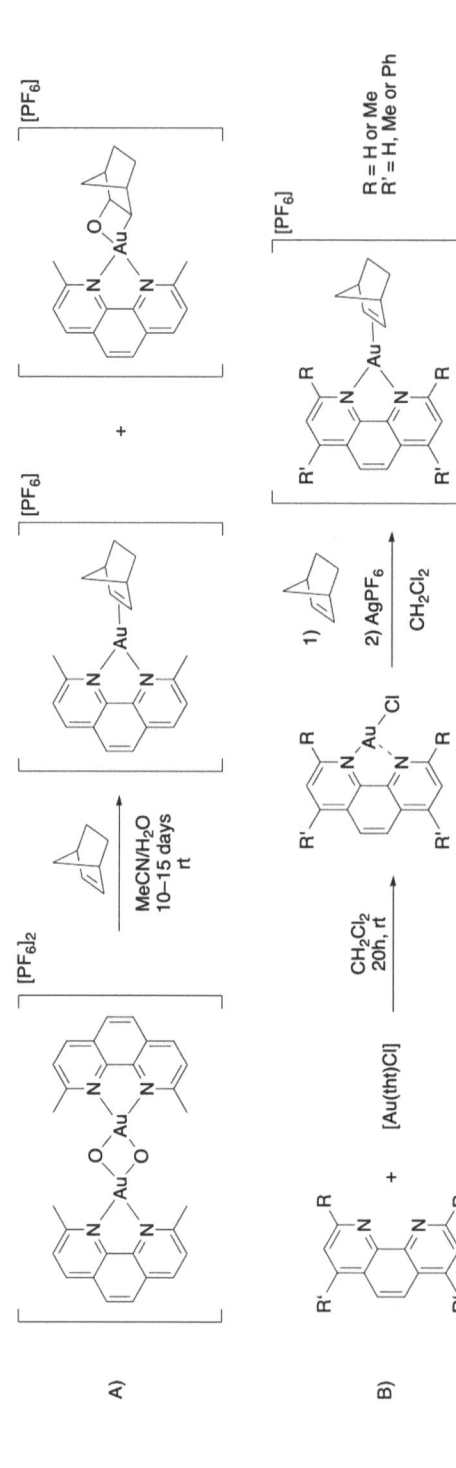

Scheme 6 Synthesis of [Au(phen)(nb)][PF$_6$] complexes by reduction of the Au(III) dioxo complexes (A) and directly by halogen abstraction and olefin coordination (B).

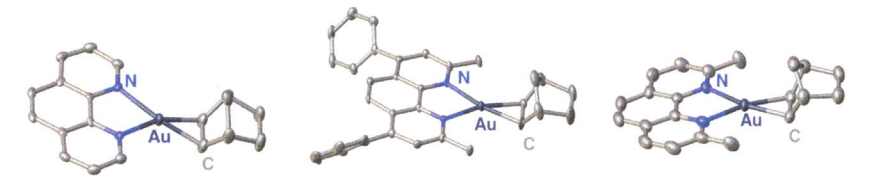

Fig. 6 Some molecular structures of phenanthroline gold(I) norbornene complexes **16**.

the nature of the molecular orbitals involved. NBO analysis indicated that σ-donation prevails over π-backdonation. While π-backdonation is practically the same for all the complexes, the σ-donation is directly related with the nature of the substituents of the phenanthroline ligands, being weaker for phenanthroline ligands featuring substituents in the position α to N, possibly due to their steric hindrance.

2.2 Gold(I) π-complexes with P^P and P^N-chelating ligands

Bidentate P-donor ligands have been less studied in gold(I) chemistry. In this respect, our group has been interested in *o*-carboranyl-bridged diphosphines (*o*-CBD) and *o*-phenylene-bridged P^N ligands. Both ligands are chelating and enforce unusual bent geometry at gold(I), which enables oxidative addition of C—X and C—C bonds to gold under mild conditions.[10–12] The hemilabile character of the P^N ligand was shown to promote Au(I)/Au(III) catalysis.[12,15–21] In addition, in recent parallel studies, our group and Patil's group have highlighted the role of tricoordinate gold(I) alkene complexes in catalytic arylation reactions merging oxidative addition and π-activation at gold.[17,19,20] Coordination of the alkene to the gold(I) center was found to occur prior the oxidative addition step, but without preventing it to take place (Scheme 7).

Reaction of the *o*-carboranyl diphosphine gold(I) chloride with AgSbF$_6$ in the presence of an excess of styrene or ethylene in dichloromethane afforded the (P^P)Au(I) π-complexes **17** and **18** in good yields (Scheme 8).[95,96] Both complexes are air stable solids. The alkene is tightly bonded to the gold(I) center, with no chemical exchange between coordinated and free alkene at the NMR time-scale. In addition, no sign of styrene or ethylene decoordination was observed under vacuum. Coordination of styrene to gold induced noticeable upfield shifts of the ^{1}H (4.75, 4.33 and 4.25 ppm) and ^{13}C NMR (65.9 and 95.0 ppm) vinylic signals with respect to free styrene (5.74, 5.24 and 6.72 ppm, 113.7 and 136.9 ppm respectively).

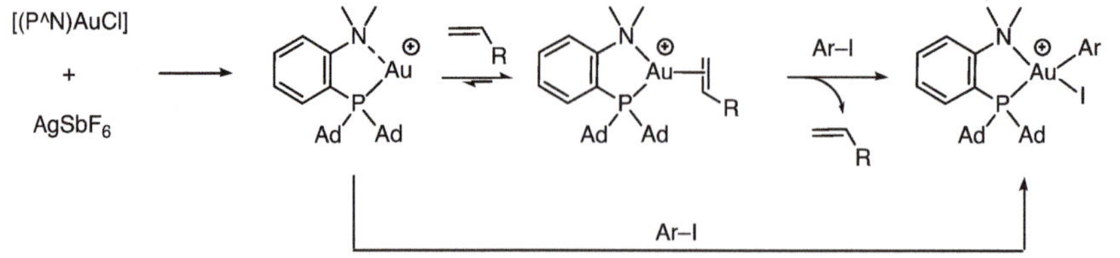

Scheme 7 Proposed mechanism of oxidative addition to gold of aryl iodides with and without the presence of alkenes.

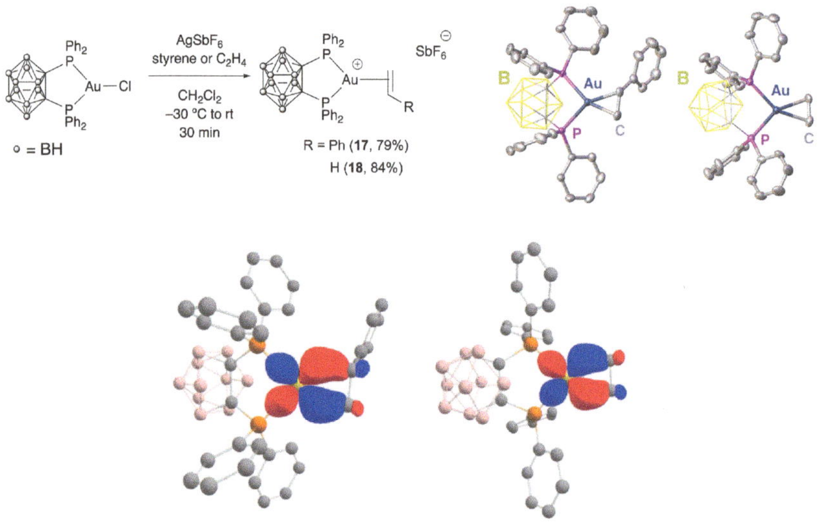

Scheme 8 Synthesis, molecular structures and plots of the NLMO associated to d(Au)→π*(C=C) backdonation of o-carboranyl diphosphine gold(I) styrene (**17**) and ethylene (**18**) complexes.

The same spectroscopic behavior was seen upon ethylene coordination. In both complexes, the o-CBD ligand is symmetrically coordinated to gold and the P–Au–P bite angle (89.13(3)° and 91.32(7)°) is similar to those previously observed in [(P⌒P)Au(CO)]$^+$ and [(P⌒P)Au=C(Ph)R]$^+$ complexes (see Section 3).[97,98] In both complexes, the gold center is in a trigonal-planar environment, and the alkene coordinates in the same plane as the P⌒P ligand. This arrangement maximizes the overlap between the HOMO of the [(P⌒P)Au]$^+$ fragment (in-plane d_{xy}-type orbital) and the π* orbital of the alkene enforcing substantial Au→alkene backdonation, as for the related N⌒N-chelating ligands. This π-backdonation is further noticed in the

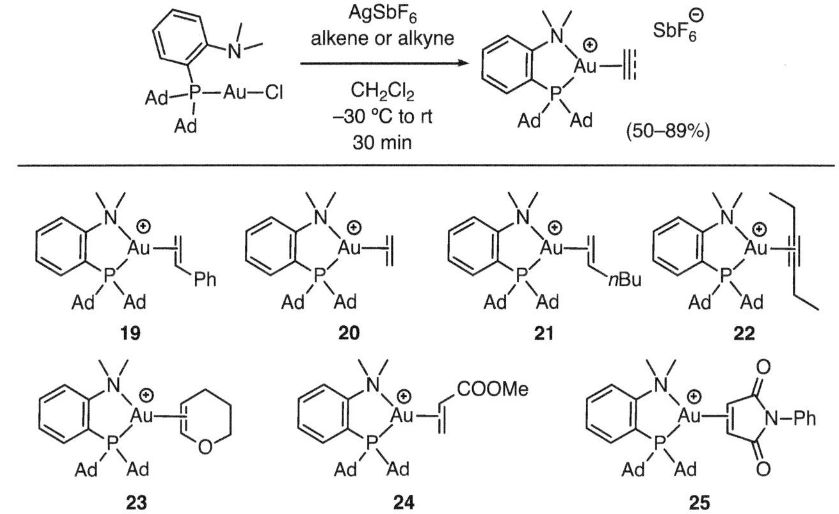

Scheme 9 Synthesis of MeDalphos gold(I) π-complexes.

metrical parameters for the styrene and ethylene molecules coordinated to gold. The C=C bonds (1.395(4) and 1.365(15) Å) are elongated in comparison to free styrene and ethylene (1.35 and 1.313 Å).

The gold(I) complex deriving from the P^N hemilabile ligand MeDalphos has also shown interesting features toward different alkenes.[95,96] Reaction of MeDalphos gold(I) chloride with unbiased alkenes and alkynes such as styrene, ethylene, 1-hexene and 3-hexyne in the presence of AgSbF$_6$ afforded the corresponding gold(I) π-complexes **19–22** (Scheme 9). The ^{1}H NMR spectra display new sets of well-defined vinylic signals (distinct from those of the free π-compounds), indicating the coordination of the alkene to gold. Also noteworthy is the presence of two distinct downfield shifted ^{1}H NMR singlets at ∼3 ppm for complexes **19** (styrene) and **21** (1-hexene) for the N(CH$_3$)$_2$ group, in comparison with one singlet at 2.57 ppm for the starting [(P^N)AuCl] complex. This splitting and the downfield shift indicate that the nitrogen atom coordinates to the Au(I) center. On the other hand and due to the symmetric nature of the ethylene and 3-hexyne, the N(CH$_3$)$_2$ group displayed in complexes **20** and **22** only one singlet at ∼3 ppm, which is downfield shifted in comparison with the [(P^N)AuCl] complex, suggesting here also *N*-coordination to gold(I). In addition, all complexes showed great stability toward air and no sign of alkene/alkyne exchange in the presence of excess of alkene/alkyne at the NMR

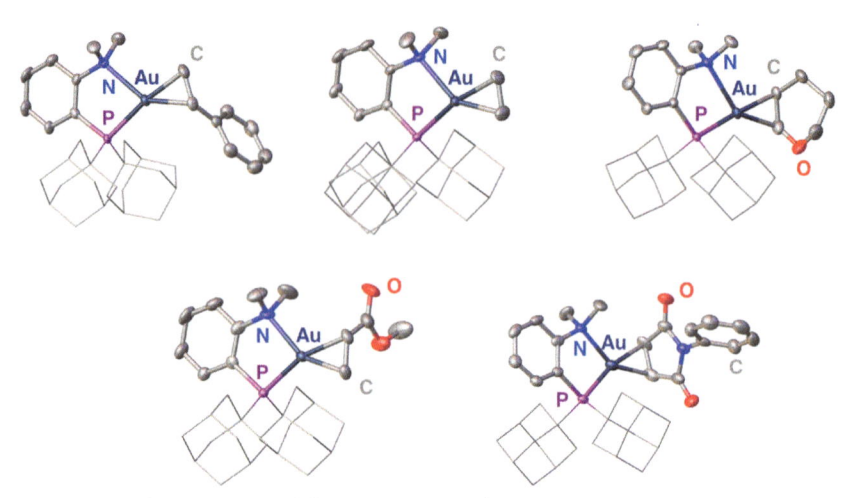

Fig. 7 Molecular structures of the [(P˄N)Au(ene)]⁺ complexes.

time-scale or dissociation under vacuum. Coordination to gold(I) induced a marked upfield shift of the ^{1}H and ^{13}C NMR signals with respect to the free alkene/alkyne. These shift differences reached up to 2.13 ppm in ^{1}H NMR and 65 ppm in ^{13}C NMR spectroscopy, suggesting substantial gold→alkene backdonation. The complexes **19** and **20** adopted similar structures in the solid state with the P˄N ligand chelating gold (Fig. 7). In both complexes, the gold center sits in a trigonal-planar environment and the alkene lies in the (P˄N)Au coordination plane. The presence of substantial d(Au)→π*(C=C) backdonation is apparent from the elongation of the C=C bond upon coordination (by ∼0.06 Å compared to the free alkene). The alkene is η²-coordinated to gold, and both complexes presented quasi-symmetric structures with almost identical Au—C bond lengths.

In contrast to N˄N (bipyridines, phenanthrolines...) and P˄P (*o*-CBD) ligands, the hemilabile MeDalphos was found to be able to form stable π-complexes not only with unbiased alkenes but also with both electron-rich and electron-poor alkenes. Following the same synthetic procedure, MeDalphos gold(I) chloride reacted with 3,4-dihydro-2*H*-pyrane (DHP), methyl acrylate (MA) and *N*-phenylmaleimide (NPM) in the presence of AgSbF$_6$ to yield the corresponding gold(I) π-complexes **23–25** (Scheme 9). The DHP and MA complexes **23** and **24** showed great stability, but the NPM complex **25** could not be isolated. Since decomposition occurred upon filtration, characterization was performed in the presence of N-phenylmaleimide without further purification. For the electron-poor MA and NPM containing complexes, ^{1}H and ^{13}C NMR spectroscopy revealed similar features to those with unbiased alkenes. The coordinated alkene signals are upfield shifted in comparison with the free alkene. In contrast, the coordination of DHP to gold was accompanied by a downfield shift of the vinylic signals.

Interestingly, the magnitude of the ^{1}H NMR downfield shift of the N(CH$_3$)$_2$ signal perfectly correlated with the electronic properties of the π-system: Δδ increases from 0.15 ppm for the DHP complex (electron-rich alkene) to 0.82 ppm for the NPM complex (electron-poor alkene), while complexes with unbiased alkenes exhibited intermediate values (Table 1). The same trends were observed in ^{13}C and ^{31}P NMR spectroscopy, suggesting that the P^N ligand adjusts its coordination to gold depending on the electronics of the π-system.

All data recorded in CD$_2$Cl$_2$. ^{1}H and ^{13}C NMR data from the free alkenes or alkyne are indicated in brackets.

Similarly to the styrene and ethylene complexes, the DHP, MA and NPM π-complexes present structures in the solid state with the P^N ligand chelating gold and a trigonal-planar environment around the gold center (Fig. 7). In all cases, the alkene is η^2-coordinated and sits in the [(P^N)Au]$^+$ coordination plane despite the ensuing steric shielding. Of note, η^2-coordination for a complex with an electron-rich DHP was unprecedented (terminal σ-coordination rather than side-on π-coordination was found in dicoordinate gold(I) π-complexes with electron-rich alkenes).[36,99–103] The three complexes showed an elongation of the C=C bond upon coordination, indicating significant gold→alkene backdonation. The Au—C and Au—P bond lengths show little changes along the series, whereas the Au—N bond length varies from 2.234(6) Å in the NPM complex (the most electron-poor alkene) to 2.505(2) Å in the DHP complex (the most electron-rich alkene) (Table 2). This variation, together with the NMR spectroscopic data of the N(CH$_3$)$_2$ group, attested the versatility and adaptive behavior of the P^N ligand to strengthen/weaken the N—Au interaction to accommodate the electronic demand at gold depending on the alkene. Theoretical calculations further supported both the strong d(Au)→π*(C=C) backdonation, and the ability of the P^N ligand to accommodate the π-coordination of a wide range of alkenes at gold. In addition, a detailed examination of the structural, electronic and spectroscopic

Table 1 Selected ^{1}H, ^{13}C and ^{31}P NMR data for the (P^N)Au(I) π-complexes **19–25**.

$H_b\!\!\underset{2}{\overset{H_a}{\diagdown}}\!\!\underset{1}{=}\!\!R$

	δ^{1}H (ppm)					δ^{13}C (ppm)				δ^{31}P (ppm)
	H$_{1a}$	H$_{1b}$	H$_2$	H$_{NMe1}$	H$_{NMe2}$	C$_1$	C$_2$	C$_{NMe1}$	C$_{NMe2}$	δ^{31}P (ppm)
(P^N)Au–Cl	–			2.57		–		47.0		53.6
(P^N)Au—⫽—⟨O⟩	7.76 (6.34)		4.78 (4.64)	2.74	2.71	144.1 (144.3)	76.7 (100.7)	49.5	49.3	56.5
(P^N)Au—⫽—Ph	4.33 (5.74)	4.17 (5.24)	6.03 (6.72)	3.09	2.82	66.4 (113.7)	99.9 (136.9)	53.0	52.8	57.3
(P^N)Au—⫽—nBu	3.94 (4.95)	3.76 (4.87)	5.66 (5.78)	3.01	2.98	75.2 (116.0)	111.1 (139.0)	52.5	52.5	58.1
(P^N)Au—⫽—	4.10 (5.43)			3.14		75.0 (116.8)		53.9		58.8
(P^N)Au—⫽—	–			3.02		97.8 (80.7)		52.9		59.4
(P^N)Au—⫽—COOMe	4.35 (6.40)	4.61 (5.82)	4.23 (6.13)	3.25	3.07	68.4 (130.6)	72.5 (128.5)	54.9	53.8	63.2
(P^N)Au—⫽—N-Ph (maleimide)	4.68 (6.81)		5.67 (6.81)	3.44	3.33	68.8 (134.1)	72.4 (134.1)	55.4	55.2	65.3

Table 2 Selected bond lengths (Å) for the π-alkene gold(I) complexes 19, 20, 23–25.

	![N-Au-P D/R/Z]	![O ring]	![Ph]	![alkene]	![COOMe]	![N-Ph maleimide]
Au–P	2.3032(6)	2.337(1)	2.3393(5)	2.326(1)	2.334(2)
Au–N	2.505(2)	2.381(4)	2.306(2)	2.275(5)	2.234(6)
Au–C	2.276(3)	2.208(5)	2.141(3)	2.114(6)	2.133(8)
	2.193(3)	2.143(5)	2.149(3)	2.155(6)	2.136(7)
C=C	1.385(5)	1.387(8)	1.387(5)	1.406(10)	1.394(12)

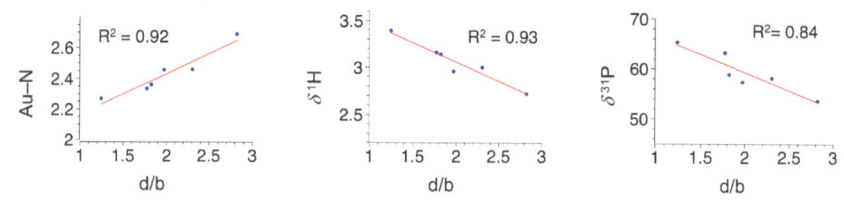

Fig. 8 Correlation graphs for the response of the P^N ligand (Au—N distance in Å, ^{1}H NMR chemical shift of the NMe$_2$ group and ^{31}P NMR chemical shift in ppm) to the alkene coordination to gold characterized by the donation/backdonation ratio d/b, as determined by Charge Decomposition Analyses.

data collected experimentally and computationally revealed a direct and linear response of the P^N MeDalphos ligand to the Au/alkene coordination along the series of gold(I) π-complexes (Fig. 8).

2.3 Comparison of gold(I) ethylene complexes with chelating and hemilabile ligands

No gold(I) ethylene complex has been reported with monodentate ancillary ligands, but as aforementioned, chelating ligands have proved powerful to stabilize such complexes. Hereafter are gathered all the known tricoordinate gold(I) ethylene complexes, showing the variety of N and P-based chelating/hemilabile ligands used so far. The ^{1}H and ^{13}C NMR chemical shifts, and the C=C bond distances of the coordinated ethylene are listed in Table 3.

Table 3 Comparison between reported Au(I) ethylene complexes and free ethylene.

Compound	Ligand type	δ ¹H (ppm)	δ ¹³C (ppm)	C=C (Å)	References
‖	–	5.43	116.8	1.313	72
[Au–ethylene cation]	–	4.94	92.7	1.371(7) 1.351(7) 1.369(7)	83
[bipyridine Au ethylene]	Neutral N N	3.09	61.6	–	66
[scorpionate Au ethylene]	Anionic N N	3.81 (**6a**) 3.69 (**6b**) 2.61 (**7a**) 3.00 (**7b**)	63.7 (**6a**) 59.3 (**6b**) 55.3 (**7a**) 56.9 (**7b**)	1.380(10) (**6a**) 1.387(9) (**6b**) 1.413(7) (**7a**) 1.410(5) (**7b**)	69,75
[fluorinated diazadiene Au ethylene]	Anionic N N	2.71	59.1	1.405(4)	77
[naphthyl diimine Au ethylene]	Neutral N N	3.31 3.28	65.4	1.455(13)	82
[difluorobipyridine Au ethylene]	Neutral N N	3.90	63.8	1.399(5)	84

Table 3 Comparison between reported Au(I) ethylene complexes and free ethylene.—cont'd

Compound	Ligand type	δ ¹H (ppm)	δ ¹³C (ppm)	C=C (Å)	References
(nBu-substituted N^N Au complex)	Neutral N^N	3.88	60.6	1.411(10) 1.383(8)	92
(Ph₂P-carborane-PPh₂ Au complex)	Neutral P^P	4.47	74.0	1.387(5)	96
(N,P(Ad)₂ Au complex)	Neutral P^N	4.10	75.0	1.365(15)	96

3. Gold(I) carbene complexes

Gold(I) carbene complexes have been known for several decades. The first Fischer-type gold(I) carbene complexes Cl–Au=C(X)Ph (X=OMe, NH₂, NHMe, NMe₂) were reported by Fischer himself as early as in 1981.[104,105] About 20 years later, a series of cationic Fischer-type gold(I) carbene complexes featuring an NHC or a phosphine as ancillary ligand were structurally characterized (Fig. 9).[106–108] The presence of a π-donating OMe group at the carbene center reduces the reactivity. These complexes feature little ordinary carbene character and are best described by their carbocationic form. The Au—C bond order is very close to one and the C—O distance is significantly contracted.

Since the first isolation of a *N*-heterocyclic carbene gold(I) complexes in 1973,[109] a number of NHC gold(I) complexes [(NHC)AuCl] have been prepared taking advantage of the many straightforward synthetic routes.[110] Notably, NHC-gold complexes have shown great potential in biological applications,[111] luminescent devices[112] and as catalysts in many organic transformations.[113] Besides the use of carbenes as ancillary ligands, gold(I) carbenoid species have been postulated as intermediate species in many transformations induced by gold catalysts. Therefore, the preparation and characterization of gold(I) carbene complexes has captured growing interest in the last years despite the difficulties of stabilizing such reactive species and

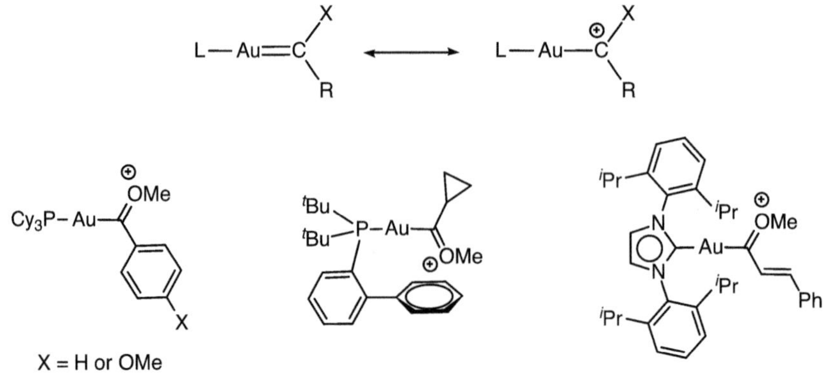

Fig. 9 Carbene and carbocationic forms of gold(I) carbenes; some structurally characterized Fischer-type gold(I) carbene complexes.

gathering useful structural information. For instance, in 2013 Widenhoefer et al. managed to observe spectroscopically at low temperature a gold carbenoid intermediate formed by a prototype enyne rearrangement for the first time.[114] Most of the gold(I) carbene complexes reported to date are stabilized by π conjugation of the electro-deficient carbenoid center with heteroatoms.[106–108] In 2014, Fürstner et al. described the first stable gold(I) carbene **26** without heteroatomic substituent (Fig. 10).[106] According to X-ray diffraction analysis, there is little Au → C$_{carbene}$ back donation and hence very modest Au—C double-bond character. The electron deficiency of the carbene center is tempered by π-delocalization of the electron-rich aryl substituents. In the same year, Widenhoefer et al. reported a gold(I) cycloheptatrienylidene complex **27**. Here, the carbene moiety is integrated into a 6π-electron aromatic ring (Fig. 10).[115] In a parallel study, Straub et al. prepared and fully characterized a stable diaryl gold(I)-carbene complex **28**. The carbene center is substituted by two Mes groups and a very bulky NHC is used as ancillary ligand, resulting in very strong steric shielding of the carbenoid moiety (Fig. 10).[116]

In 2014, our group took advantage of the chelating properties of the o-carboranyl diphosphine ligand to synthetize and fully characterize the first classical carbonyl complex of gold.[97] The bent PAuP$^+$ fragment is crucial to raise the energetic level of the occupied d$_{xz}$(Au) orbital and enhance π-backdonation. Inspired by the unusual electronic structure of the gold carbonyl complex, we targeted the first carbene complex stabilized by the gold fragment rather than the carbene substituents. Reaction of [(P P)AuNTf$_2$] with diphenyldiazomethane at low temperature (−40 °C) afforded the

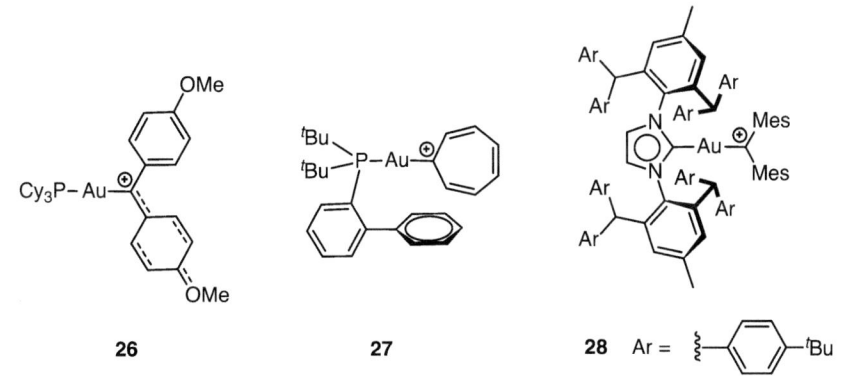

Fig. 10 Non-heteroatom substituted gold(I) carbene complexes featuring phosphine/NHC ancillary ligands.

gold(I) carbene complex [(P P)Au(CPh$_2$)]$^+$ **29** in excellent yield as a highly thermally stable solid (Scheme 10).[97] The gold center is arranged in a trigonal-planar environment, in which the DPCb ligand chelates the gold center with a bite angle of 90.26(4)°, similar to those observed in the π-complexes **17** and **18** (89.12(2)° and 91.32(7)°). The carbene center is perfectly planar and almost perpendicular to the [(P P)Au]$^+$ coordination plane (with a twist angle of 85.2°). This orientation minimizes steric repulsion and enables optimal interaction between the high-energy occupied d$_{xz}$(Au) orbital in the plane of the [(P P)Au]$^+$ fragment and the vacant 2p orbital at the carbene center. The Au–C$_{carbene}$ bond is slightly shorter (1.984 Å) to those of complexes **26–28**. Optimization of the geometry by DFT calculations and NBO analysis corroborated the bonding interaction between the d$_{xz}$(Au) and 2p^π(C) orbitals in the coordination plane of gold, indicating significant carbene π-backdonation and some degree of Au–C π bonding.

The easy generation and versatile reactivity of α-oxo gold carbene complexes was recognized in the early 2000s, giving rise to numerous applications in gold catalysis.[117–119] In a pioneering work of Pérez and Nolan, the reaction of α-diazo carbonyl compounds with phosphine and NHC gold(I) precursors proved to be a very efficient and general route to generate highly electrophilic α-oxo carbene intermediates[120,121] of many carbene transfer[122–126] and coupling[127,128] catalytic reactions. Despite their role as key intermediates, gold(I) α-oxo carbene complexes have remained elusive for many years.[129–131] Remarkably, the group of Zhang developed a Au(I)-catalyzed modular synthesis of 2,4-disubstituted oxazoles via [3 + 2]

$\delta\ ^{13}C = 316.2$ ppm
($J_{CP} = 75.2$ Hz)

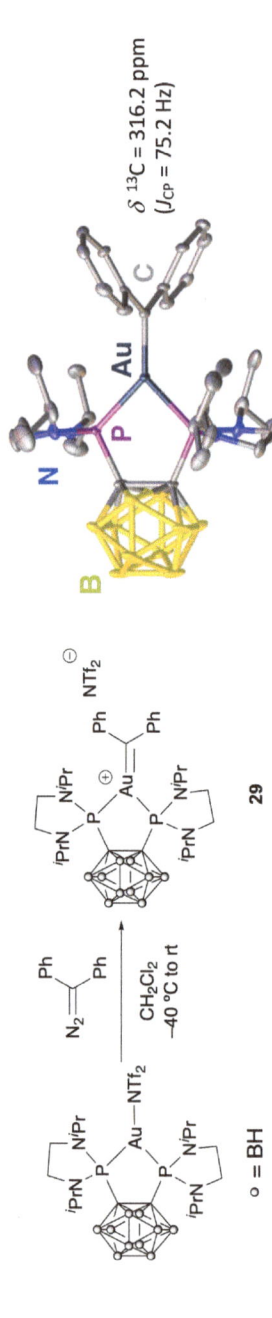

○ = BH

29

Scheme 10 Synthesis and structure of the P^P gold carbene complex **29**.

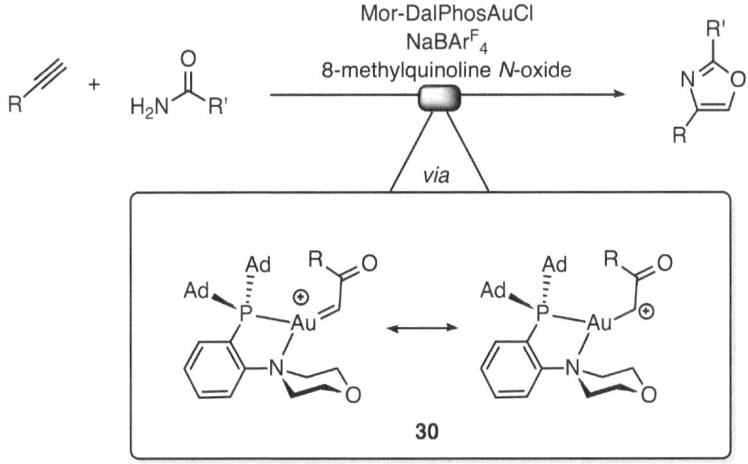

Scheme 11 Gold(I)-catalyzed synthesis of 2,4-disubstituted oxazoles through the generation of a (P⌒N)-chelated α-oxo gold carbene **30**.

annulations between alkynes and aromatic/alkenic carboxamides under mild conditions.[132–135] A highly electrophilic α-oxo gold(I) carbene generated *via* gold-promoted oxidation of a terminal alkyne was postulated as the key reaction intermediate (Scheme 11). However, the use of P⌒N or P⌒S bidentate ligands, especially MorDalphos-type ligands,[136] significantly tempered its reactivity, permitting its efficient trapping by a carboxamide en route to the formation of the oxazole ring. The formation of a tricoordinate gold(I) carbene species **30** by coordination of the N or S atom was proposed to reduce the electrophilicity of the carbene center.[137–139]

In view of the studies on gold(I) α-oxo carbene complexes stabilized by P⌒N and P⌒S chelating ligands, together with the unprecedented reactivity and bonding situation induced by the P⌒N bidentate ligand MeDalphos,[15–17] our group envisioned that a tricoordinate α-oxo carbene gold(I) complex could be generated by reacting the α-diazo ester PhC(=N$_2$)CO$_2$Et with the [(MeDalphos)AuCl] precursor in the presence of AgSbF$_6$ (Scheme 12).[98] However, despite *in situ* and low-temperature NMR monitoring, the proposed [(P⌒N)Au=C(Ph)CO$_2$Et]$^+$ **31** was not detected. Its geometry was optimized by DFT (Scheme 12) and an NBO analysis was carried out. This revealed a Au...N distance of 2.53 Å, which is relatively long [46% shorter than the Σr_{VdW} (3.07 Å)[74] but 22% longer than the Σr_{cov} (2.07 Å)].[140] In addition, the P–Au–C$_{\text{carbene}}$ arrangement is almost linear (163.3°), and thus with the absence of a significant N to Au coordination the gold center becomes electron-deficient with a significant transfer of

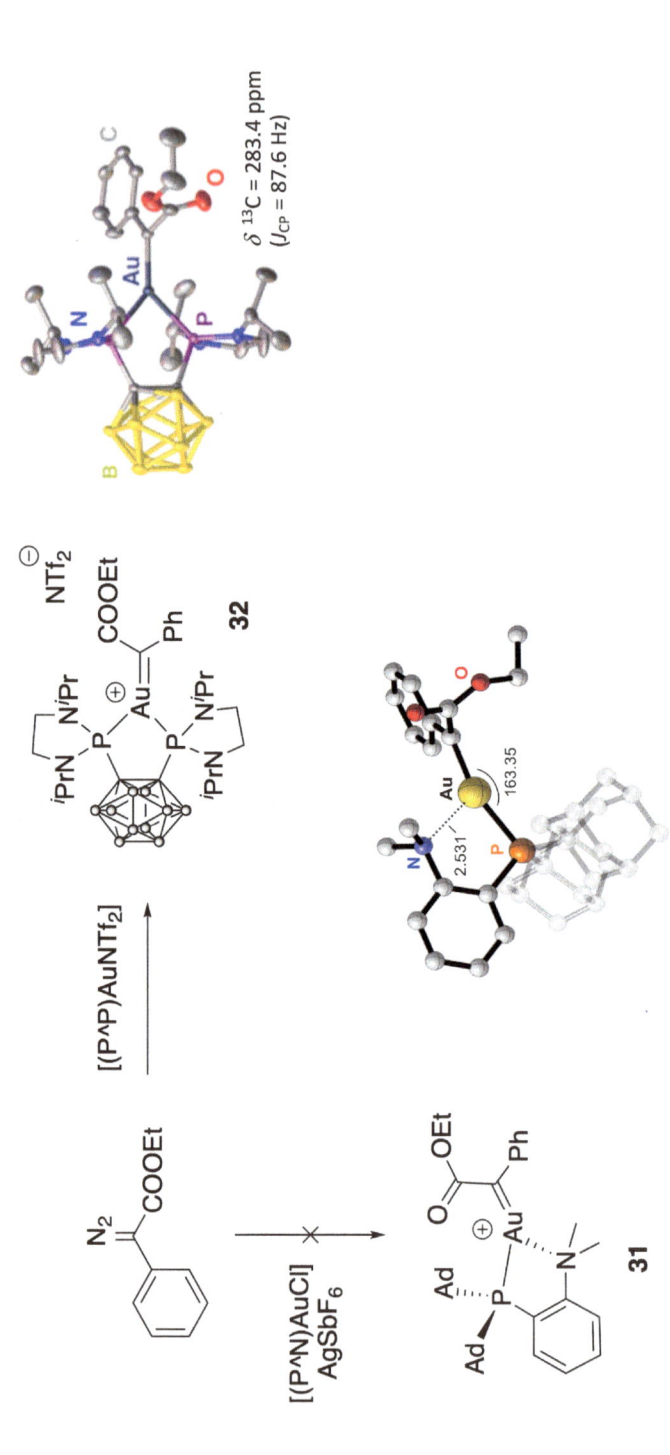

Scheme 12 α-Oxo gold(I) carbene complexes with chelating P^P and P^N ligands: (attempted) synthesis, DFT-optimized and cristallographically-determined structures.

electron density from the carbene to the metal (Charge Transfer = 0.8e). This bonding situation explains the high electrophilicity and instability of the proposed carbene complex.

On the other hand, reaction of the *o*-carboranyl diphosphine gold(I) precursor with PhC(=N$_2$)CO$_2$Et at −40 °C was accompanied by immediate dinitrogen evolution, and generated the α-oxo carbene gold(I) complex **32** (Scheme 12). The [(P P)Au=C(Ph)CO$_2$Et]$^+$ complex was found to be thermally unstable (decomposition is quite fast above 0 °C). It was characterized spectroscopically at low temperature and crystals suitable for X-ray diffraction analysis could be obtained. ^{31}P NMR spectroscopy showed a single resonance signal at δ = 138.9 ppm, indicative of symmetric coordination of the phosphorus atoms to gold. In addition, ^{13}C NMR spectroscopy showed a characteristic carbene resonance signal appearing as a triplet (*J*$_{CP}$ = 87.6 Hz) at 283.4 ppm, which is in the range of previously characterized gold(I) carbenoid complexes. As apparent from the solid-state structure, the gold center is arranged in a trigonal-planar environment with the DPCb ligand coordinated in a κ^2-coordination fashion. The P–Au–P bite angle of 90.26(4)° is similar to those observed in related [(P P)Au–L]$^+$ complexes (L = styrene, ethylene, CO or CPh$_2$). The Au—C bond distance (1.961(2) Å) is in the shortest range of those reported for gold carbene complexes. Of note, the phenyl ring is nearly co-planar with the carbene center and the C$_{carbene}$–C$_{ipso}$ bond is slightly contracted (1.448(4) Å) indicating a certain degree of π-delocalization between the phenyl ring and the carbene center, thus stabilizing the carbene species. NBO analysis revealed π$_{C=C}$(Ph) → 2p$^\pi$(C) and n$_C^\sigma$ → π*$_{C=O}$ donor-acceptor interactions indicating push-pull stabilization of the α-oxo carbene. In addition, a significant d$_{xz}$(Au) → 2p$^\pi$(C) backdonation was also apparent. Charge transfer (CT) and charge-decomposition analysis (CDA) indicated the relative contributions of C$_{carbene}$ → Au donation and Au → C$_{carbene}$ backdonation with a d/b ration of 1.76, which is significantly smaller than that computed for the [(P N)Au=C(Ph)CO$_2$Et]$^+$ analog (d/b = 2.16), supporting a lower electrophilicity and higher stability of the P P-chelated complex (Fig. 11).

In 2009, Toste et al. reported a detailed investigation of the bonding situation in gold(I) carbenoid complexes. The influence of the carbene substituents and ancillary ligand at gold on the structure (carbene *versus* carbocationic form) and reactivity was analyzed.[141] A few years later, Kästner et al. also studied stabilizing effects in gold(I) carbene complexes by Intrinsic Bond Orbital (IBO) calculations.[142] Again, π-stabilization through delocalization or by the presence of heteroatoms in α position of the carbene center

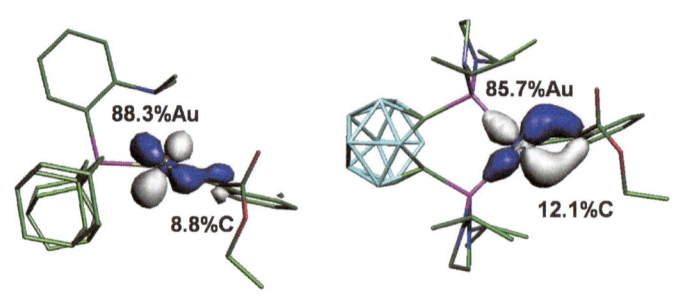

Fig. 11 Plots of the NLMO associated with the $d_{xz}(Au) \rightarrow 2p^{\pi}(C)$ backdonation in the α-oxo carbene gold(I) complexes featuring either P^N or P^P chelating ligand (with the contributions of the gold and $C_{carbene}$ atoms in percent).

appeared crucial for the stabilization of gold(I) carbene complexes. In addition, $C_{carbene}$ to gold σ-donation was found to prevail in the Au—C bonding. It is similar for di and tricoordinate gold(I) carbene complexes, however π-backdonation is significantly higher in tricoordinate complexes (0.175 versus 0.074), in line with the electronic properties of the bent P–Au–P$^+$ fragment.

To demonstrate that the P^P-Au α-oxo carbene **32** complex can be indeed considered as a mimic of transient species, different representative transformations were tested, first in stoichiometric and then in catalytic conditions. Reaction of styrene with the α-diazo ester in the presence of 5 mol% of [(P^P)AuNTf$_2$] afforded the cyclopropane derivative **33** in 58% yield after 64 h at room temperature (Scheme 13). The Au(I) carbene complex also reacted with Ph$_3$P→BH$_3$ to give the B—H insertion product **34** in 60% yield after 30 min. In addition, the Au(I) carbene complexes reacted with phenol (60% yield after 2 h) to give the O—H insertion aryl ether product **35** (Scheme 13). This O—H insertion differs from the reported monoligated phosphine gold(I) complexes.[125,143] Gold carbene complexes possess unique propensity to insert into the *para* C—H bond, enabling the direct and selective functionalization of unprotected phenols. Chemoselectivity for C—H versus O—H insertion is favored with electron-deprived phosphine ligands, which increase the electrophilicity of the α-oxo carbene intermediate. However, the P^P ligand strengthens the Au → $C_{carbene}$ backdonation and decreases the electrophilicity of the carbene switching the chemoselectivity toward O—H insertion leading to the formation of the aryl ether.[144]

The use of an α-diazo ester with electron-withdrawing groups (CF$_3$ groups) in remote positions was evaluated to confirm the critical role of the gold carbene electrophilicity. In this case, the functionalized phenol

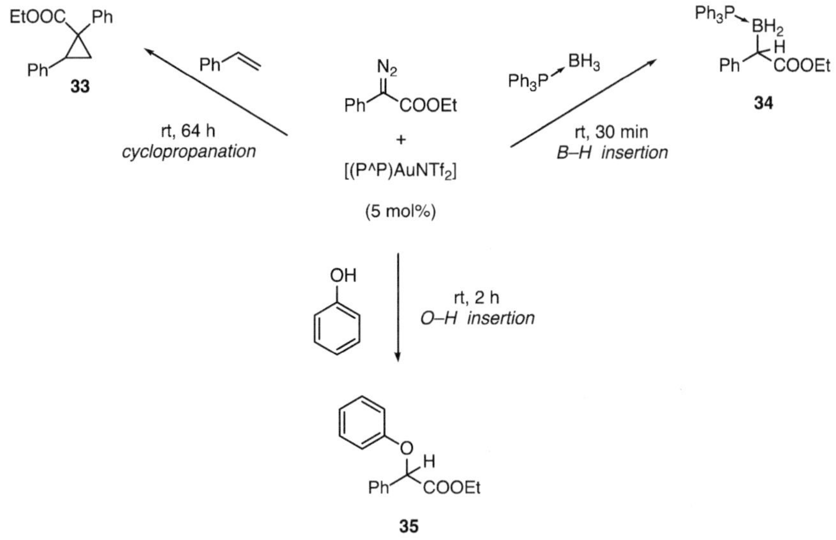

Scheme 13 Gold carbene reactivity upon reaction of a-diazo ester with styrene, triphenylphosphine-borane, and phenol catalyzed by the [(P⌃P)AuNTf₂] complex.

36 derived from C—H insertion was obtained as major product along with some aryl ether **37** (Scheme 14). According to DFT calculations, the frontier orbitals of the (F₃C)₂Ph-substituted gold carbene are noticeably lower in energy than those of the corresponding Ph-substituted gold carbene. The resulting P P-chelated gold carbene actually resembles the [(P⌃N)AuC(Ph)CO₂Et]⁺ complex in carbene electrophilicity, which explains why C—H insertion prevails in this case.

Extending the variety of stable reactive gold(I) carbene complexes is an obvious outlook of the recent advances, in particular with respect to the substitution pattern at the carbene. In a recent study, Echavarren et al. generated and spectroscopically characterized for the first time a monosubstituted gold(I) carbene in solution.[145] The reaction of the gold(I) complexes bearing JohnPhos-type ligands **38** with mesityl diazomethane first afforded the corresponding gold(I) carbenoids **39** (Scheme 15). Subsequently, addition of GaCl₃ at −90 °C generated the gold(I) carbenes **40**, which are stable only below −70 °C. The monosubstituted gold(I) carbene complexes were characterized spectroscopically at low temperature, presenting a characteristic signal in ¹H NMR spectroscopy at 11.74–12.67 ppm corresponding to the proton in α position to the carbene. The carbenic carbon centers appear as doublets at 284.6–290.0 ppm with ²J_{CP} of 96.8–99.8 Hz, in line with previously characterized gold(I) carbenes.[97,98,106–108,115,116] NBO analyses

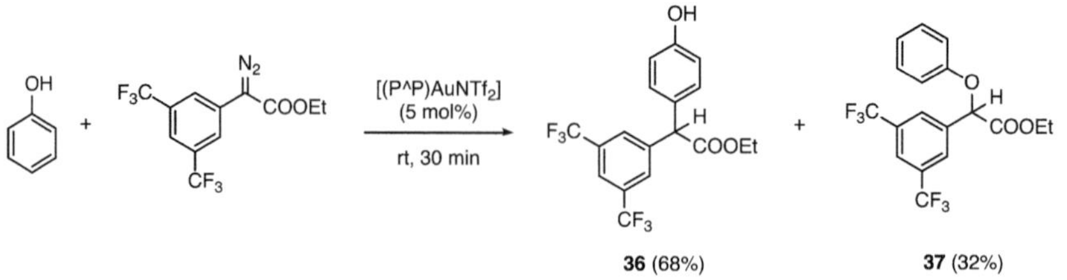

Scheme 14 C—H versus O—H functionalization of phenol with the α-diazo ester catalyzed by the [(P P)AuNTf$_2$] complex.

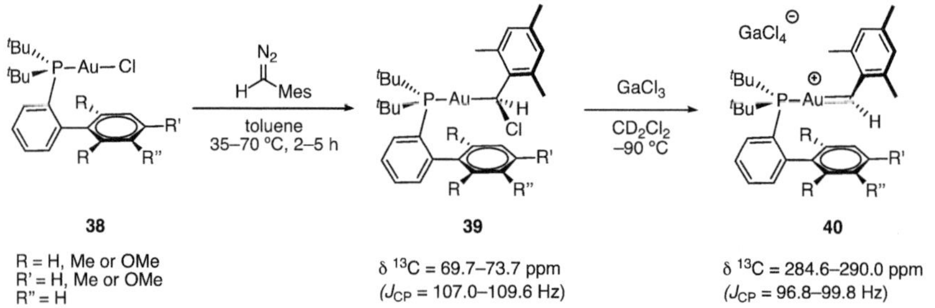

Scheme 15 Synthesis of the chloro(mesityl)methylgold(I) carbenoids **39** and generation of the corresponding monosubstituted gold(I) carbenes **40**.

revealed weak π-backdonation from the d$_{xz}$(Au) orbital to the 2p$^\pi$(C) orbital, but significant C–C(Ar) π-bonding.

The monosubstituted gold(I) carbenes **40** display rich reactivity and undergo typical carbene-type transformations such as cyclopropanation, oxidation, and C—H insertion reactions (Scheme 16). For instance, treatment with pyridine N-oxide or cyclohexane generated 2,4,6-trimethylbenzaldehyde and the C—H insertion product, respectively. Small amounts of the E-configured alkene dimerization product were observed in both cases. The alkene as well as the corresponding cyclopropane derivative were actually obtained when the gold carbene was warmed up and let to evolve without external reagent. It is nevertheless possible to trap the carbene with alkenes. Mono, di and tri-substituted substrates work well, and the corresponding cyclopropanes were all obtained in high yields.

Recently, Fürstner et al. have spectroscopically characterized gold(I) difluorocarbenoid complexes for the first time.[146] These species were generated by Lewis-acid mediated α-fluoride elimination from dicoordinate L–Au–CF$_3$ complexes. Reaction of complex **41** with TMSOTf or TMSNTf$_2$ resulted in the formation of the gold(I) carbenoid complexes

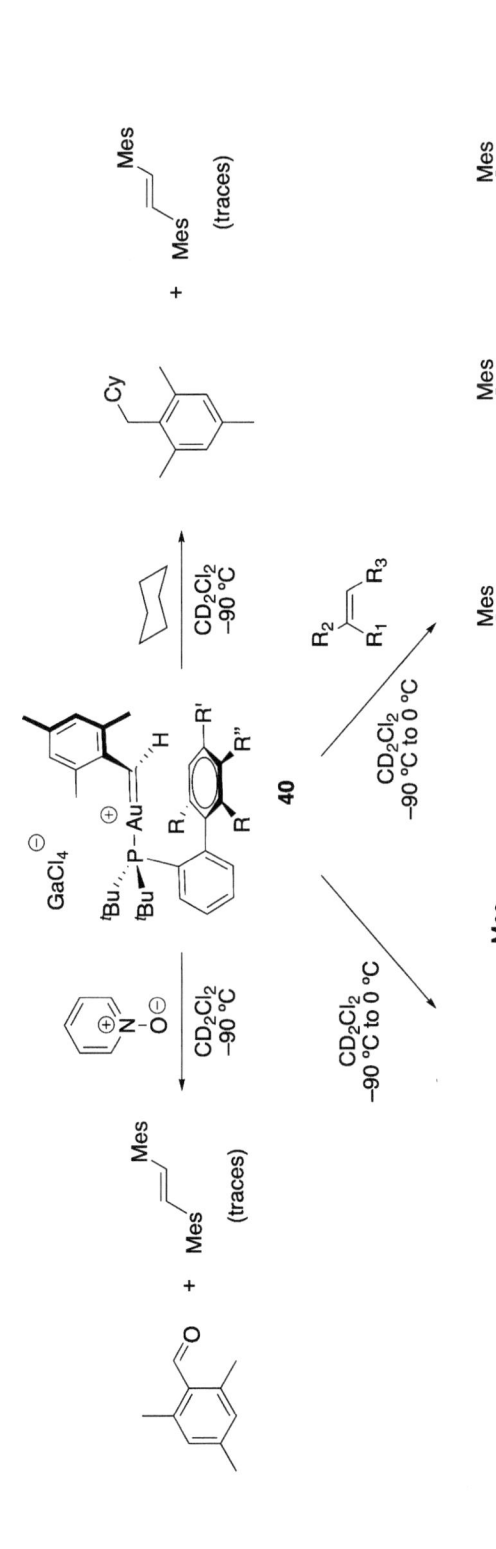

Scheme 16 Reactivity of monosubstituted gold(I) carbenes **40**.

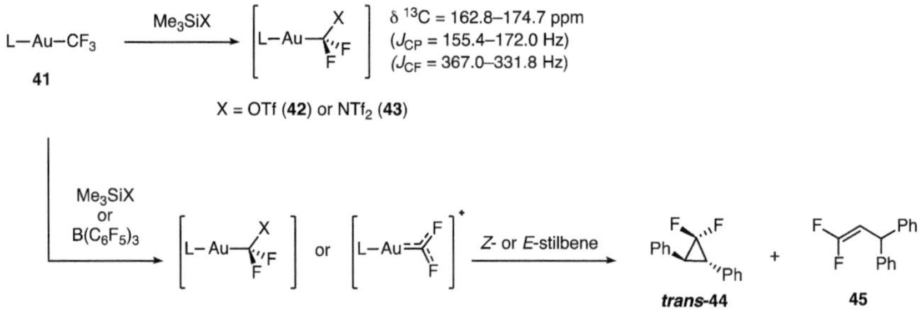

Scheme 17 Generation of gold(I) difluorocarbenoids (L=PPh$_3$, PCy$_3$, XPhos, [2,4-(tBu)$_2$C$_6$H$_3$O]$_3$P or (F$_3$CC$_6$H$_4$)$_3$P) complexes and their reactivity with stilbenes.

42 and **43**, which were only stable below −50 °C (Scheme 17). ^{19}F and ^{31}P NMR spectroscopy showed characteristic signals for one phosphorus center and two fluorine atoms mutually coupled. The carbon atom bonded to gold appeared as a doublet of triplets at ∼170 ppm in the ^{13}C NMR spectrum, in line with a carbenoid structure rather than a true "metal carbene" which would resonate at much lower field.[147] Consistently, ^{19}F COSY NMR showed cross peaks between the signals of the carbenoid Au–CF$_2$–X and of the OTf group. π-Donation from the fluorine atoms is insufficient to compensate for the high electrophilicity of the carbon atom, and therefore the weakly coordinating OTf/NTf$_2$ counteranions remain covalently bonded. In the presence of TMSOTf, TMSNTf$_2$ or B(C$_6$F$_5$)$_3$, the L–Au–CF$_3$ complexes react with stilbenes to give the corresponding difluorocyclopropane. The obtention of the *trans* cyclopropane ***trans*-44** from both *E* and *Z*-stilbene, along the formation of some difluoroalkene **45**, argue in favor of a stepwise reaction involving a metal-carbocation as intermediate.

Chelating ligands enhance Au→carbene π-backdonation and are therefore attractive to stabilize both mono-substituted carbenes (with an aryl or an alkyl group), fluorocarbenes and/or fluoroalkyl carbenes at gold(I). *o*-Carboranyl diphosphine derivatives together with other chelating ancillary ligands are certainly to be tested to this end. Besides P-based chelating ligands, rigid and strongly donation N^N ligands such of those used for gold(I) π-complexes are also worthwhile to be explored.

4. Concluding remarks

In a few years of time, chelating ligands have proved extremely versatile and powerful in gold(I) chemistry. In particular, N^N, P^P and

P^N ligands have enabled to prepare a variety of stable π-alkene/alkyne and carbene complexes mimicking key transient species involved in important catalytic transformations. The field is still in its infancy and offers many perspectives. The variety of such complexes can certainly be extended further, using other bidentate ligands (including cyclometalated ones) and varying the nature and substitution pattern of the π-substrate and carbene moiety. Exploring the reactivity of these species is also clearly of high interest. It will for sure advance our knowledge in gold chemistry and open new avenues in gold catalysis.

Acknowledgments

Financial support from the Centre National de la Recherche Scientifique, the Université de Toulouse and the Agence Nationale de la Recherche is gratefully acknowledged. M.N. thanks the Fonds National Suisse de la Recherche Scientifique for an Early Postdoc Mobility fellowship. Special gratitude is expressed to all the coworkers (whose names appear in the references) for their invaluable contribution. K. Miqueu (CNRS/Université de Pau et des Pays de l'Adour, IPREM) is warmly thanked for long-standing, very fruitful and stimulating collaboration. M. Rigoulet is acknowledged for the calculations on the model phosphine and diphosphine complexes.

References

1. Hashmi ASK, Hutchings GJ. Gold catalysis. *Angew Chem Int Ed*. 2006;45(47):7896–7936. https://doi.org/10.1002/anie.200602454.
2. Hashmi ASK, Toste FD. *Modern Gold Catalyzed Synthesis*. Wiley-VCH; 2012. https://doi.org/10.1002/anie.201207733.
3. Carvajal MA, Novoa JJ, Alvarez S. Choice of coordination number in d^{10} complexes of group 11 metals. *J Am Chem Soc*. 2004;126(5):1465–1477. https://doi.org/10.1021/ja038416a.
4. Crespo O, Gimeno MC, Laguna A, Jones PG. Two-, three- and four-coordinate gold(I) complexes of 1,2-bis(diphenylphosphino)1,2-dicarba-*closo*-dodecarborane. *J Chem Soc Dalton Trans*. 1992;(10):1601–1605. https://doi.org/10.1039/DT9920001601.
5. Otsuka S. Chemistry of platinum and palladium compounds of bulky phosphines. *J Organomet Chem*. 1980;200(1):191–205. https://doi.org/10.1016/S0022-328X(00)88646-5.
6. Gorin DJ, Toste FD. Relativistic effects in homogenous gold catalysis. *Nature*. 2007;446:395–403. https://doi.org/10.1038/nature05592.
7. Lauterbach T, Livendahl M, Rosellón A, Espinet P, Echavarren AM. Unlikeliness of Pd-free gold(I)-catalyzed sonogashira coupling reactions. *Org Lett*. 2010;12(13):3006–3009. https://doi.org/10.1021/ol101012n.
8. Livendahl M, Goehry C, Maseras F, Echavarren AM. Rationale for the sluggish oxidative addition of aryl halides to Au(I). *Chem Commun*. 2014;50(13):1533–1536. https://doi.org/10.1039/C3CC48914K.
9. Teles JH. Oxidative addition to gold(I): a new avenue in homogeneous catalysis with Au. *Angew Chem Int Ed*. 2015;54(19):5556–5558. https://doi.org/10.1002/anie.201501966.

10. Joost M, Zeineddine A, Estévez L, et al. Facile oxidative addition of aryl iodides to gold(I) by ligand design: bending turns on reactivity. *J Am Chem Soc*. 2014;136(42):14654–14657. https://doi.org/10.1021/ja506978c.
11. Joost M, Estévez L, Miqueu K, Amgoune A, Bourissou D. Oxidative addition of carbon-carbon bonds to gold. *Angew Chem Int Ed*. 2015;54(17):5236–5240. https://doi.org/10.1002/anie.201500458.
12. Zeineddine A, Estévez L, Mallet-Ladeira S, Miqueu K, Amgoune A, Bourissou D. Rational development of catalytic Au(I)/Au(III) arylation involving mild oxidative addition of aryl halides. *Nat Commun*. 2018;8:565. https://doi.org/10.1038/s41467-017-00672-8.
13. Fernandez I, Wolters LP, Bickelhaupt FM. Controlling the oxidative addition of aryl halides to Au(I). *J Comput Chem*. 2014;35(29):2140–2145. https://doi.org/10.1002/jcc.23734.
14. Huang B, Hu M, Toste FD. Homogenous gold redox chemistry: organometallics, catalysis, and beyond. *Trends Chem*. 2020;2(8):707–720. https://doi.org/10.1016/j.trechm.2020.04.012.
15. Rodriguez J, Zeineddine A, Sosa-Carrizo ED, et al. Catalytic Au(I)/Au(III) arylation with the hemilabile MeDalphos ligand: unusual selectivity for electron-rich iodoarenes and efficient application to indoles. *Chem Sci*. 2019;10(30):7183–7192. https://doi.org/10.1039/C9SC01954E.
16. Rodriguez R, Adet N, Saffon-Merceron N, Bourissou D. Au(I)/Au(III)-catalyzed C–N coupling. *Chem Commun*. 2020;56(1):94–97. https://doi.org/10.1039/C9CC07666B.
17. Rigoulet M, Thillaye du Boullay O, Amgoune A, Bourissou D. Gold(I)/Gold(III) catalysis that merges oxidative addition and π-alkene activation. *Angew Chem Int Ed*. 2020;59(38):16625–16630. https://doi.org/10.1002/anie.202006074.
18. Akram MO, Das A, Chakravarty I, Patil NT. Ligand-enabled gold-catalyzed C(sp^2)–N cross coupling reactions of aryl iodides with amines. *Org Lett*. 2019;21(19):8101–8105. https://doi.org/10.1021/acs.orglett.9b03082.
19. Chintawar CC, Yadav AK, Patil NT. Gold catalyzed 1,2-diarylation of alkenes. *Angew Chem Int Ed*. 2020;59(29):11808–11813. https://doi.org/10.1002/anie.202002141.
20. Tathe AG, Chintawar CC, Bhoyare VW, Patil NT. Ligand-enabled gold-catalyzed 1,2-heteroarylation of alkenes. *Chem Commun*. 2020;56(65):9304–9307. https://doi.org/10.1039/D0CC03707A.
21. Zhang S, Wang C, Ye X, Shi X. Intermolecular alkene difunctionalization *via* gold-catalyzed oxyarylation. *Angew Chem Int Ed*. 2020;59(46):20470–20474. https://doi.org/10.1002/anie.202009636.
22. Dyker G. An eldorado for homogeneous catalysis? *Angew Chem Int Ed*. 2000;39(23):4237–4239. https://doi.org/10.1002/1521-3773(20001201)39:23<4237::AID-ANIE4237>3.0.CO;2-A.
23. Hashmi ASK. Homogeneous catalysis. *Gold Bull*. 2004;37:51–65. https://doi.org/10.1007/BF03215517.
24. Hoffmann-Röder A, Krause N. The golden gate to catalysis. *Org Biomol Chem*. 2005;3:387–391. https://doi.org/10.1039/B416516K.
25. Toste FD, Michelet V. *Gold Catalysis: An Homogeneous Approach*. Imperial College Press; 2014.
26. Slaughter LM. *Homogeneous Gold Catalysis*. Springer; 2015. https://doi.org/10.1007/978-3-319-13722-3.
27. Winter C, Krause N. Gold-catalyzed nucleophilic cyclization of functionalized allenes: a powerful access to carbo- and heterocycles. *Chem Rev*. 2011;111(3):1994–2009. https://doi.org/10.1021/cr1004088.
28. Chiarucci M, Bandini M. New developments in gold-catalyzed manipulation of inactivated alkenes. *Beilstein J Org Chem*. 2013;9:2586–2614. https://doi.org/10.3762/bjoc.9.294.

29. Obradors C, Echavarren AM. Intriguing mechanistic laberynths in gold(I) catalysis. *Chem Commun*. 2014;50(1):16–28. https://doi.org/10.1039/C3CC45518A.
30. Dorel R, Echavarren AM. Gold(I)-catalyzed activation of alkynes for the construction of molecular complexity. *Chem Rev*. 2015;115(17):9028–9072. https://doi.org/10.1021/cr500691k.
31. Halliday CJV, Lynam M. Gold-alkynyls in catalysis: alkyne activation, gold cumulenes and nuclearity. *Dalton Trans*. 2016;45(32):12611–12626. https://doi.org/10.1039/C6DT01641C.
32. Mascareñas JL, Varela I, López F. Allenes and derivatives in gold(I)- and platinum(II)-catalyzed formal cycloadditions. *Acc Chem Res*. 2019;52(2):465–479. https://doi.org/10.1021/acs.accounts.8b00567.
33. Schmidbaur H, Schier A. Gold η^2-coordination to unsaturated and aromatic hydrocarbons: the key step in gold-catalyzed organic transformations. *Organometallics*. 2010;29(1):2–23. https://doi.org/10.1021/om900900u.
34. Cinellu MA. *Modern Gold Catalyzed Synthesis*. Wiley-VCH; 2012:175–199.
35. Brooner REM, Widenhoefer RA. Cationic, two-coordinate gold π complexes. *Angew Chem Int Ed*. 2013;52(45):11714–11724. https://doi.org/10.1002/anie.201303468.
36. Jones AC. Gold π-complexes as model intermediates in gold catalysis. *Top Curr Chem*. 2015;357:133–166. https://doi.org/10.1007/128_2014_593.
37. Blons C, Amgoune A, Bourissou D. Gold(III) π complexes. *Dalton Trans*. 2018;47(31):10388–10393. https://doi.org/10.1039/C8DT01457D.
38. Zeise WC. Von der Wirkung zwischen Platinchlorid und Alkohol, und von den dabei entstehenden neuen Substanzen. *Ann Phys Chem*. 1831;97(4):497–541. https://doi.org/10.1002/andp.18310970402.
39. Dell'Amico DB, Calderazzo F, Dantona R, Straehle J, Weiss H. Olefin complexes of gold(I) by carbonyl displacement from carbonylgold(I) chloride. *Organometallics*. 1987;6(6):1207–1210. https://doi.org/10.1021/om00149a014.
40. Davila RM, Staples RJ, Fackler Jr JP. Synthesis and structural characterization of Au$_4$(MNT)(dppe)$_2$(Cl)$_2$·1/4CH$_2$Cl$_2$ (MNT = 1,2-dicyanoetheen-1,2-dithiolate-*S*, *S*'; dppe = *cis*-bis(diphenylphosphino)ethylene): a gold(I) metal-olefin complex in which the olefin orientation relative to the coordination plane involving the metal is defined. *Organometallics*. 1994;13(2):418–420. https://doi.org/10.1021/om00014a007.
41. Mingos DMP, Yau J, Menzer S, Williams DJ. A gold(I) [2]catene. *Angew Chem Int Ed*. 1995;34(17):1894–1895. https://doi.org/10.1002/anie.199518941.
42. Lang H, Kohler K, Zsolnai L. Unusual coordination mode of organogold(I) compounds: trigonal-planar complexation of gold(I) centers by alkynes. *Chem Commun*. 1996;17:2043–2044. https://doi.org/10.1039/CC9960002043.
43. Köhler K, Silverio SJ, Hyla-Krypsin I, et al. Trigonal-planar-coordinated organogold(I) complexes stabilized by organometallic 1,4-diynes: reaction behavior structure, and bonding. *Organometallics*. 1997;16(23):4970–4979. https://doi.org/10.1021/om970302i.
44. Schulte P, Behrens U. Strong coordination of cycloheptynes by gold(I) chloride: synthesis and structure of two complexes of the type [(alkyne)AuCl]. *Chem Commun*. 1998;16:1633–1634. https://doi.org/10.1039/A803791D.
45. McIntosh DF, Ozin GA. Direct synthesis using gold atoms: synthesis, infrared and ultraviolet-visible spectra and molecular orbital investigation of monoethylene gold(0), (C$_2$H$_4$)Au. *J Organomet Chem*. 1976;121(1):127–136. https://doi.org/10.1016/S0022-328X(00)85515-1.
46. McIntosh DF, Ga O, Messmer RP. Optical and SCF-X$_a$-SW investigations of M(π-C$_2$H$_4$), where M = Cu, Ag, and Au. *Inorg Chem*. 1980;19(11):3321–3327. https://doi.org/10.1021/ic50213a023.
47. Kasai PH. Acetylene and ethylene complexes of gold atoms: matrix isolation ESR study. *J Am Chem Soc*. 1983;105(22):6704–6710. https://doi.org/10.1021/ja00360a026.

48. Kasai PH. Propylene complexes of copper, silver and gold atoms: matrix isolation ESR study. *J Am Chem Soc.* 1984;106(11):3069–3075. https://doi.org/10.1021/ja00323a001.
49. Kasai PH. Bis(ethylene)gold(0) and bis(propylene)gold(0) complexes: gold nuclear hyperfine and quadrupole coupling tensors. *J Phys Chem.* 1988;92(8):2161–2165. https://doi.org/10.1021/j100319a016.
50. Nicolas G, Spiegelmann F. Theoretical study of ethylene-noble metal complexes. *J Am Chem Soc.* 1999;112(14):5410–5419. https://doi.org/10.1021/ja00170a003.
51. Bond GC, Serman PA. Gold catalysis for olefin hydrogenation. *Gold Bull.* 1973;6:102–105. https://doi.org/10.1007/BF03215018.
52. Savjani N, Rosca D, Schormann M, Bochmann M. Gold(III) olefin complexes. *Angew Chem Int Ed.* 2013;52(3):874–877. https://doi.org/10.1002/anie.201208356.
53. Langseth E, Scheuermann ML, Balcells D, et al. Generation and structural characterization of a gold(III) alkene complex. *Angew Chem Int Ed.* 2013;52(6):1660–1663. https://doi.org/10.1002/anie.201209140.
54. Rocchigiani L, Fernandez-Cestau J, Agonigi G, Chambrier I, Budzelaar PHM, Bochmann M. Gold(III) alkyne complexes: bonding and reaction pathways. *Angew Chem Int Ed.* 2017;56(44):13861–13865. https://doi.org/10.1002/anie.201708640.
55. Rekhroukh F, Blons C, Estévez L, et al. Gold(III)-arene complexes by insertion of olefins into gold-aryl bonds. *Chem Sci.* 2017;8(6):4539–4545. https://doi.org/10.1039/C7SC00145B.
56. Rodríguez J, Szalóki G, Sosa-Carrizo ED, Saffon-Merceron N, Miqueu K, Bourissou D. Gold(III) π-allyl complexes. *Angew Chem Int Ed.* 2020;59(4):1511–1515. https://doi.org/10.1002/anie.201912314.
57. Holmsen MSM, Nova A, Oien-Odegaard S, Heyn RH, Tilset M. A highly asymmetric gold(III) η^3-allyl complex. *Angew Chem Int Ed.* 2020;59(4):1516–1520. https://doi.org/10.1002/anie.201912315.
58. Zuccaccia D, Belpassi L, Tarantelli F, Macchioni A. Ion pairing in cationic olefin-gold(I) complexes. *J Am Chem Soc.* 2009;131(9):3170–3171. https://doi.org/10.1021/ja809998y.
59. Brown TJ, Dickens MG, Widenhoefer RA. Synthesis, X-ray crystal structures, and solution behavior of monomeric, cationic, two-coordinate gold(I) π-alkene complexes. *J Am Chem Soc.* 2009;131(18):6350–6351. https://doi.org/10.1021/ja9015827.
60. de Frémont P, Marion N, Nolan SP. Cationic NHC–gold(I) complexes: synthesis, isolation, and catalytic activity. *J Organomet Chem.* 2009;694(4):441–560. https://doi.org/10.1016/j.jorganchem.2008.10.047.
61. Hooper TN, Green M, McGrady JE, Patel JR, Russell CA. Synthesis and structural characterisation of stable cationic gold(I) alkene complexes. *Chem Commun.* 2009;(26):3877–3879. https://doi.org/10.1039/B908109G.
62. Brown TJ, Dichens MG, Widenhoefer RA. Synthesis and X-ray crystal structures of cationic, two-coordinate gold(I) π-alkene complexes that contain a sterically hindered *o*-biphenylphosphine ligand. *Chem Commun.* 2009;(42):6451–6453. https://doi.org/10.1039/B914632F.
63. Motloch P, Blahut J, Cisorova I, Roithova J. X-ray characterization of triphenylphosphine-gold(I) olefin π-complexes and the revision of their stability in solution. *J Organomet Chem.* 2017;848:114–117. https://doi.org/10.1016/j.jorganchem.2017.07.011.
64. Griebel C, Hodges DD, Yager BR, et al. Bisphenyl phosphines: structure and synthesis of gold(I) alkene π-complexes with variable phosphine donicity and enhanced stability. *Organometallics.* 2020;39(14):2664–2671. https://doi.org/10.1021/acs.organomet.0c00278.

65. Cinellu MA, Minghetti G, Stoccoro S, Zucca A, Manassero M. Reaction of gold(III) oxo complexes with alkenes. Synthesis of unprecedented gold alkene complexes, [Au(N,N)(alkene)][PF$_6$]. crystal structure of [Au(bipy^{ip})(η^2-CH$_2$=CHPh)][PF$_6$] (bipy^{ip} = 6-ispropyl-2,2'bipyridine). *Chem Commun*. 2004;10(14):1618–1619. https://doi.org/10.1039/B404890C.
66. Cinellu MA, Minghetti G, Cocco F, et al. Synthesis and properties of gold alkene complexes. Crystal sructure of [Au(bipy^{oXyl})(η^2-CH$_2$=CHPh)](PF$_6$) and DFT calculations on the model cation [Au(bipy)(η^2-CH$_2$=CHPh)]$^+$. *Dalton Trans*. 2006;(48):5703–5716. https://doi.org/10.1039/B610657A.
67. Cinellu MA, Minghetti G, Cocco F, Stoccoro S, Zucca A, Manassero M. Reactions of gold(III) oxo complexes with cyclic alkenes. *Angew Chem Int Ed*. 2005;44(42): 6892–6895. https://doi.org/10.1002/anie.200501754.
68. Dias HVR, Wu J. Structurally characterized coinage-metal-ethylene complexes. *Eur J Inorg Chem*. 2008;(12):2113. https://doi.org/10.1002/ejic.200800231.
69. Dias HVR, Wu J. Thermally stable gold(I) ethylene adducts: [(HB{3,5 (CF$_3$)$_2$Pz}$_3$)Au(CH$_2$=CH$_2$)]. *Angew Chem Int Ed*. 2007;46(41):7814–7816. https://doi.org/10.1002/anie.200703328.
70. Dias HVR, Wu J. Structurally similar, thermally stable copper(I), silver(I), and gold(I) ethylene complexes supported by a fluorinated scorpionate. *Organometallics*. 2012;31(4):1511–1517. https://doi.org/10.1021/om201185v.
71. Ridlen SG, Wu J, Kulkarni NV, Dias HVR. Isolable ethylene complexes of copper(I), silver(I), and gold(I) supported by fluorinated scorpionates [HB{3-(CF$_3$),5-(CH$_3$)Pz}$_3$]$^-$ and [HB{3-(CF$_3$),5-(Ph)Pz}$_3$]$^-$. *Eur J Inorg Chem*. 2016;2016(15–16):2573–2580. https://doi.org/10.1002/ejic.201501365.
72. Nes GJH, Vos A. Single-crystal structures and electron density distributions of ethane, ethylene and acetylene. III. Single-crystal X-ray structure determination of ethylene at 85 K. *Acta Crystallogr B*. 1979;35(11):2593–2601. https://doi.org/10.1107/S0567740879009961.
73. Krossing I, Reisinger A. A stable salt of the Tris(ethene)silver cation: structure and characterization of [ag(η_2-C$_2$H$_4$)$_3$]$^+$[Al{OC(CF$_3$)$_3$}$_4$]$^-$. *Angew Chem Int Ed*. 2003;42(46): 5725–5728. https://doi.org/10.1002/anie.200352080.
74. Batsanov SS. Van der Waals radii of elements. *Inorg Mater*. 2001;37:871–885. https://doi.org/10.1023/A:1011625728803.
75. Wu J, Noonikara-Poyil A, Muñoz-Castro A, Dias HVR. Gold(I) ethylene complexes supported by electron-rich scorpionates. *Chem Commun*. 2021;57(8):978–981. https://doi.org/10.1039/D0CC07717H.
76. Flores JA, Dias HVR. Gold(I) ethylene and copper(I) ethylene complexes supported by a polyhalogenated triazapentadienyl ligand. *Inorg Chem*. 2008;47(11):4448–4450. https://doi.org/10.1021/ic800373u.
77. Dias HVR, Flores JA, Wu J, Kroll P. Monomeric copper(I), silver(I), and gold(I) alkyne complexes and the coinage metal family group trends. *J Am Chem Soc*. 2009;131 (31):11249–11255. https://doi.org/10.1021/ja904232v.
78. Hooper TN, Green M, Russell CA. Cationic Au(I) alkyne complexes: synthesis, structure and reactivity. *Chem Commun*. 2010;46(13):2313–2315. https://doi.org/10.1039/B923900F.
79. Zuccaccia D, Belpassi L, Rocchigiani L, Tarantelli F, Macchioni A. A phosphine gold(I) π-alkyne complex: tuning the metal-alkyne bond character and counterion position by the choice of the ancillary ligand. *Inorg Chem*. 2010;49(7):3080–3082. https://doi.org/10.1021/ic100093n.
80. Brown TJ, Widenhoefer RA. Cationic gold(I) π-complexes of terminal alkynes and their conversion to dinuclear σ,π-acetylide complexes. *Organometallics*. 2011;30(21): 6003–6009. https://doi.org/10.1021/om200840g.

81. Ciancaleoni G, Biasiolo L, Bistoni G, et al. NHC-gold-alkyne complexes: influence of the carbene backbone on the ion pair structure. *Organometallics*. 2013;32(15): 4444–4447. https://doi.org/10.1021/om4005912.
82. Klimovica K, Krischbaum K, Daugulis O. Synthesis and properties of "Sandwich" diimine-coinage metal ethylene complexes. *Organometallics*. 2016;35(17):2938–2943. https://doi.org/10.1021/acs.organomet.6b00487.
83. Dias HVR, Fianchini M, Cundari TR, Campana CF. Synthesis and characterization of the gold(I) Tris(ethylene) complex [Au(C$_2$H$_4$)$_3$][SbF$_6$]. *Angew Chem Int Ed*. 2007; 47(3):556–559. https://doi.org/10.1002/anie.200703515.
84. Harper MJ, Arthur CJ, Crosby J, et al. Oxidative addition, transmetalation, and reductive elimination at a 2,2′-bipyridyl-ligated gold center. *J Am Chem Soc*. 2018;140 (12):4440–4445. https://doi.org/10.1021/jacs.8b01411.
85. Cadge JA, Sparkes HA, Bower JF, Russel CA. Oxidative addition of alkenyl and alkynyl iodides to a Au^I complex. *Angew Chem Int Ed*. 2020;59(16):6617–6621. https://doi.org/10.1002/anie.202000473.
86. Brand JP, Charpentier J, Waser J. Direct alkynylation of indole and pyrrole heterocycles. *Angew Chem Int Ed*. 2009;48(49):9346–9349. https://doi.org/10.1002/anie. 200905419.
87. Brand JP, Waser J. Direct alkynylation of thiophenes: cooperative activation of TIPS-EBX with gold and bronsted acids. *Angew Chem Int Ed*. 2010;49(40): 7304–7307. https://doi.org/10.1002/anie.201003179.
88. Li Y, Brand JP, Waser J. Gold-catalyzed regioselective synthesis of 2- and 3-alkynyl furans. *Angew Chem Int Ed*. 2013;52(26):6743–6747. https://doi.org/10.1002/anie. 201302210.
89. Peng H, Xi Y, Ronaghi N, Dong B, Akhmedov NG, Shi X. Gold-catalyzed oxidative cross-coupling of terminal alkynes: selective synthesis of unsymmetrical 1,3-diynes. *J Am Chem Soc*. 2014;136(38):13174–13177. https://doi.org/10.1021/ja5078365.
90. Cai R, Lu M, Aguilera EY, et al. Ligand-assisted gold-catalyzed cross-coupling with arylazonium salts: redox gold catalysis without an external oxidant. *Angew Chem Int Ed*. 2015;54(30):8772–8776. https://doi.org/10.1002/anie.201503546.
91. Li X, Xie X, Sun N, Liu Y. Gold-catalyzed cadiot-chodkiewicz-type cross-coupling of terminal alkynes with alkynyl hypervalent iodine reagents: highly selective synthesis of unsymmetrical 1,3-diynes. *Angew Chem Int Ed*. 2017;56(24):6994–6998. https://doi.org/10.1002/anie.201702833.
92. Yang Y, Antoni P, Zimmer M, et al. Dual gold/silver catalysis involving alkynylgold(III) intermediates formed by oxidative addition and silver-catalyzed C–H activation for the direct alkynylation of cyclopropenes. *Angew Chem Int Ed*. 2019;58(15):5129–5133. https://doi.org/10.1002/anie.201812577.
93. Yang Y, Eberle L, Mulks FF, et al. Trans influence of ligands on the oxidation of gold(I) complexes. *J Am Chem Soc*. 2019;141(43):17414–17420. https://doi.org/10. 1021/jacs.9b09363.
94. Cinellu MA, Arca M, Ortu F, et al. Structural, theoretical and spectroscopic characterisation of a series of novel gold(I)-norbornene complexes supported by phenantrolines: effects of the supporting ligand. *Eur J Inorg Chem*. 2019;2019(44):4784–4795. https://doi.org/10.1002/ejic.201901116.
95. Navarro M, Toledo A, Joost M, Amgoune A, Mallet-Ladeira S, Bourissou D. π-complexes of P^P and P^N chelated gold(I). *Chem Commun*. 2019;55(55): 7974–7977. https://doi.org/10.1039/C9CC04266K.
96. Navarro M, Toledo A, Mallet-Ladeira S, Sosa-Carrizo ED, Miqueu K, Bourissou D. Versatility and adaptative behaviour of the P^N chelating ligand MeDalphos within gold(I) π complexes. *Chem Sci*. 2020;11(10):2750–2758. https://doi.org/10.1039/C9SC06398F.

97. Joost M, Estévez L, Mallet-Ladeira S, Miqueu K, Amgoune A, Bourissou D. Enhanced π-backdonation from gold(I): isolation of original carbonyl and carbene complexes. *Angew Chem Int Ed*. 2014;53(52):14512–14516. https://doi.org/10.1002/anie.201407684.
98. Zeineddine A, Rekhroukh F, Sosa-Carrizo ED, et al. Isolation of a reactive tricoordinate α-oxo gold carbene complex. *Angew Chem Int Ed*. 2018;57(5): 1306–1310. https://doi.org/10.1002/anie.201711647.
99. Fürstner A, Alcarazo M, Goddard R, Lehmann CW. Coordination chemistry of Ene-1,1-diamines and a prototype "carbodicarbene". *Angew Chem Int Ed*. 2008;47(17): 3210–3214. https://doi.org/10.1002/anie.200705798.
100. Sanguramath RA, Hooper TN, Butts CP, Green M, McGrady JE, Russell CA. The interaction of gold(I) cations with 1,3-dienes. *Angew Chem Int Ed*. 2011;50(33): 7592–7595. https://doi.org/10.1002/anie.201102750.
101. Zhu Y, Day CS, Jones AC. Synthesis and structure of cationic phosphine gold(I) enol ether complexes. *Organometallics*. 2012;31(21):7332–7335. https://doi.org/10.1021/om300893q.
102. Sriram M, Zhu Y, Camp AM, Day CS, Jones AC. Structure and dynamic behavior of phosphine gold(I)-coordinated enamines: characterization of α-metalated iminium ions. *Organometallics*. 2014;33(16):4157–4164. https://doi.org/10.1021/om500670z.
103. Zhdanko A, Maier M. Mechanistic study of gold(I)-catalyzed hydroamination of alkynes: outer or inner sphere mechanism? *Angew Chem Int Ed*. 2014;53(30): 7760–7764. https://doi.org/10.1002/anie.201402557.
104. Aumann R, Fischer EO. Neue Gold-Carbene-Komplexe durch Carbenübertragung. *Chem Ber*. 1981;114(5):1853–1857. https://doi.org/10.1002/cber.19811140523.
105. Schubert U, Ackermann K, Aumann R. Chloro[dimethylamino(phenyl)carbene] gold(I), ClAuC(C$_6$H$_5$)N(CH$_3$)$_2$. *Cryst Struct Commun*. 1982;11(2):591–594. For X-ray diffraction analysis of the Cl–Au=C(NMe$_2$)Ph complex, see.
106. Seidel G, Fürstner A. Structure of a reactive gold carbenoid. *Angew Chem Int Ed*. 2014;53(19):4807–4811. https://doi.org/10.1002/anie.201402080.
107. Fañanas-Mastral M, Aznar F. Carbene transfer reaction from chromium(0) to gold(I): synthesis and reactivity of new Fischer-type gold(I) alkenyl carbene complexes. *Organometallics*. 2009;28(3):666–668. https://doi.org/10.1021/om801146z.
108. Brooner REM, Widenhoefer RA. Experimental evaluation of the electron donor ability of a gold phosphine fragment in a gold carbene complex. *Chem Commun*. 2014; 50(19):2420–2423. https://doi.org/10.1039/C3CC48869A.
109. Minghetti G, Bonati F. Bis(carbene) complexes of gold(I) and gold(III). *J Organomet Chem*. 1973;54:C62–C63. https://doi.org/10.1016/S0022-328X(00)84984-0.
110. Raubenheimer HG, Cronje S. Carbene complexes of gold: preparation, medical application and bonding. *Chem Soc Rev*. 2008;37(9):1998–2011. https://doi.org/10.1039/B708636A.
111. Mora M, Gimeno MC, Visbal R. Recent advances in gold-NHC complexes with biological properties. *Chem Soc Rev*. 2018;48(2):447–462. https://doi.org/10.1039/C8CS00570B.
112. Visbal R, Gimeno MC. N-heterocylic carbene metal complexes: photoluminescence and applications. *Chem Soc Rev*. 2014;43(10), 35513574. https://doi.org/10.1039/C3CS60466G.
113. Marion N, Nolan SP. N-heterocylic carbenes in gold catalysis. *Chem Soc Rev*. 2008; 37(9):1776–1782. https://doi.org/10.1039/B711132K.
114. Brooner REM, Brown TJ, Widenhoefer RA. Direct observation of a cationic gold(I)-bicyclo[3.2.0]hept-1(7)-ene complex generated in the cycloisomerization of a 7-phenyl-1,6-enyne. *Angew Chem Int Ed*. 2013;52(24):6250–6261. https://doi.org/10.1002/anie.201301640.

115. Harris RJ, Widenhoefer RA. Synthesis, structure, and reactivity of a gold carbenoid complex that lacks heteroatom stabilization. *Angew Chem Int Ed.* 2014;53(35): 9369–9371. https://doi.org/10.1002/anie.201404882.
116. Hussong MW, Rominger F, Krämer P, Straub BF. Isolation of a non-heteroatom-stabilized gold-carbene complex. *Angew Chem Int Ed.* 2014;53(35):9372–9375. https://doi.org/10.1002/anie.201404032.
117. Qian D, Zhang J. Gold-catalyzed cyclopropanation reactions using a carbenoid precursor toolbox. *Chem Soc Rev.* 2015;44(3):677–698. https://doi.org/10.1039/C4CS 00304G.
118. Liu L, Zhang J. Gold-catalyzed transformations of α-diazocarbonyl compounds: selectivity and diversity. *Chem Soc Rev.* 2016;45(3):506–516. https://doi.org/10.1039/C5CS00821B.
119. Fructos MR, Díaz-Requejo MM, Pérez PJ. Gold and diazo reagents: a fruitful tool for developing molecular complexity. *Chem Commun.* 2016;46(52):7326–7335. https://doi.org/10.1039/C6CC01958G.
120. Fructos MR, Belderrain TR, de Frémont P, et al. A gold catalyst for carbene-transfer reactions from ethyl diazoacetate. *Angew Chem Int Ed.* 2005;44(33):5285–5288. https://doi.org/10.1002/anie.200501056.
121. de Frémont P, Stevens ED, Fructos MR, Díaz-Requejo MM, Pérez PJ, Nolan SP. Synthesis, isolation and characterization of a cationic gold(I) N-heterocyclic carbene (NHC) complexes. *Chem Commun.* 2006;(19):2045–2047. https://doi.org/10.1039/B601547F.
122. Fructos MR, de Frémont P, Nolan SP, Díaz-Requejo MM, Pérez PJ. Alkene carbon-hydrogen bond functionalization with (NHC)MCl precatalysts (M = Cu, Au; NHC = N-heterocylic carbene). *Organometallics.* 2006;25(9):2237–2241. https://doi.org/10.1021/om0507474.
123. Barluenga J, Lonzi G, Tomás M, López LA. Reactivity of stabilized vinyl diazo derivatives toward unsaturated hydrocarbons: regioselective gold-catalyzed carbon-carbon bond formation. *Chem—Eur J.* 2013;19(5):1573–1576. https://doi.org/10.1002/chem.201203217.
124. López E, Lonzi G, López LA. Gold-catalyzed C–H bond functionalization of metallocenes: synthesis of densely functionalized ferrocene derivatives. *Organometallics.* 2014;33(21):5924–5927. https://doi.org/10.1021/om500638t.
125. Yu Z, Ma B, Chen M, Wu H-H, Liu L, Zhang J. Highly site-selective direct C–H bond functionalization of phenols with α-aryl-α-diazoacetates and diazooxindoles via gold catalysis. *J Am Chem Soc.* 2014;136(19):6904–6907. https://doi.org/10.1021/ja503163k.
126. Pagar VV, Liu R-S. Gold-catalyzed cycloaddition reactions of ethyl diazoacetate, nitrosoarenes, and vinyldiazo carbonyl compounds synthesis of isoxazolidine and benzo[b]azepine derivatives. *Angew Chem Int Ed.* 2015;54(16):4923–4926. https://doi.org/10.1002/anie.201500340.
127. Zhang D, Xu G, Ding D, Zhu C, Li J, Sun J. Gold(I)-catalyzed diazo coupling: strategy towards alkene formation and tandem benzannulation. *Angew Chem Int Ed.* 2014; 53(41):11070–11074. https://doi.org/10.1002/anie.201406712.
128. Xu G, Zhu C, Gu W, Li J, Sun J. Gold(I)-catalyzed diazo cross-coupling: a selective and ligand-controlled denitrogenation/cyclization cascade. *Angew Chem Int Ed.* 2015; 54(3):883–887. https://doi.org/10.1002/anie.201409845.
129. Schulz J, Jasiková L, Skríba A, Roithová K. Role of gold(I) α-oxo carbenes in the oxidation reactions of alkynes catalyzed by gold(I) complexes. *J Am Chem Soc.* 2014; 136(32):11513–11523. https://doi.org/10.1021/ja505945d.
130. Schulz J, Jasik J, Gray A, Roithová J. Formation of oxazoles from elusive gold(I) α-oxocarbenes: a mechanistic study. *Chem—Eur J.* 2016;22(28):9827–9834. https://doi.org/10.1002/chem.201601634.

131. Straub BF, Hofmann P. Copper(I) carbenes: the synthesis of active intermediates in copper-catalyzed cyclopropanation. *Angew Chem Int Ed.* 2001;40(7):1288–1290. α-Oxo copper(I) carbene complex were earlier authenticated, see: https://doi.org/10.1002/1521-3773(20010401)40:7 < 1288::AID-ANIE1288 > 3.0.CO;2-6.
132. Luo Y, Ji K, Li Y, Zhang L. Tempering the reactivities of postulated α-oxo gold carbenes using bidentate ligands: implication of tricoordinated gold intermediates and the development of an expedient bimolecular assembly of 2,4-disubstituted oxazoles. *J Am Chem Soc.* 2012;134(42):17412–17415. https://doi.org/10.1021/ja307948m.
133. Wu G, Zheng R, Nelson J, Zhang L. One-step synthesis of methanesulfonyloxymethyl ketones via gold-catalyzed oxidation of terminal alkynes. A combination of ligand and counter anion enales high efficiency and a one-pot synthesis of 2,4-disubstituted thiazoles. *Adv Synth Catal.* 2014;356(6):1229–1234. https://doi.org/10.1002/adsc.201300855.
134. Ji K, D'Souza B, Nelson J, Zhang L. Gold-catalyzed oxidation of propargylic ethers with internal C–C triple bonds: impressive regioselectivity enabled by inductive effect. *J Organomet Chem.* 2014;770:142–145. https://doi.org/10.1016/j.jorganchem.2014.08.005.
135. Li J, Ji K, Zheng R, Nelson J, Zhang L. Expanding the horizon of intermolecular trapping of *in situ* generated α-oxo gold carbenes: efficient oxidative union of allylic sulfides and terminal alkynes *via* C–C bond formation. *Chem Commun.* 2014;50(31):4130–4133. https://doi.org/10.1039/c4CC00739E.
136. Ji K, Zhao Y, Zhang L. Optimizing P,N-bidentate ligands for oxidative gold catalysis: efficient intermolecular trapping of α-oxo gold carbenes by carboxylic acids. *Angew Chem Int Ed.* 2013;52(25):6508–6512. https://doi.org/10.1002/anie.201301601.
137. Ji K, Zhang L. A non-diazo strategy to cyclopropanation *via* oxidatively generated gold carbene: the benefit of a conformationally rigid *P,N*-bidentate ligand. *Org Chem Front.* 2014;1(1):34–38. https://doi.org/10.1039/C3QO00080J.
138. Wang Y, Zheng Z, Zhang L. Intramolecular insertions into unactivated C(sp³)–H bonds by oxidatively generated β-diketone-α-gold carbenes: synthesis of cyclopentanones. *J Am Chem Soc.* 2015;137(16):5316–5419. https://doi.org/10.1021/jacs.5b02280.
139. Ji K, Zhang Z, Wang Z, Zhang L. Enantioselective oxidative gold catalysis enabled by a designed chiral P,N-bidentate ligand. *Angew Chem Int Ed.* 2015;54(4):1245–1249. https://doi.org/10.1002/anie.201409300.
140. Cordero B, Gómez V, Platero-Prats AE, et al. Covalent radii revisited. *Dalton Trans.* 2008;21(21):2832–2838. https://doi.org/10.1039/B801115J.
141. Benitez D, Shapiro ND, Tkatchouk E, Wang Y, Goddard III WA, Toste FD. A bonding model for gold(I) carbene complexes. *Nat Chem.* 2009;1:482–486. https://doi.org/10.1038/nchem.331.
142. Nunes dos Santos Comprido L, JEMN K, Knizia G, Kästner J, ASK H. The stabilizing effects in gold carbene complexes. *Angew Chem Int Ed.* 2015;54(35):10336–10340. https://doi.org/10.1002/anie.201412401.
143. Xi Y, Su Y, Yu Z, et al. Chemoselective carbophilic addition of α-diazoesters through ligand-controlled gold catalysis. *Angew Chem Int Ed.* 2014;53(37):9817–9821. https://doi.org/10.1002/anie.201404946.
144. Liu Y, Yu Z, Zhang JZ, Liu L, Xia F, Zhang J. Origins of unique gold-catalysed chemo- and site-selective C–H functionalization of phenols with diazo compounds. *Chem Sci.* 2016;7(3):1988–1995. https://doi.org/10.1039/C5SC04319K.
145. García-Morales C, Pei X-L, Sarria Toro JM, Echavarren AM. Direct observation of aryl gold(I) carbenes that undergo cyclopropanation, C–H insertion, and dimerization reactions. *Angew Chem Int Ed.* 2019;58(12):3957–3961. https://doi.org/10.1002/anie.201814577.

146. Tskhovreboc AG, Lingnau JB, Fürstner A. Gold difluorocarbenoid complexes: spectroscopic and chemical profiling. *Angew Chem Int Ed*. 2019;58(26):8834–8838. https://doi.org/10.1002/anie.201903957.
147. Crabtree RH. *The Organometallic Chemistry of the Transition Metal*. 4th ed. Hoboken: Wiley; 2004.

CHAPTER THREE

Imido complexes of groups 8–10 active in nitrene transfer reactions

Caterina Damiano, Paolo Sonzini, Alessandro Caselli, and Emma Gallo*

Department of Chemistry, University of Milan, Milan, Italy
*Corresponding author: e-mail address: emma.gallo@unimi.it

Contents

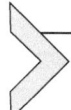

1. Introduction: Involvement of imido complexes in the formation of C—N bonds   145
2. Imido complexes of group 8   148
   2.1 Iron   148
   2.2 Ruthenium   166
   2.3 Concluding remarks on group 8   171
3. Imido complexes of groups 9 and 10   172
   3.1 Cobalt   173
   3.2 Nickel and palladium   175
4. Summary and outlook   179
References   180

1. Introduction: Involvement of imido complexes in the formation of C—N bonds

The presence of C—N bonds in several biologically and/or pharmaceutically active molecules has always encouraged the scientific community to develop new selective and efficient procedures for the synthesis of nitrogen containing compounds. Considering the growing demand for eco-sustainable and atom-economically methodologies, metal-catalyzed one-pot nitrene transfer reactions have replaced multiple-steps, time-consuming synthetic procedures and have become one of the most employed strategies for the insertion of a *N*-moiety in organic substrates.

Nitrene ligands are neutral, electron deficient and very reactive intermediates in which the four non-bonding electrons can be arranged either in two lone pairs or in one lone pair and an unpaired biradical, forming an electrophilic singlet or a triplet nitrene, respectively.[1] Both free species are very reactive and consequently their reactions with organic substrates usually proceed with poor chemo- and stereo-selectivities.[2] Even if nitrene functionalities can be generated from several sources, such as iminophenyliodinanes[3] and haloamine-T,[4] one of the most successful and sustainable class of reagents are organic azides[5,6] due to their high atom-economy and the formation of benign molecular nitrogen as the only by-product. The reaction of one of the above-mentioned nitrene precursors with transition metal complexes allows the formation of imido complexes, generally indicated as $L_nM=NR$, containing a more stable nitrene fragment (NR) whose transfer to organic molecules can be better driven and controlled. Imido complexes are active catalysts in several nitrene transfer reactions such as alkene aziridinations,[7] C—H aminations[8,9] and sulfamidation reactions[10] (Scheme 1).

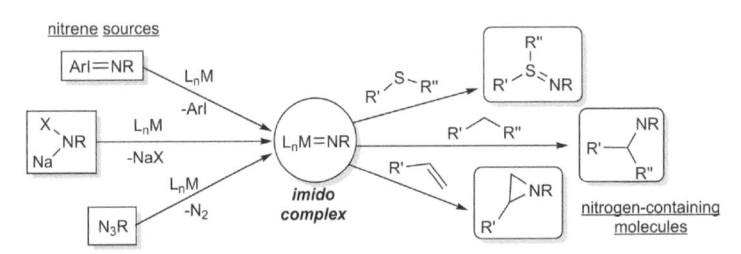

Scheme 1 Imido complex formation and nitrene transfer to organic substrates.

Imido complexes can be classified as Fischer and Schrock-type species thanks to the analogies between nitrene and carbene ligands.[11] Even if carbenes are formally defined as neutral species, nitrene ligands can be better envisaged as dianionic ones (NR^{2-}) due to the stronger electronegativity of the nitrogen atom with respect to the carbon one. The "NR" ligand also presents one lone pair, which can be involved in a π-donation from the ligand to the metal with the consequent formation of a triple $L_nM\equiv NR$ bond.[12] The metal-nitrogen bond order, the geometry of the complex and the catalyst stability are strongly influenced by the nitrogen hybridization as well as the metal electronic configuration. Electropositive metals, which possess empty d_π-type orbitals, stabilize Schrock type imido complexes in which the sp-hybridized nitrogen atom establishes two π-interactions with the metal, favoring the linear geometry. On the other hand, in Fischer type complexes the sp^2-hybridized nitrogen atom preferably coordinates late transition metals in a bent geometry, avoiding the unfavorable π-interactions between the nitrogen lone pair and the filled d_π-type orbitals of the metal.

Additionally, the one-electron reduction of Fischer type complexes, or the one-electron oxidation of Schrock type, affords L_nM—NR radical species[13] that are also active in nitrene transfer reactions (Scheme 2).

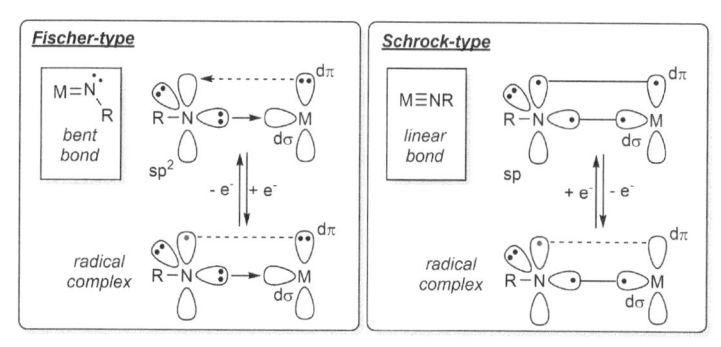

Scheme 2 Fischer and Schrock type metal imido complexes.

In view of the strong dependence of the imido complex reactivity on the nature of the M—NR bond, it is fundamental modulating the π-interaction between the metal and the NR ligand by changing the nature and oxidation state of the metal as well as the properties of the ligands (L_n) and R group of the nitrogen atom. It should be noted that high stable imido complexes, which can be easily isolated and characterized, often show lower reactivity in transferring the nitrene functionality with respect to less stable ones, which are suggested to be transient reactive intermediates in metal-catalyzed nitrene transfer reactions.

As well established by previously published reviews,[14–18] late transition metal complexes represent the most active catalysts in these reactions, even if the majority of their corresponding imido complexes have only been supposed as active species without being either isolated or detected. Bearing in mind this statement, this Chapter will only discuss the catalytic activity of *isolated and characterized* imido complexes of late transition metals that have been published since 2010, focusing the attention on the characteristics that can affect the catalyst stability and reactivity in nitrene transfer reactions. Considering that no stable imido complexes of group 11 have been reported in the analyzed period (2010 – 20), the catalytic activity of groups 8–10 derivatives will be only discussed. Older references will be only briefly described to clarify significant concepts and we sincerely apologize in advance if some important contributions have been unintentionally omitted.

2. Imido complexes of group 8
2.1 Iron

Since first examples of iron imido heme[19] and non-heme complexes[20] that were reported in the early 1980s, the iron-catalyzed nitrene transfer reactions have received high interest becoming the topic of several papers and reviews.[7,21–23] Despite the widespread use of iron complexes as active catalysts for C—H amination and alkene aziridination reactions, the isolation and characterization of active iron imido species (L_nFe=NR) remains rare. Up to now several monometallic and multi-metallic L_nFe=NR complexes have been synthesized and characterized, showing different iron oxidation states (+2, +3, +4 and +5) and spin states, but for many of them applications have been limited to stoichiometric reactions[24–27] in which the imido complex instead being involved in amination reactions, acts as a precursor for the synthesis of other iron complexes.

The chemical characteristic of iron-imido complexes is strongly dependent on the nature of the ancillary ligands, whose ligand field can stabilize the metal-nitrogen multiple bond allowing the isolation and characterization of imido species. Among all the ligands employed for synthetizing iron complexes active in promoting nitrene transfer reactions, nitrogen containing ligands represent one of the most employed classes that are here classified depending on their coordination modes.

Bidentate nitrogen ligands. Betley and co-authors[28] reported the use of different bidentate dipyrromethene ligands for the synthesis of iron (II) complexes active in C—H bond amination and aziridination reactions. The solid-state molecular structure of the synthesized complexes showed that the iron coordination sphere was strongly influenced by the nature of substituents present on the dipyrromethene skeleton. Alkyl substituents, such as *tert*-butyl groups (^{t}Bu) in complex **1** or adamantyl groups (Ad) in complex **2**, favored the trigonal-pyramidal geometry in which the solvent molecule occupied the axial position of the pyramid. On the other hand, the presence of bulky groups, such as 2,4,6-$Ph_3C_6H_2$ (complex **3**), afforded a trigonal-planar geometry, where no solvent was coordinated to the metal center. Independently from the observed four- or three-coordinated geometries, the obtained complexes were tested in the model C—H amination reaction of toluene by 1-azidoadamantane (AdN$_3$). Obtained data showed the catalytic superiority of **2**, which can efficiently catalyze at room temperature the synthesis of the desired benzyl adamantly amine (95% yield) with 6.5 turnovers number (TON). An improvement of TON values was observed by increasing the reaction temperature, and simultaneously, the radical coupling of two PhCH$_2$ molecules was responsible for the formation of the undesired 1,2-diphenylethane. This result suggested that a radical reaction mechanism occurred in which, after the formation of the active imido complex, an H-atom abstraction and then a radical recombination afforded the desired C—H aminated product (Scheme 3). It should be noted that the hydrogen atom abstraction step is well supported by the high primary kinetic isotope effect (12.8) which was observed in the reaction of complex **2** with adamantyl azide in an equimolar toluene/toluene-d_8 mixture.

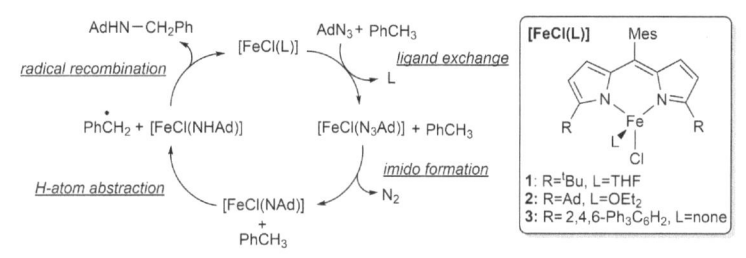

Scheme 3 Proposed mechanism for toluene amination catalyzed by iron dipyrromethene complexes.

In order to isolate the active imido intermediate [FeCl(NR)], the less active complex **3** was reacted with a stoichiometric amount of phenyl azide (N$_3$Ph) and the bimolecular species **4** was obtained probably by the radical coupling of two monomeric precursors (*path a*, Scheme 4). To overcome *path a*, complex **3** was reacted with a stoichiometric amount of the more hindered para-*tert*-butyl phenyl azide achieving the mononuclear iron complex **5**, whose Mössbauer and X-ray diffraction analyses were in accordance with the presence of a terminal imido ligand (*path b*, Scheme 4). The imido complex **5** transferred in a stoichiometric reaction the nitrogen-containing moiety to substrates such as toluene and styrene, affording the desired benzylic amine (42% yield) and aziridine (76% yield), respectively. The primary kinetic isotopic value of 24, which was registered in the reaction of **5** with an equimolar toluene/toluene-d_8 mixture, confirmed the mechanism proposed above for the catalytic reaction.

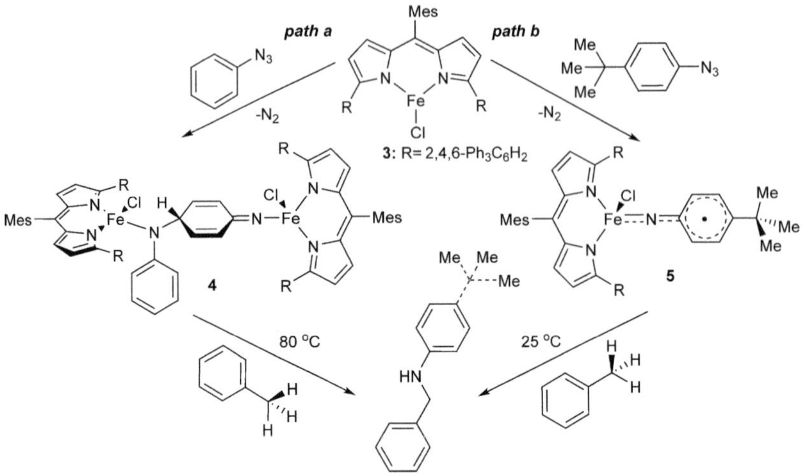

Scheme 4 Synthetic strategy adopted for the isolation of the iron imido intermediate **5**.

Iron dipyrromethene complex **3** was also used as a promoter for C—H amination and aziridination reactions. Catalytic data, together with the obtained Mössbauer parameters, indicated the formation of iron imido species containing a high spin Fe (III) (d^5, $S=5/2$) center that is antiferromagnetically coupled with an imido-based radical ($S=1/2$) fragment.

The electronic structure of the active imido intermediate prompted the authors to investigate the catalytic activity of other iron dipyrromethene complexes [Fe(II)Cl(L)] in several nitrene transfer reactions such as intramolecular C—H aminations,[29] aziridinations, and benzylic/allylic aminations.[30] The dipyrromethene skeleton was opportunely modified by replacing the mesityl group (Mes) in the *meso* position with 2,6-Cl$_2$C$_6$H$_3$ substituent in order to avoid the side-amination of accessible benzylic C—H bonds of the dipyrromethene ligand itself. Complex **6** (Scheme 5) efficiently catalyzed the C—H functionalization and subsequent cyclization[29] of several alkyl azides containing allylic, benzylic, primary and secondary C—H bonds affording the desired Boc-protected pyrrolidines (**7a–7o**, Scheme 5) in yields up to 98%. Additionally, catalyst **6** promoted C—H benzylic aminations and alkene aziridinations[30] yielding the desired amines **8a–8e** and aziridines **9a–9h** in yields up to 60% and 86%, respectively.

According to the reaction mechanism previously described for the toluene amination (Scheme 3), it was proposed that C—H benzylic aminations and alkene aziridinations generally proceed *via* one-electron reaction

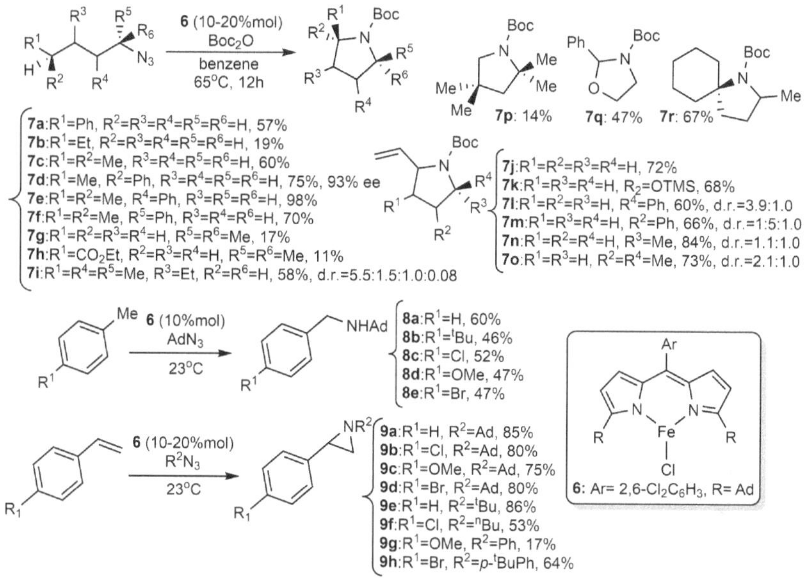

Scheme 5 Intramolecular C—H aminations, aziridinations and C—H benzylic aminations catalyzed by iron complex **6**.

pathways in which a radical imido complex is the active intermediate. The latter is also the putative species suggested in the reaction mechanism of intramolecular C—H aminations catalyzed by **6** (Scheme 6). For this reaction the authors proposed a multistep process in which the oxidation of iron (II) complex (**A**) forms iron (III) radical imido intermediate (**B**), which can be involved in two different reaction pathways. In *path* a, the formation of iron radical imido **B** is followed by an H-atom abstraction yielding intermediate **C** that evolves in compound **D** by a radical recombination process.

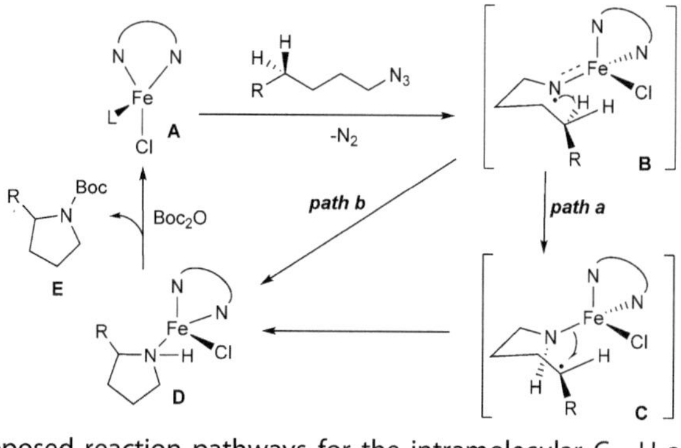

Scheme 6 Proposed reaction pathways for the intramolecular C—H amination catalyzed by iron complex **6**.

The final reaction of **D** with Boc$_2$O (Boc = *tert*-butoxycarbonyl) affords the desired product **E** by reforming the starting catalyst **A**. The alternative *path b* provides the formation of intermediate **D** by a direct C—H insertion on the radical imido **B**. Due to the spatial arrangement of the iron dipyrromethene complex and the steric hindrance generated by the adamantyl groups, the azide substrate approaches in both cases from the opposite side with respect to the chloride atom forming an imido intermediate in which the ring-closing process is geometrically favored.

Betley and co-authors also tested the activity of iron (II) dipyrromethene complexes in allylic amination reactions to discriminate eventual preference for aziridinations over allylic C—H aminations.[30] Collected data revealed that only allylic amine products **11a–11c** (Scheme 7) were obtained by reacting non-styrenyl alkenes with AdN$_3$ in the presence of catalyst **10**. Under the same experimental conditions, an analogous result was observed by using cyclic alkenes and products **11d** and **11e** were obtained in the low yields of 19% and 18%, respectively.

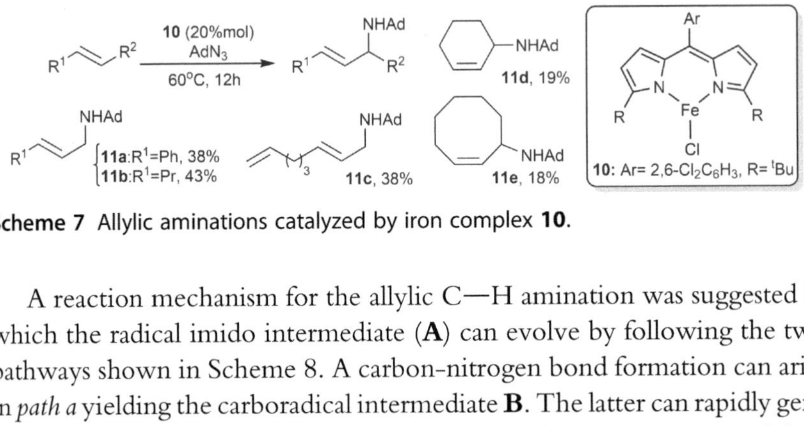

Scheme 7 Allylic aminations catalyzed by iron complex **10**.

A reaction mechanism for the allylic C—H amination was suggested in which the radical imido intermediate (**A**) can evolve by following the two pathways shown in Scheme 8. A carbon-nitrogen bond formation can arise in *path a* yielding the carboradical intermediate **B**. The latter can rapidly generate the aziridine ring **C** through a radical recombination process and then the desired allyl amine **E** could be obtained by the subsequent homolytic (or heterolytic) aziridine ring-opening reaction. In the alternative *path b*, the allylic radical compound was obtained by an H-atom abstraction process and reacts with the amino functionality of the iron intermediate **D** forming the final product **E**. Bearing in mind that allylic linear amines were obtained as the sole products as well as allylic radical species usually recombine through the less-hindered terminal carbon atom, the authors suggested the *path b* as the operative one for the allylic amination promoted by iron (II) dipyrromethene catalysts.

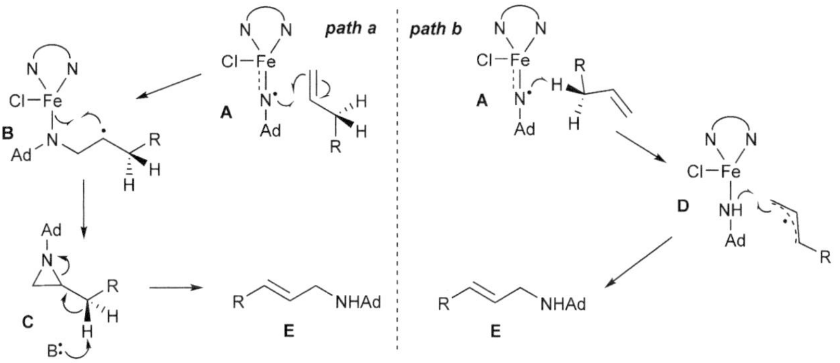

Scheme 8 Proposed reaction pathways for the allylic amination catalyzed by iron dipyrromethene complexes.

To shade some light onto the nature of the imido intermediate involved in nitrene transfer reactions catalyzed by complex **10**, the imido complex **12** (Scheme 9), synthesized by reacting the iron bimetallic precursor with the 2,6-diisopropylphenyl azide, was isolated and characterized.[31] Crystallographic data and Mössbauer parameters of **12** confirmed the formation of an iron (III) imido complex in which the metal is antiferromagnetically coupled with an imido-radical ligand. It should be noted that observed results are in accordance with the electronic structure already proposed for the previously isolated imido complex **5** (Scheme 4). Complex **12**, instead transferring the nitrene functionality to the C—H bond (as observed with complex **5**), was involved in a H-atom abstraction process, which forms the amino species

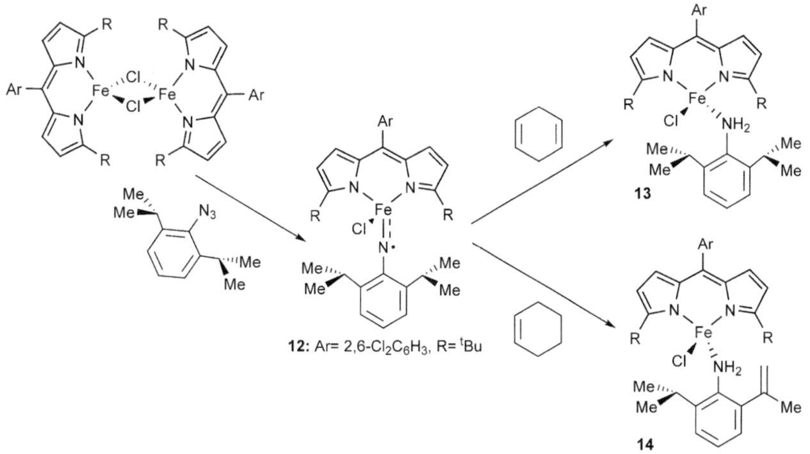

Scheme 9 Synthesis of imido complex **12** and its reactivity toward cyclohexene and 1,4-cyclohexadiene.

corresponding to the employed azide. As reported in Scheme 9, the reaction of complex **12** with an excess of cyclohexene or 1,4-cyclohexadiene yielded the corresponding amine derivatives and in both cases the C—H amination was not observed. This tendency was evident in the formation of complex **14** in which a double H-atom abstraction was responsible for the dehydrogenation of one isopropyl group of the amine fragment.

The unexpected reactivity of complex **12** encouraged a deeper investigation of all the factors that can influence the reactivity of both radical and non-radical imido species derived from iron (II) dipyrromethene complexes.[32,33] Thus, N-alkyl- and N-arylimido complexes (Scheme 10) were synthesized and characterized. Mössbauer and X-ray diffraction analyses confirmed that both radical (**5, 18**) and non-radical (**15, 17**) complexes contained a high spin iron (III) center. Note that the oxidation state of iron did not change during the reduction of **5** to **15** as well as of **17** to **18** because in both cases the added electron provoked a periphery reduction. While in the first case the aromaticity of the N-aryl imido ligand was restored, in the second one the added electron was responsible for the formation of the N-centered radical imido moiety.

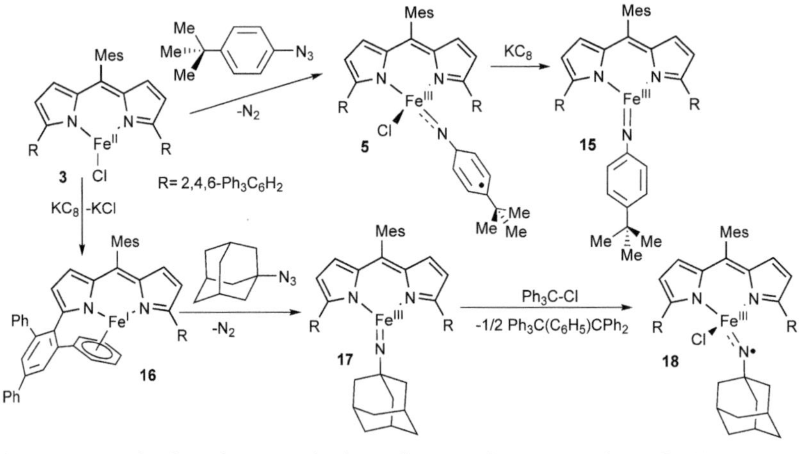

Scheme 10 Radical and non-radical imido complexes tested in C—H amination reactions.

All the obtained imido complexes were tested as catalysts of C—H aminations and comparative kinetic analysis of the toluene C—H amination established the higher reactivity of **5** and **18** with respect to that of the non-radical molecules **15** and **17**.

Complex **18** performed better than **5** due to the different radical density, which was *N*-centered in **18** and delocalized on the aromatic ring in complex **5**. The radical delocalization on the aromatic ring of **5** reduces the C—H amination activity by favoring the side H-atom abstraction yielding the NH$_2$ functionality (see Scheme 9).

Considering the negative effect associated to the presence of the aromatic substituent, molecular orbital considerations and thermodynamic studies were performed on *N*-alkyl complexes **17** and **18**. Both non-radical **17** and radical **18** complexes reacted with toluene forming the corresponding C—H aminated product through a H-atom abstraction and subsequent radical recombination (Scheme 11). In view of the frontier molecular orbitals involved in this process, the authors proposed that the reaction was favored using **18** due to the lower energy required for the orbital interaction between the toluene C—H σ bond and the *N*-radical (2p$_y$)^1 of **18** with respect to that was necessary when the Fe—N π bond of **17** was involved.

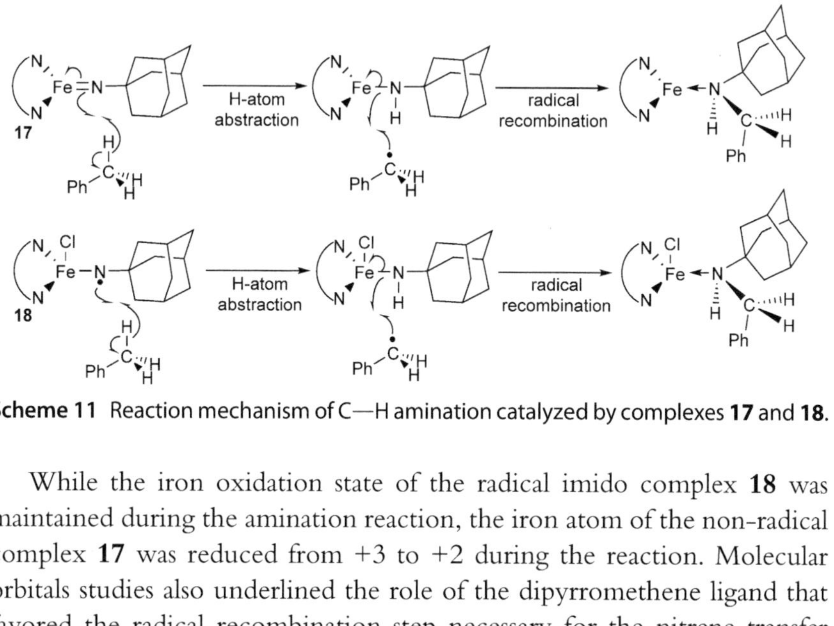

Scheme 11 Reaction mechanism of C—H amination catalyzed by complexes **17** and **18**.

While the iron oxidation state of the radical imido complex **18** was maintained during the amination reaction, the iron atom of the non-radical complex **17** was reduced from +3 to +2 during the reaction. Molecular orbitals studies also underlined the role of the dipyrromethene ligand that favored the radical recombination step necessary for the nitrene transfer process.

Tetradentate and pentadentate ligands. Porphyrins and porphyrinoids are nitrogen tetradentate ligands largely used for the synthesis of bio-inspired iron catalysts for nitrene transfer reactions. Over the years, very active iron complexes have been obtained by using these ligands but up to now the formation of imido species as active catalytic intermediates has only been supposed. In this regard, Goldberg and co-authors, inspired by the high activity of heme enzymes, investigated the use of a heme-like corrolazine (Cz) as a ligand for the synthesis of stable iron imido complexes able to transfer the N-moiety to organic substrates.[34] The usual strategy employed to synthesize iron imido complexes, by reacting an iron precursor with alkyl or aryl azides, was ineffective in presence of iron corrolazine **19** and the desired imido species **20** was efficiently achieved by using chloramine-T (Na$^+$TsNCl$^-$·3H$_2$O) as the alternative nitrene source (Scheme 12).

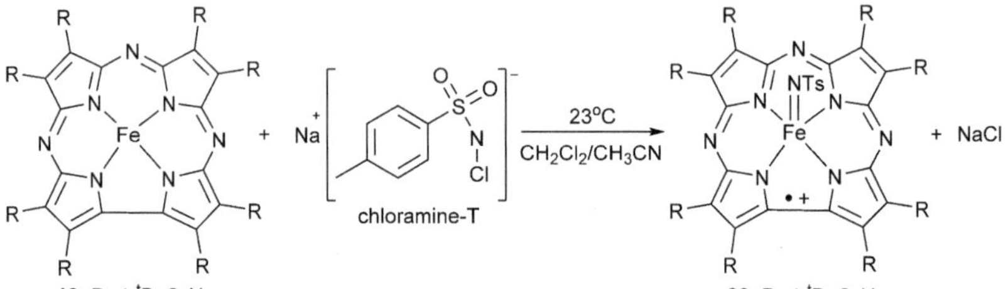

Scheme 12 Synthesis of iron imido complex **20**.

UV–visible spectra acquired during the reaction confirmed the high stability of **20** in solution at room temperature due to the ability of corrolazines in stabilizing high-valent iron species. On the other hand, the lower stability of **20** in the solid state prevented to obtain it in a pure crystalline form for performing a structural investigation. The desired imido complex was isolated in a solid state alongside 25% of the starting iron (III) precursor (**19**) due to the auto-reduction of the imido complex during the precipitation process. Although **20** was not obtained in a pure form, Mössbauer and EPR spectroscopies (EPR = electron spin resonance) confirmed the formation of an imido complex with an iron (IV) center antiferromagnetically coupled with the π-radical cation delocalized on the corrolazine ligand.

The ability of **20** to transfer the N-tosyl (NTs) group to triphenylphosphine was investigated. The formation of the desired Ph$_3$P=NTs was accompanied by the reduction of the imido complex to the precursor **19**. This observation prompted the authors to test complex **19** in the same reaction under catalytic conditions. UV–visible and NMR data confirmed the formation of the imido intermediate **20** during the reaction and Ph$_3$P=NTs was rapidly obtained with no other phosphorane by-products. To acquire more information on the reaction mechanism, the **19**-catalyzed N-Ts transfer to triphenylphosphine was monitored by ^{31}P NMR and collected data indicated the formation of Ph$_3$P=NH and Ph$_3$P=O as side-products of the reaction. The contemporary formation of the two different iron complexes **21** and **22** was in accordance with the reaction mechanism reported in Scheme 13. It was proposed that the desired Ph$_3$P=NTs compound reacts with complex **19** forming Ph$_3$P=NH by the hydrolytic cleavage of the tosyl group. The so-formed phosphinimine evolves by following two different pathways. In *path a*, the coordination of Ph$_3$P=NH to complex **19** yields the isolated complex **21**, while in *path b* the hydrolyzation of Ph$_3$P=NH formed Ph$_3$PO that coordinates **19** giving complex **22**.

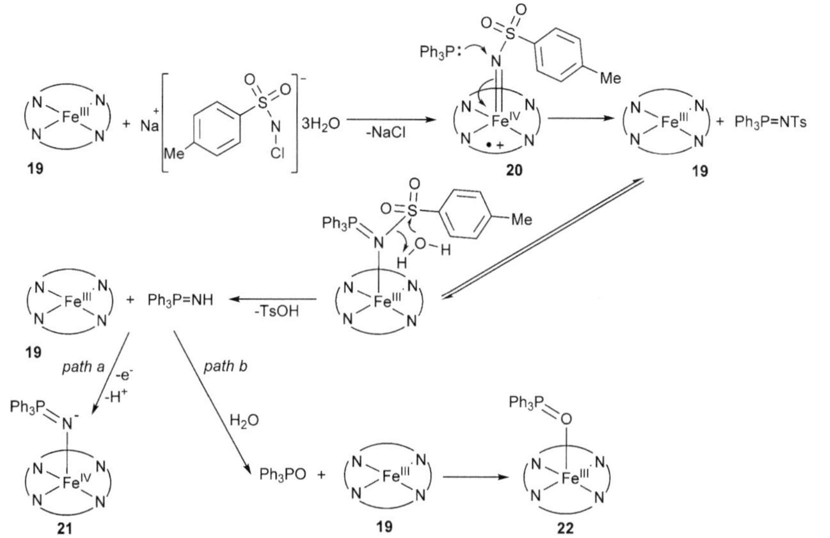

Scheme 13 Proposed mechanism for the NTs transfer catalyzed by complex **19**.

The use of non-heme tetradentate ligands was investigated by Nam and co-authors[35,36] who studied the nitrene transfer activity of iron(V) imido complexes containing to the so-called TAML ligand (TAML = tetraamido macrocyclic ligand). As reported in Scheme 14, imido complex [Fe(V)(TAML)(NTs)]Na (**24**) can be efficiently synthesized from its iron (III) precursor [Fe(III)(TAML)]Na (**23**) by employing *N*-tosyliminophenyliodinane (PhINTs) as the nitrene source.

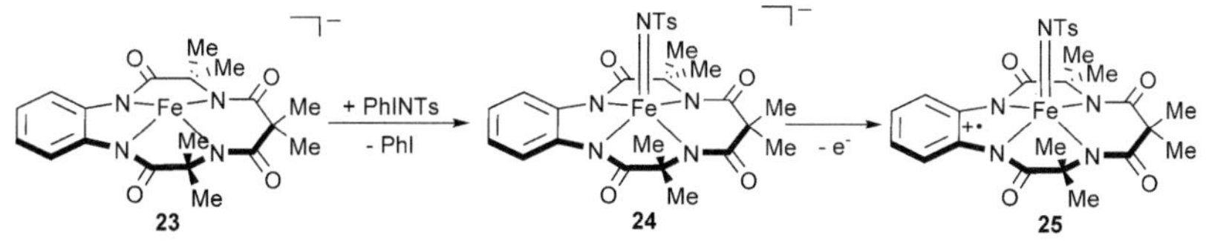

Scheme 14 Synthesis of the iron imido complexes **24** and **25** containing TAML ligand.

The oxidation of **24** with one-electron oxidants, such as [Fe(bpy)$_3$]$^{3+}$ (bpy=2,2′-bipyridyl), [Ru(bpy)$_3$]$^{3+}$ or tris(4-bromophenyl)-ammoniumyl hexachloroantimonate [(4-BrC$_6$H$_4$)$_3$N]SbCl$_6$, formed the metastable complex **25**. Despite complex **25** was less stable than its precursor **24**, both complexes were fully characterized by spectroscopic techniques including Mössbauer, X-ray absorption spectroscopy (XAS) and extended X-ray absorption fine structure (EXAFS). In addition, DFT calculations (DFT=density functional theory) elucidated their electronic structures. Based on acquired data, while complex **24** was recognized as a low-spin iron (V) imido complex ($S = 1/2$), complex **25** was identified as a diamagnetic complex ($S = 0$) in which the low-spin iron (V) is antiferromagnetically coupled with the TAML radical cation which is formed during the one-electron oxidation step.

Due to the different electronic structures, both complexes were tested as catalysts in nitrene transfer reactions using fluorene and thioanisole as model substrates for the C—H bond activation and sulfimidation, respectively.

As reported in Scheme 15, while complexes **24** and **25** had comparable reactivities toward fluorene affording the desired product **26a** in similar yields (~90%), a higher activity of **25** was observed in the sulfimidation forming **27a**. The authors proposed two different mechanisms for these reactions suggesting that while the rate-determining step of the C—H amination is the H-atom abstraction, during which the aminated compound **26a** and Fe(IV)(TAML) complex are directly formed in a single step, the sulfimidation reaction proceeds in two consecutive steps. It was proposed that the first step of sulfimidation is an electron transfer from thioanisole substrate PhS(CH$_3$) to the iron (V) catalyst, which is followed by the nitrene transfer process yielding Fe(IV)(TAML) species together with the desired Ph(CH$_3$)S=NTs compound.

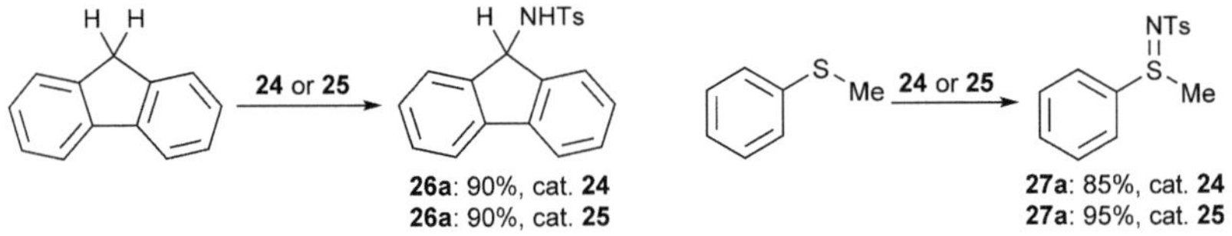

Scheme 15 Nitrene transfer reactions promoted by imido complexes **24** and **25**.

In order to better support these hypotheses, H atom affinity (HAA) in C—H aminations and electron affinity (EA) in sulfimidations were calculated for both catalysts. The very similar HAA values calculated for **24** and **25** (81.9 and 85.2 kcal mol^{-1}, respectively) complexes are in accordance with the comparable reactivity of these two catalysts observed in the fluorene C—H amination. Conversely, the EA value obtained for **25** (122.8 kcal mol^{-1}) was higher than that of **24** (103.9 kcal mol^{-1}) to support the pivotal role of the electron transfer process in the thioanisole sulfimidation and explained the superiority of **25** over **24** in promoting the synthesis of **27a**.

In order to investigate the general involvement of electron transfer processes in sulfimidations, Nam's research group investigated the electrochemical properties of complexes **24** and **25** (Scheme 14) and their catalytic propensity in promoting the amination of several para-X-substituted thioanisoles (X = OMe, Me, H, Cl, CN) showing different one-electron oxidation potentials.[37]

Experimental studies and DFT calculations supported a direct nitrene transfer when complex **24** was employed to promote the thioanisole sulfimidation, which occurred with rate constants higher than those estimated if an out-sphere electron transfer process was the active catalytic pathway. On the other hand, when complex **25** was tested as the catalyst, the transfer of the NTs group can follow both mechanisms and the occurrence of one or the other depends on the oxidation potential of the chosen thioanisole substrate. When substrates showing low oxidation potentials, such as p-methoxythioanisole, were tested, the sulfimidation reaction proceeds through the one-electron reduction of **25**. Conversely, if para-X-substituted thioanisoles showing higher oxidation potentials were used, the sulfimidation reaction produces the desired organic product together with an iron(III) complex presenting the TAML ligand as a radical cation.

The NTs transfer to organic substrates was also investigated by Maldivi and co-authors[38] who evaluated, from the theoretical point of view, the activity and electronic properties of the different iron(IV) imido complexes **28–33** (Fig. 1), which were obtained by using tetradentate ligands (L$_n$) containing phenolate and nitrogen donor groups. Acquired data showed the pivotal role of phenolate substituents in stabilizing the high valent iron imido

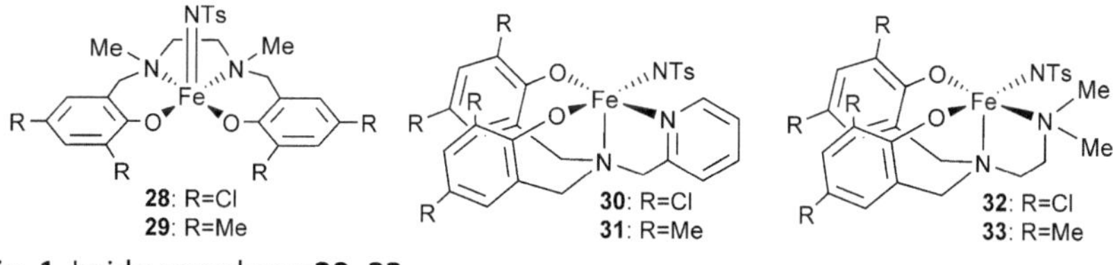

Fig. 1 Imido complexes **28–33**.

complexes in the $S=2$ iron spin state thanks to their basicity and strong π-donation capability. Bearing in mind that the nitrene transfer from high valent iron complexes to an organic substrate can occur either by the H-atom abstraction or in two steps by the initial electron transfer from the substrate to the iron complex, the authors calculated the EA of L_nFe(IV)NTs catalysts and the BDE (bond dissociation energy) of the N—H bond of L_nFe(III) NHTs complexes in order to better rationalize their reactivity.

Comparing the EA energies of all the examined complexes, higher EA values were found for **28, 30** and **32** in which the methyl groups of the ligands were replaced by chloride atoms to indicate that the presence of electron withdrawing substituents on the ligand periphery can increase the iron complex electrophilicity and in turn their catalytic activity. Higher values of BDE energies were found for **32**- and **33**-deriving L_nFe(III)NHTs complexes in view of the higher basicity of the employed ligands with respect to that of ligands used for synthesizing catalysts **28–31**. These data suggested the propensity of **32** and **33** to promote reactions involving the direct nitrene transfer into C—H bonds by a H-atom abstraction mechanism.

Despite L_nFe=NR complexes formed by tetradentate ligands are the most studied systems due to their similarities with natural enzymes, iron (IV) tosyl imido complexes presenting pentadentate ligands have been also studied. de Visser and co-authors reported the synthesis and complete characterization of complexes **34** and **35**, showed in Fig. 2.[39] The mass and ^{1}H NMR characterization of complex **35**, synthesized from its iron (II) precursor by the UV–Vis activation of PhINTs, revealed the formation of an iron (IV) ion in $S=1$ spin state. The reactivity of **35** was initially evaluated in the model NTs transfer to thioanisole and then compared to that of complex **34**. Lower reaction rates were observed when **35** was used in place of **34** suggesting that probably the **35**-catalyzed sulfimidation reaction was governed by a direct NTs transfer mechanism rather than by the more efficient electron transfer process. The study of the **35**-catalyzed sulfimidation of different para-X-substituted thioanisoles (X=OMe, Me, H, Cl) disclosed

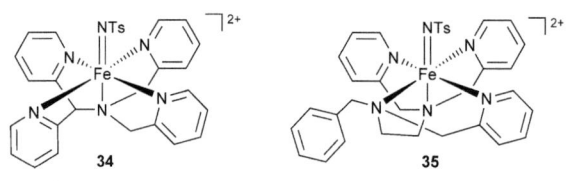

Fig. 2 Imido complexes **34** and **35**.

an improvement of the reaction rate when electron-donating substituents, such as methoxy and methyl groups, were present on the substrate. Kinetic data and the resulting Hammett plot, collected by employing substituted thioanisoles as substrates, supported a reaction mechanism in which the electron transfer process is the rate-determining step and the positive effect of electron-donating groups due to the stabilization of the transition state positive charge. In addition, DFT calculations and the high BDE values of L_nFe(III)NHTs complexes suggested that L_nFe(IV)NTs imido complex abstracts a H atom from the substrate forming a strong N—H bond, which are less prone to be involved in the nitrene transfer to organic substrates.

The nitrene transfer ability of complex **34** was also investigated by Latour and co-authors[40] who tested the activity of **34** and the structurally similar compound **38** (Scheme 16) in the styrene aziridination. Both complexes **34** and **38** were synthesized from their iron (II) precursors **36** and **37**, respectively (Scheme 16) and characterized by using several spectroscopic techniques. Spectroscopic and spin densities data revealed the formation of two iron (IV) imido complexes ($S = 1$) in which a strong radical character was located on the nitrogen imido atom. Preliminary Mössbauer experiments, performed by reacting **38** with thioanisole, showed the formation of the desired sulfimidated product together with a high spin iron (II) species, which was supposed to be the starting complex **37**. This observation suggested that the imido complex **38** was the active intermediate in the NTs transfer to thioanisole forming the desired product by restoring the starting catalyst **37**. Thus, the catalytic activities of complexes **36** and **37** were compared in the model styrene aziridination to underline the ligand effect on the catalytic performances. Complex **37** showed a better catalytic performance than **36** in forming **39a** that was obtained in higher yield (70% *vs* 59%). Collected data are in accordance with the higher EA energy estimated for the imido complex **38** than that calculated for **34**. The authors attributed the enhanced reactivity of **37** and the higher EA of **38** to the less basic and electron-donating character of the employed ligand. In order to better investigate the reactivity of **36** and **37**, Hammett correlation plot was performed by analyzing competitive experiments between styrene and different *para*-X-substituted styrenes (Scheme 16). A quite similar behavior was observed for both catalysts and the observed negative slope of the Hammett plot was in agreement with the involvement of an electrophilic active species and the occurrence of a partial charge transfer during the rate-determining step.

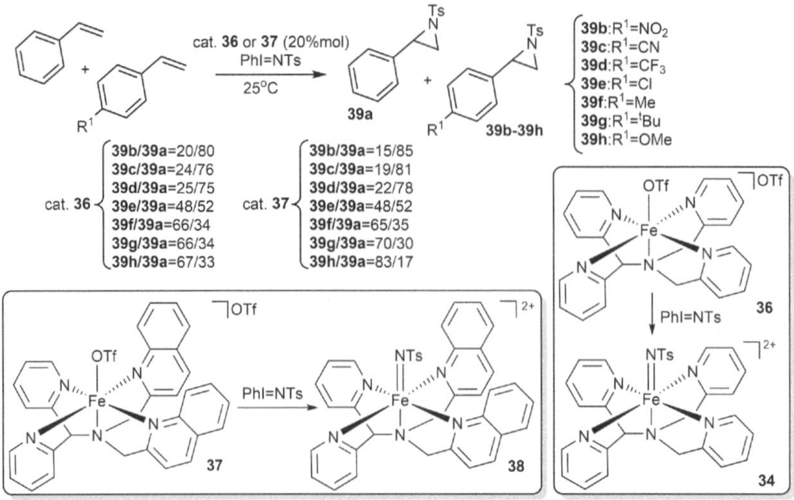

Scheme 16 Competitive styrene aziridinations catalyzed by complexes **37** and **36**.

In view of the Hammett correlation and DFT calculations performed on the aziridination reaction promoted by **37** and **36**, the authors proposed a two-step reaction mechanism (Scheme 17) in which the electron-transfer from styrene to the tosylimido group of the iron complex represents the rate-determining step.

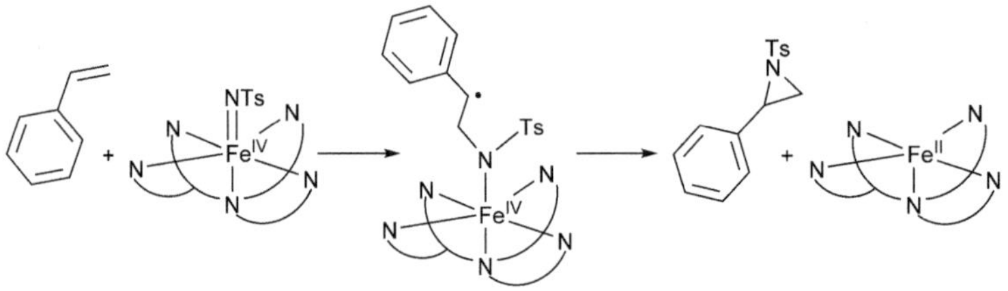

Scheme 17 Stepwise reaction mechanism proposed for styrene aziridination catalyzed by **34** and **38**.

Steric factors do not affect the reaction rate because the proposed electron-transfer does not require the close contact of the involved regents. On the other hand, the electronic nature of the R group on the imido moiety (NR) together with the overall charge of the active imido intermediate ($L_nFe=NR$) have a great influence on the kinetics of the catalytic reaction. The charge of the imido complex is strictly dependent from the employed ligand and in fact the use of neutral penta-nitrogen ligands cause higher EA values than those registered by testing imido complexes derived from dianionic ligands, such as phenolate-containing ones. In addition, the presence of electron-withdrawing substituents on the nitrogen imido atom increases the EA value (EA of $L_nFe=NTs$ was 20 kcal mol^{-1} higher than that of Fe= NTol (Tol=tolyl)), and in turn enhances catalytic performances.

Multimetallic iron imido complexes. In addition to the above discussed monometallic iron systems, multimetallic imido complexes have recently demonstrated to be very efficient in nitrene transfer reactions. As already reported for monometallic species, the active imido compounds, derived from multinuclear precursors, resulted extremely reactive and not stable enough to be characterized by standard spectroscopic techniques. Thus, the chemical structures of the majority of these imido intermediates were only supposed and modeled by theoretical calculations.[27,41,42] Nevertheless, Betley and co-authors reported the synthesis and catalytic applications of a di-iron bridging imido complex **41**,[31] which was proposed as the active intermediate in C—H amination and aziridination reactions. Di-iron complex **41**, which bound two bidentate nitrogen-containing ligands, was obtained by reacting the iron precursor **40** with 3,5-bis(trifluoromethyl)phenyl azide in *n*-hexane at −40 °C for 6 h (Scheme 18). Experimental and theoretical characterization of **41** revealed the formation of a bimetallic complex in which the two anti-ferromagnetically coupled high spin iron (III) atoms gave rise to five possible dimer states with a total spin that is in the 0–5 range. In addition, DFT calculations revealed the dependence of the antiferromagnetic coupling value on the thermal energy and physical state of the complex. The theoretical study indicated that the geometry of the Fe-N-Fe linkage can be disrupted in solution by molecular motions and the consequent reduction of the orbital overlap causes the diminution of the antiferromagnetic interaction. As a result, the

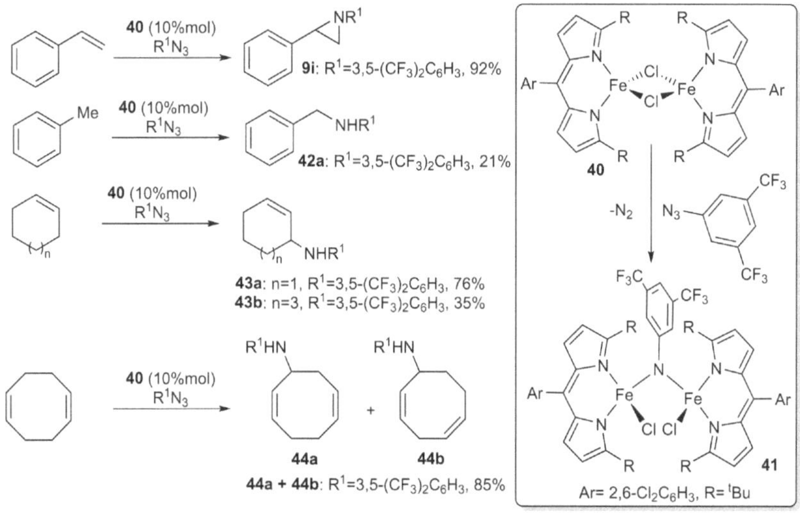

Scheme 18 Nitrene transfer reactions catalyzed by complex **40**.

thermal energy produces a dimeric system presenting two non-interacting iron (III) ions in which the unpaired electron density along the Fe—N bond is responsible for the C—H bond activation through H-atom abstraction processes. Due to the above-mentioned electron density, imido complex **41** was very efficient in transferring the nitrene functionality to compounds containing allylic and benzylic C—H bonds. These results encouraged the authors to evaluate the catalytic activity of the precursor **40** in different nitrene transfer reactions. Good results were achieved in the C—H allylic amination of cyclohexene, cyclooctene and cyclooctadiene affording the desired products (**43a**, **43b** and **44a** + **44b**) in yields up to 85%. Lower yield (21%) was instead obtained in the synthesis of product **42a** due to the stronger C—H bond of toluene. In addition to amination reactions, **40** efficiently promoted the styrene aziridination affording **9i** in 92% yield (Scheme 18).

The good achieved catalytic results prompted the authors to investigate the reaction mechanism by monitoring *in situ* the cyclohexene amination as the model reaction. Time-course NMR spectra indicated that the imido complex **41** is formed together with a monomeric iron imido radical species. It was proposed a reaction mechanism for the C—H amination in which complex **40** reacts with the aryl azide forming first a monomeric imido radical intermediate which can be involved in two different pathways. The radical intermediate can either react with cyclohexene yielding a carboradical by a H-atom abstraction or coupling with another molecule of **40** forming the bridging imido **41** complex. The so-formed **41** promotes the H-atom abstraction from cyclohexene to yield the intermediate **45** that finally affords the desired product by a radical recombination with the cyclohexene carboradical (Scheme 19).

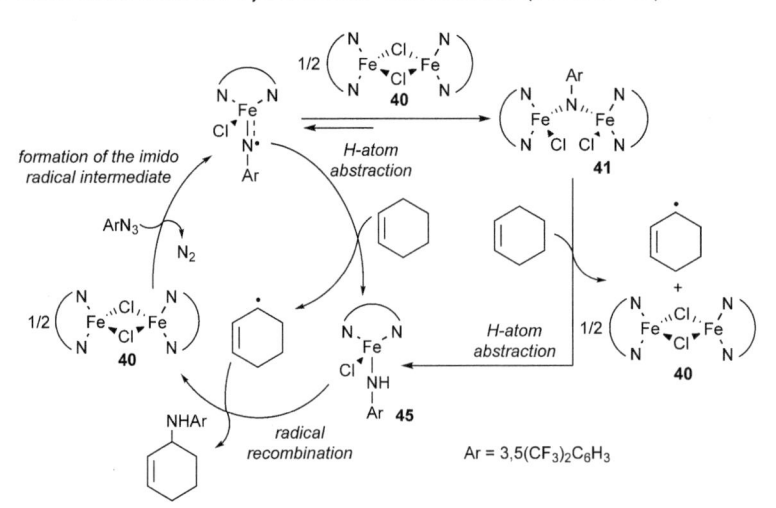

Scheme 19 Reaction mechanism proposed for the C—H amination catalyzed by complex **40**.

Considering the short lifetime of diiron imido species active in nitrene transfer reactions, Latour and co-authors studied the thioanisole sulfimidation promoted by the diiron (III,II) complex **46** (Scheme 20) by using a combination of desorption electrospray ionization mass spectrometry (DESI-MS), quantitative chemical quench experiments and DFT calculations.[43] The DESI-MS study of the reaction between thioanisole and tosyl aryliodinane (ArI=NTs) in the presence of complex **46** revealed the formation of the imido complex **48**. Additional stopped-flow and Mössbauer experiments confirmed the presence of other two species, which were identified as the amido derivative **49** and the iodinane adduct **47**. Competitive sulfimidation reactions performed by using thioanisole and para-X-substituted thioanisoles (X = NO$_2$, COMe, OMe, Me, Cl) confirmed the electrophilic nature of the imido intermediate. The linear correlation between the kinetic constants and the redox potential of the investigated para-X-substituted thioanisoles suggested the implication of a direct nitrene transfer reaction rather than a one-electron oxidation process. Based on these results, the authors proposed a reaction mechanism in which complex **46** reacts with the nitrene source affording the iodinane adduct **47**. The decomposition of the latter complex yields the reactive imido intermediate **48**, which either rapidly transfers the NTs group to thioanisole or forms the amido complex **49** through a H-atom abstraction (Scheme 20). Due to the short lifetime and very high reactivity of **48**, its electronic structure was only suggested by DFT calculations which supported the formation of a diiron (III,IV) imido complex with a radical character located on the NTs moiety. The possibility of the iron (IV) ion of complex **48** to assume both $S = 1$ and $S = 2$ spin states can explain the observed high chemical reactivity.

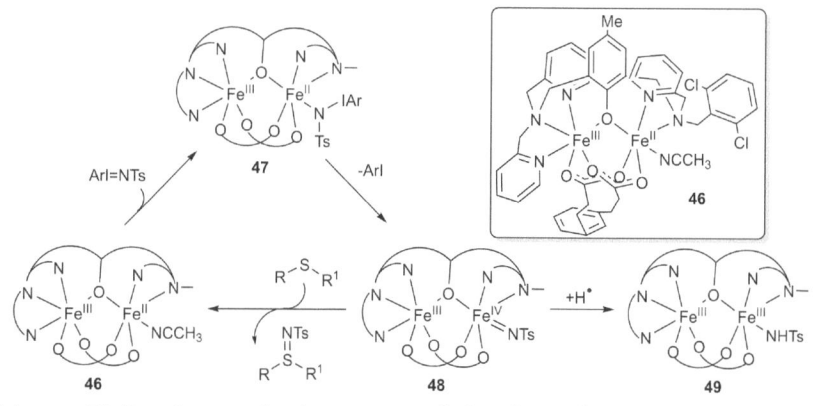

Scheme 20 Reaction mechanism proposed for the sulfimidation catalyzed by complex **46**.

2.2 Ruthenium

Despite the lower abundance and the higher cost of ruthenium than iron, ruthenium complexes have been extensively employed in nitrene transfer reactions showing excellent performances in aziridination,[7] sulfimidation[10] and C—H amination reactions.[9,44] Several ligands and nitrene sources have been investigated to maximize the nitrene transfer ability of these complexes and, similarly to what has been described for iron-catalyzed reactions, imido intermediates have been proposed as active intermediates. A general isolation of putative imido species, involved in ruthenium-mediated nitrene transfers, has been prevented due to their high chemical instability and their molecular structures have often been supposed or suggested by theoretical calculations.[45–49]

Some examples of active ruthenium bis-imido porphyrin complexes have been reported and they are generally obtained by reacting ruthenium porphyrin precursors with organic azides (RN$_3$) as nitrene sources. Che and co-authors investigated the synthesis, chemical stability and reactivity of various ruthenium bis-imido complexes of general formula Ru(VI)(porphyrin)(NY)$_2$ in which π-conjugated arylimides were employed as axial ligands to rise the complex stability by increasing the electron delocalization along the metal-imido multiple bond.[50] The additional use of sterically encumbered porphyrins and electron withdrawing substituents on the arylimido ligands allowed to efficiently synthesize complexes **51–54** (Scheme 21) by reacting

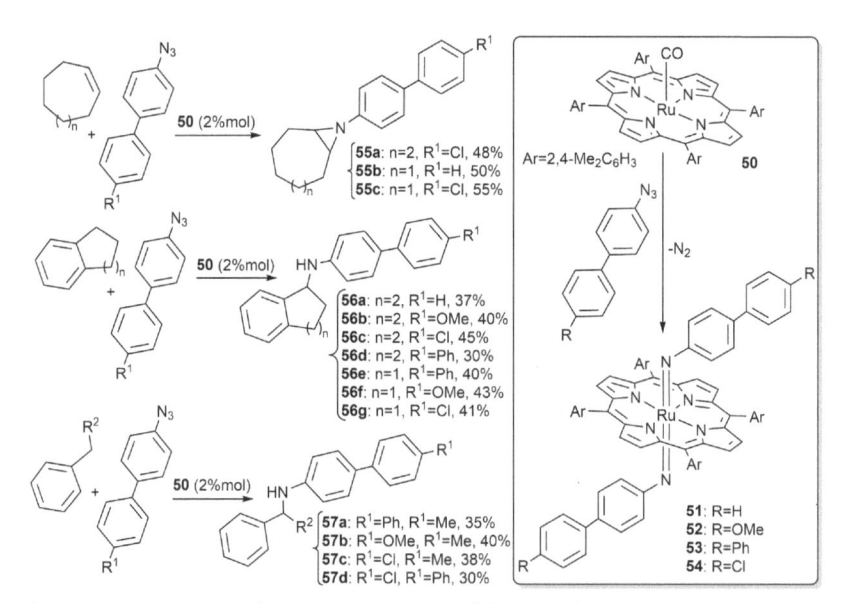

Scheme 21 Nitrene transfer reactions promoted by complex **50**.

the ruthenium (II) porphyrin **50** with different aryl azides. All the obtained bis-imido complexes were fully characterized by NMR, UV–Vis and IR spectroscopies thanks to their high stability and, in the case of complex **52**, the molecular structure was also determined by X-ray crystallography.

Thus, the reactivity of these complexes was firstly investigated in the stoichiometric nitrene transfer to alkenes and hydrocarbons forming in moderate yields (up to 58%) the desired aziridines and C—H aminated products, respectively. In view of these positive results, the authors tested the nitrene transfer activity of ruthenium (II) complex **50** in the catalytic amination of different organic substrates (Scheme 21). The desired aza-compounds were obtained in yields (up to 55%) similar to those obtained by using ruthenium bis-imido complexes under stoichiometric conditions. The modest catalytic performance of **50** was attributed to the employed porphyrin ligand whose steric hindrance could hamper the efficient transfer of the nitrene functionality to the approaching organic substrate.

A less encumbered porphyrin was instead used by E. Gallo and co-authors to study the role of the ruthenium bis-imido complex **59** in allylic amination reactions promoted by complex **58**.[51] The bis-imido ruthenium complexes **59** was obtained by reacting ruthenium (II) porphyrin **58** with an excess of 3,5-bis(trifluoromethyl)phenyl azide (Scheme 22) and fully characterized by standard spectroscopic techniques and X-ray diffraction. Complex **59** was stable for several days if stored under nitrogen in the solid state and decomposed only in presence of coordinative solvents able to replace the aryl imido ligands. Considering the nature of **59**,

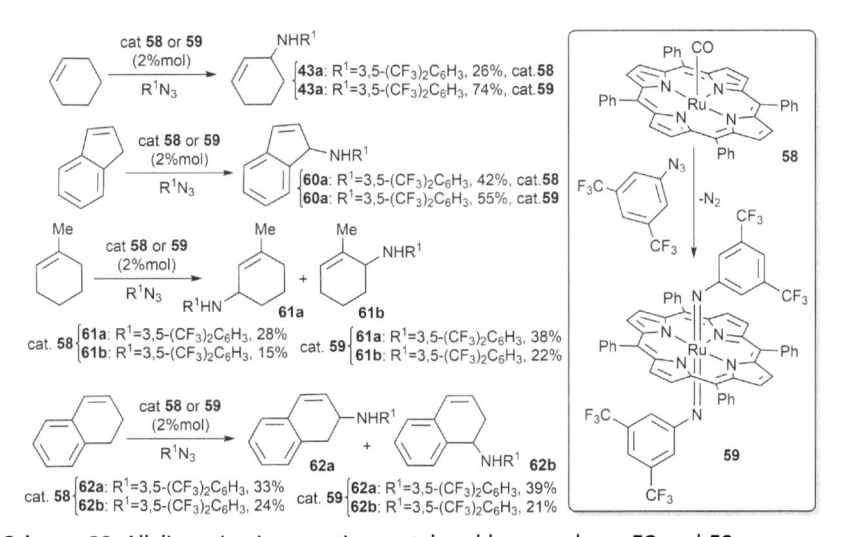

Scheme 22 Allylic amination reactions catalyzed by complexes **58** and **59**.

the authors explored its reactivity toward different allylic substrates, both in stoichiometric and catalytic conditions and collected data confirmed the imido complex ability to efficiently transfer the nitrene functionality to organic substrates. The catalytic performance of **59** was compared to that of **58** (Scheme 22) and achieved results highlighted a comparable, and even higher efficiency, of the bis-imido complex with respect to that of its precursor **58**.

These results suggested the role of bis-imido complexes as active intermediates in nitrene transfer reactions promoted by ruthenium (II) porphyrins. Experimental and kinetic studies performed on the model amination of cyclohexene revealed that the allylic amination can occur following two different pathways depending on the employed aryl azide and the substrate concentration. As shown in Scheme 23, in both pathways was proposed the initial formation of the mono-imido species **B**, which can either directly transfer the nitrene functionality to the substrate affording the desired product (*path a*) or react with another molecule of aryl azide forming the bis-imido complex **C**. The latter compound is then involved in *path b* for generating the allylic amine. It was proposed that *path a* was the favorite mechanism when less electron-withdrawing substituents on the aryl moiety were present and high concentrations of cyclohexene were employed. Conversely, using strong electron-withdrawing substituents and low concentrations of substrate, complex **B** is more prone to react with another azide molecule forming the bis-imido complex **C** that becomes the reaction catalyst of *path b*. The proposed reaction mechanism was also supported by the isolation of ruthenium degradation products **F** and **D**, which were formed by using 4-trifluoromethylphenyl azide (less electron-withdrawing substituent, *path a*) and 3,5-bis(trifluoromethyl)phenyl azide (strong electron-

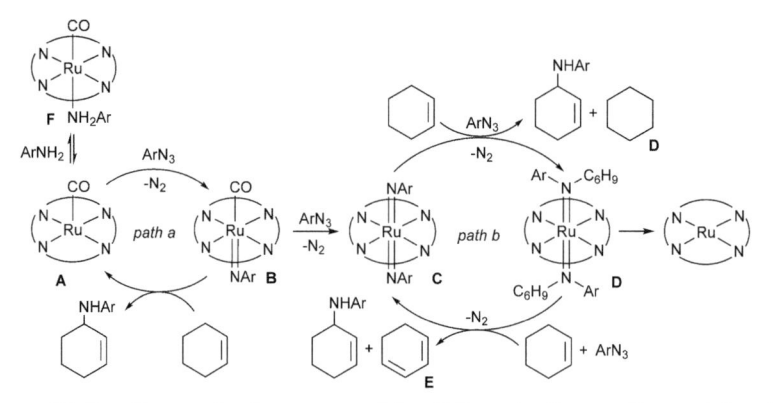

Scheme 23 Reaction mechanism proposed for allylic amination reactions catalyzed by ruthenium porphyrins.

withdrawing substituent, *path b*), respectively. Unfortunately, the proposed mono-imido **B** has not been isolated up to now and its formation was only supported by Raman analysis which revealed the presence of a Ru–CO band at low wavelength number probably due to the coordination of the imido moiety on the *trans*-axial position with respect to the CO ligand.[52]

In order to investigate the role of the ruthenium oxidation state in the formation of bis-imido complexes, E. Gallo and co-authors studied the reactivity of μ-oxo dimeric ruthenium (IV) porphyrin **63** in different nitrene transfers by using aryl azides as nitrene sources.[53] Complex **63** efficiently catalyzed benzylic/allylic amination and aziridination reactions (Scheme 24) affording the desired products in yields comparable to those previously reported by using ruthenium (II) porphyrin catalysts. The analogous catalytic performances which were observed by using either ruthenium (II) or (IV) complexes suggested the involvement of similar intermediates and thus, the formation of bis-imido complex **59** also in catalytic reactions promoted by the dimeric complex **63** was verified. The stoichiometric reaction of **63** with 3,5-bis(trifluoromethyl)phenyl azide afforded **59** in good yields suggesting that bis-imido complexes should be considered the active species in nitrene transfer reactions independently from the oxidation state (II or IV) of the ruthenium complex employed as the reaction promoter.

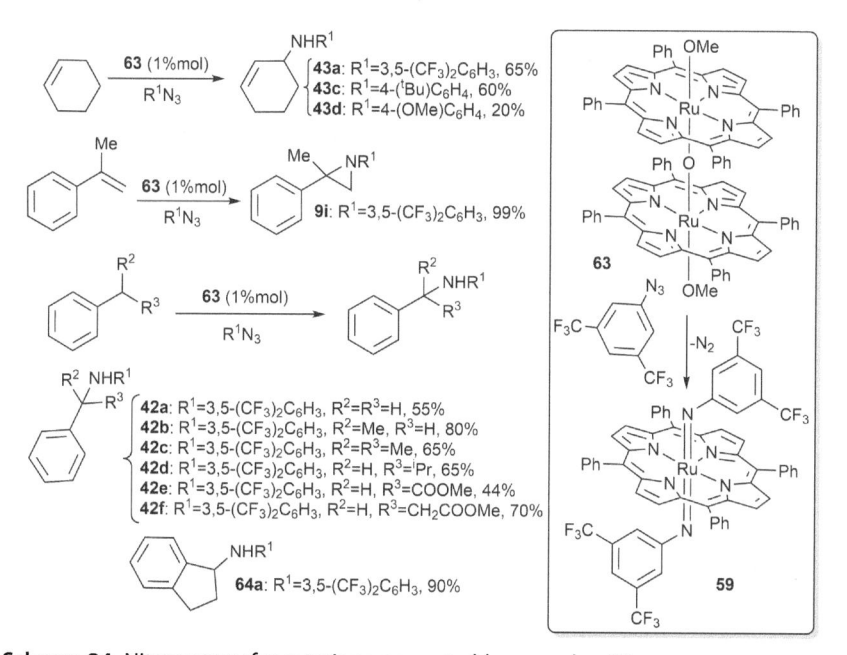

Scheme 24 Nitrene transfer reactions promoted by complex **63**.

Is important to note that bis-imido complex **59** was also very active in promoting the synthesis of C$_3$-substituted indoles by the nitrene transfer from aryl azides to alkynes.[54,55] As reported in Scheme 25, complex **59** catalyzed the regioselective synthesis of several substituted indoles affording the expected products in yields up to 95%. High yields were achieved by using aryl azides substituted with two equal electron withdrawing groups on the *meta* positions of the aryl ring due to the presence of two equivalent *ortho* C—H bonds whose activation forms the same indole product (compounds **65a–65q**). On the other hand, a mixture of the two indole isomers was obtained by employing aryl azides substituted with different groups (compounds **66a–66h**), due to the presence of chemically different C—H bonds where the reaction can take place.

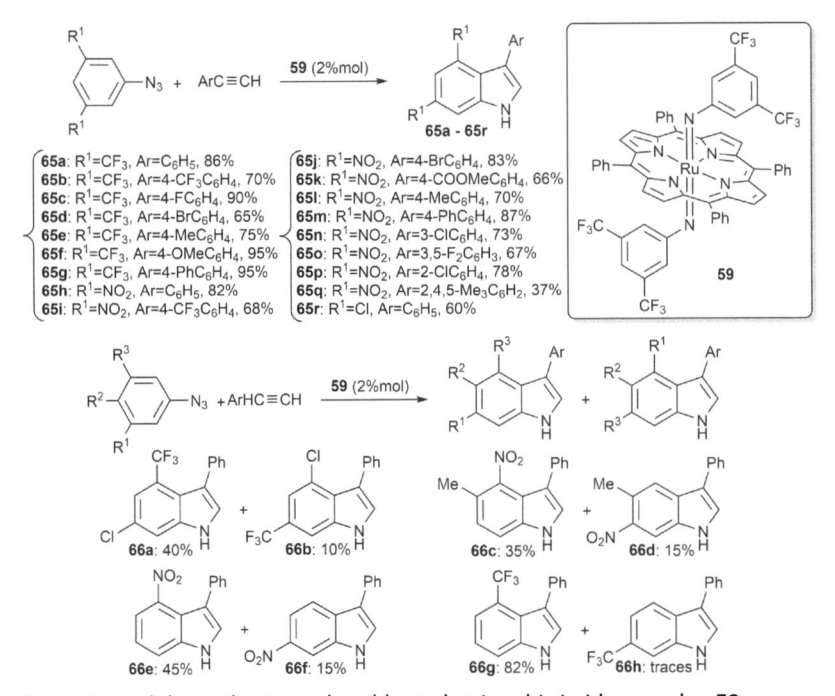

Scheme 25 Indole synthesis catalyzed by ruthenium bis-imido complex **59**.

Experimental and computational studies of the **59**-catalyzed synthesis of indole **65a** allowed proposing the reaction mechanism reported in Scheme 26. Note that in order to simplify DFT calculations, complex **A**, in which one of the aryl imido ligands of **59** was replaced by a methyl imido group, was modeled. The catalytic cycle starts from the reaction of bis-imido

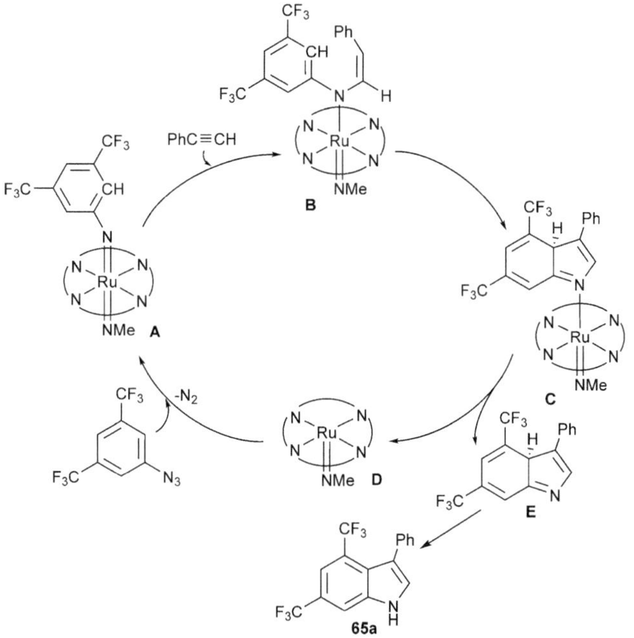

Scheme 26 Reaction mechanism proposed for the indole synthesis catalyzed by ruthenium bis-imido complexes.

A with alkyne to form the intermediate **B**. The latter is transformed into **C** through a stereochemical rearrangement which is permitted by a double intersystem crossing of singlet and triplet ground states. At this point, an inner redox process can occur forming compound **E** and the mono-imido intermediate **D** that reacts with another azide molecule restoring the starting bis-imido **A**. Finally, two subsequent H migrations on compound **E** are responsible for the formation of the final product **65a**.

2.3 Concluding remarks on group 8

In this chapter recent advances on imido complexes of group 8, active in nitrene transfer reactions, were described by pointing out the influence of substituents and ligands on the imido stability and reactivity.

High spin iron (III) and iron (IV) imido complexes, containing nitrogen ligands, resulted to be stable enough to be isolated and, at the same time, they showed a very good ability to transfer the imido functionality to organic substrates. These complexes are described as electrophilic species with a radical character whose delocalization on the whole molecule plays an important role for the catalytic efficiency. In fact, N-centered radical $L_nFe=NR$

complexes, formed by using bidentate dipyrromethene ligands and aliphatic R groups, were more active than complexes where the radical is delocalized either on aromatic R substituents or on π-conjugated L ligands.

Due to the presence of a radical character, iron imido complexes can transfer the nitrene functionality following two different mechanisms: (i) a direct nitrene transfer through a H-atom abstraction, (ii) an electron transfer followed by the nitrene transfer. Thus, the combination of experimental and theoretical studies, fundamental to evaluate the electron affinity of $L_nFe=NR$ species and the N—H bond dissociation energy of the corresponding $L_nFe-NHR$ complex, was extremely important to propose plausible reaction mechanisms and design active catalytic systems. In the case of ruthenium, no examples of isolated mono-imido complexes, active in nitrene transfer reactions, have been reported and they remain only supposed as putative intermediates of catalytic processes. Nevertheless, the use of porphyrin ligands allowed the isolation of some examples of active bis-imido ruthenium complexes of the general formula $Ru(VI)(porphyrin)(NY)_2$. These complexes can be prepared both from Ru (II) and Ru (IV) precursors and they are well stabilized by the presence of aromatic Y groups. Theoretical calculations performed on the most catalytically active bis-imido systems underlined that ruthenium bis-imido complexes can generate the mono-imido counterpart during the catalytic cycle and that both of them can act as active intermediates in nitrene transfer reactions.

3. Imido complexes of groups 9 and 10

Transition metal complexes of groups 9 and 10 are recognized very efficient catalysts in nitrene transfer reactions such as C—H aminations and alkene aziridinations.[18,56–60] As previously described for group 8-catalyzed reactions, imido species have also been proposed as key transient intermediates of catalytic processes mediated by group 9 and 10 metal complexes. Unfortunately, the characterization of molecular and electronic structures of catalytically active imido complexes remains rare due to their very high reactivity and instability. The stability of imido complexes containing metals of groups 9 and 10 are strongly influenced by the oxidation state and electronic configuration of the employed metal. These metals possess $d\pi$ orbitals more filled than those of group 8 and this is responsible for a worst energetic overlap between the metal orbitals and the N valence orbitals of the imido fragment. Consequently, the multiple bond order of the $L_nM=NR$ complex, where M is a late transition metal, is better described as a single bond and the d-electrons populate the M-L$_n$ antibonding orbitals forming an unstable imido complex with electrophilic reactivity.[17,61–63] Nevertheless, the use of metals with less filled dπ orbitals and low-coordinate scaffolds, such as bidentate ligands, allowed the isolation of some cobalt[64,65] and nickel[66–68] imido complexes, which were tested in stoichiometric nitrene transfer reactions. The involvement of imido species in catalytic reactions promoted by metal complexes of groups 9 and 10 has been mostly proposed by DFT calculations[69–76] because very few imido intermediates were detected and/or isolated. It should be noted that the high difficulty in isolating and characterizing imido active complexes of groups 9 and 10 has often been ascribed to their radical character that increases their chemical reactivity.

3.1 Cobalt

Nitrene transfer reactions promoted by group 9 metal complexes are mainly restricted to cobalt-catalyzed reactions which were extensively studied by de Bruin and co-authors. These studies on the reactivity of cobalt (II) porphyrin complexes toward different nitrene sources highlighted the general formation of radical imido intermediates during the catalytic cycle.[77] Bearing in mind previous DFT calculations performed on aziridinations and C—H amination reactions, de Bruin and co-authors proposed the imido radical complex **B**, formed by the reaction of azide with cobalt (II) porphyrin **A**, as the putative intermediate for both catalytic processes (Scheme 27).

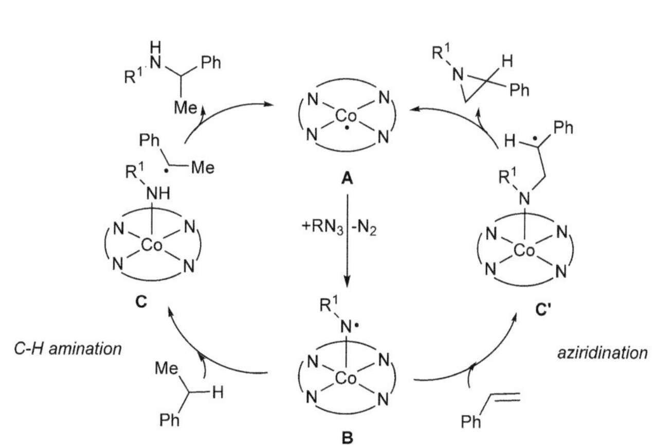

Scheme 27 Reaction mechanism proposed for cobalt porphyrin-catalyzed nitrene transfer reactions.

In order to detect the formation of imido radical intermediates, the reactions of porphyrin **67** (Scheme 28) with N-nosyl azide (NsN$_3$) and N-nosyl iminoiodane (NsI=IPh) were monitored by several techniques such as UV–Vis, IR, EPR, VCD (vibrational circular dichroism), XAFS and X-ray absorption near edge spectroscopic (XANES). Depending on the employed nitrene precursor, collected data revealed the formation of two different radical imido complexes. Reaction of the organic azide with **67** afforded the cobalt (III) mono-radical imido **68** through the single electron transfer from the cobalt (II) porphyrin to the nitrene fragment where is located the unpaired electron. Conversely, the use of the iminoiodane as the nitrene source was responsible for the formation of cobalt (III) triple-radical imido **69** in which two unpaired electrons are located on the two imido moieties and the other on the porphyrin skeleton (Scheme 28).

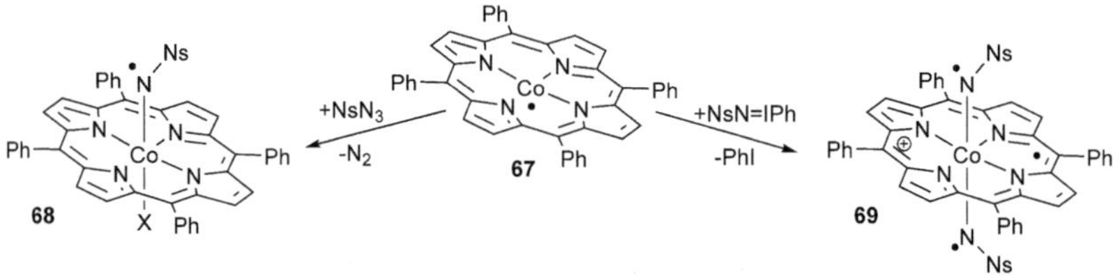

Scheme 28 Radical imido complexes formed in nitrene transfer promoted by cobalt (II) porphyrins.

The formation of the triple-radical imido species was ascribed to an additional one-electron transfer from the porphyrin ring to the nitrene moiety. Further experiments showed that bis-imido complexes were not formed by reacting iminoiodanes with sterically hindered porphyrin ligands to indicate the pivotal role of the macrocycle, which can drive the formation of either mono- or triple-radical imido intermediates independently from the employed nitrene precursor. Preliminary studies revealed good catalytic performances of **68** and **69** in promoting styrene aziridinations and supported their active involvement in cobalt (II) porphyrin-catalyzed nitrene transfer processes. It should be noted that even if XANES data revealed that both **68** and **69** were six-coordinate species, the nature of the X ligand of **68** remained uncertain. This result underlined that an additive can play a crucial catalytic role in nitrene transfer reactions by coordinating the axial position of the mono-radical intermediate **68**.

3.2 Nickel and palladium

The influence of the nitrene sources in driving the catalytic mechanism and the radical character of imido intermediates have also been reported for C—H amination reactions promoted by metal complexes of group 10. Recently, Betley and co-authors investigated the synthesis and characterization of imido complexes formed by the reaction of nickel (I) dipyrromethene complex with alkyl or aryl azides.[78] As previously described in Section 2.1 of this chapter for analogous iron complexes, the reaction of **70** with adamantly or mesityl azides afforded two imido complexes **71** and **72**, which show different electronic structures and reactivity (Scheme 29).

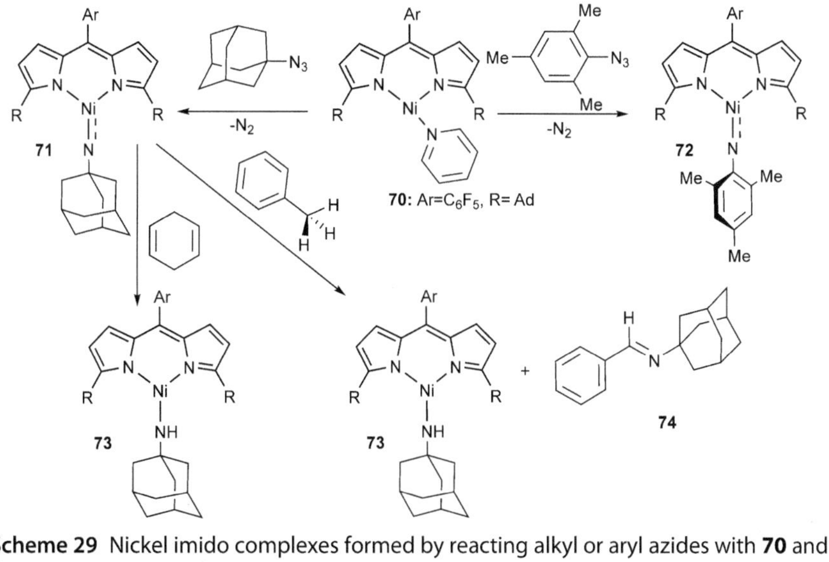

Scheme 29 Nickel imido complexes formed by reacting alkyl or aryl azides with **70** and their reactivity toward organic substrates.

The complete characterization of these compounds by NMR, EPR, solid-state magnetometry and X-ray crystallography analyses provided a detailed description of their molecular structure and electronic characteristics. While complex **71** was recognized as a formal low spin ($S = 1/2$) nickel (II) radical imido complex with the electron density located on the nitrogen imido atom, complex **72** was identified as a low spin ($S = 1/2$) nickel (III) imido complex with the electron density located on the aryl moiety of the imido fragment. In view of the N-centered electron density, complex **71** was tested in the nitrene transfer reaction to cyclohexadiene and toluene.

Even if, in both cases the expected compounds were not formed and **71** decomposed to the corresponding amido complex **73** (Scheme 29), the reaction of **71** with toluene also afforded the imine compound **74**, which derives from the desired amino product. The authors proposed that the competition between **71** and **73** in capturing the tolyl radical allowed the formation of another amido intermediate which forms the imine product **74** by a β-H elimination process. Despite the desired benzylic amine was not isolated, the formation of **74** proved the activity of **71** in efficiently transferring the nitrene functionality in nickel-catalyzed amination reactions. In view of the achieved results, the efficiency of complex **70** in promoting the ring-closing C—H amination of substrates, whose structure prevented the α-hydride migration, was investigated.[58] Complex **70** showed a very high activity (yields up to 95%) and chemoselectivity in favoring the amination of C—H bonds showing low bond dissociation energy. In addition, the presence of a wide range of functional groups was well tolerated (Scheme 30).

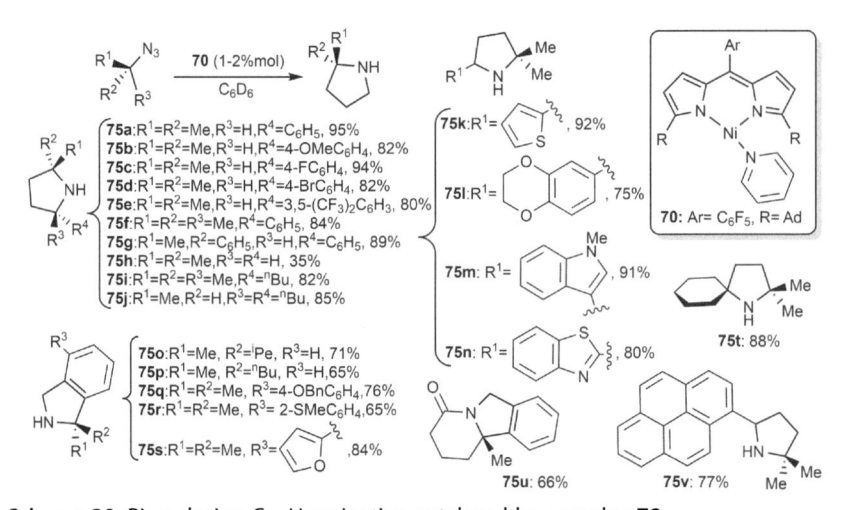

Scheme 30 Ring-closing C—H amination catalyzed by complex **70**.

Considering the proposed electronic structure of imido complex **71**, collected kinetic data and DFT calculations, the authors proposed a reaction mechanism of the **70**-catalyzed ring-closing C—H amination in which the imido **D** is the key intermediate (Scheme 31). The latter complex should be formed by the replacement of the pyridine ligand of **A** with the radical imido fragment presenting the radical character on the nitrogen atom. Then, the radical intermediate **E** is formed by a H-atom abstraction and finally

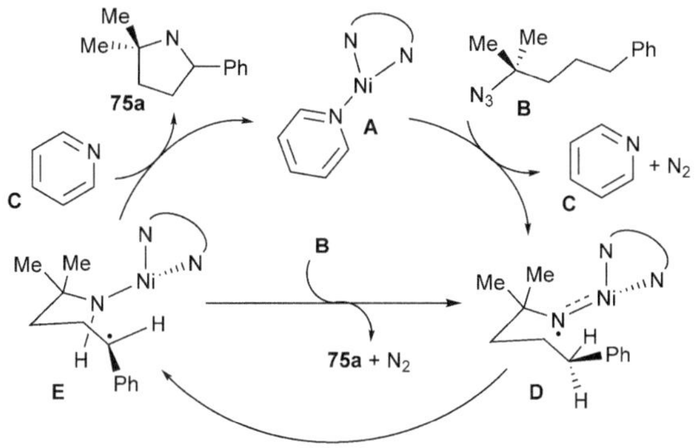

Scheme 31 Reaction mechanism proposed for the ring-closing C—H amination catalyzed by complex **70**.

the desired product **75a** is formed by a radical recombination process. Intermediate **E** can also react with another molecule of the starting azide **B** affording the product **75a** and imido **D** that restarts the catalytic cycle.

A nickel imido complex involved in nitrene transfers was also isolated by Warren and co-authors who studied the catalytic activity of a nickel (I) β-diketiminato complex (**76**) in mediating the carbodiimide synthesis from organic azides and isocyanides.[79] Complex **76** (Scheme 32) promoted the conversion of a wide range of aryl azides into corresponding carbodiimides and very high yields (up to 99%) were reached by using electron-rich azides as starting materials. To shade some light on the reaction mechanism, the starting catalyst **76** was reacted either with two different isocyanides or aryl azide forming the isocyanide adducts **77** and **78** as well as the di-nickel imido complex **79**, respectively (Scheme 32). The latter compound was fully characterized by NMR, IR and UV–Vis spectroscopies but, albeit numerous attempts, **79** was not isolated as a single crystal for performing a X-ray analysis. On the other hand, the molecular structure of an analogous complex, with Ar=2,6-Me$_2$C$_6$H$_3$, was determined by X-ray spectroscopy and used to model the structure of **79** by DFT calculations. Theoretical calculations predicted the two spin states $S=0$ and $S=1$ of nickel with the $S=0$ state more stable than $S=1$ state by only 0.7 kcal mol^{-1}. The reaction of **79** with *tert*-butyl isocyanide afforded the corresponding carbodiimide in 88% yield to suggest the possible role of **79** in the catalytic process. This result supports the formation of di-nickel imido complexes as key intermediates of nitrene transfer reactions from azides to isocyanides promoted by nickel (I) β-diketiminato complexes.

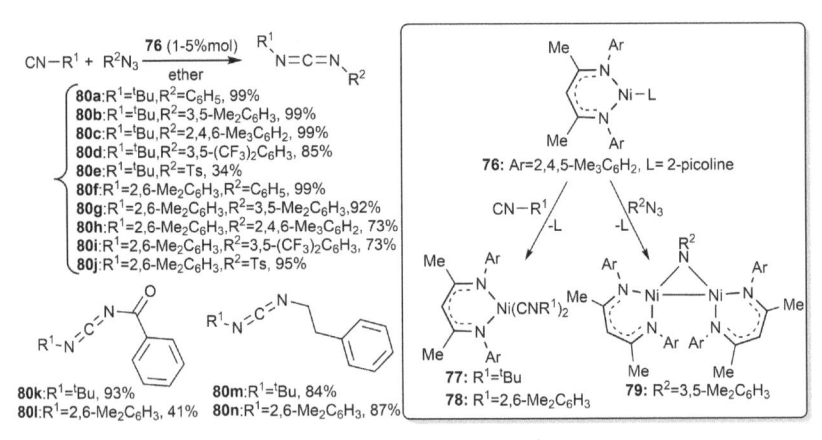

Scheme 32 Carbodiimide synthesis catalyzed by complex **76**.

Not only nickel but also palladium was employed to catalyze coupling reactions of azides with isocyanides. In fact, Pardasani and co-authors investigated the catalytic use of palladium acetate Pd(OAc)$_2$ for the synthesis of nitrogen containing heterocycles through the azide–isocyanide cross-coupling/cyclization protocol.[80] As reported in Scheme 33, Pd(OAc)$_2$ efficiently catalyzed the nitrene transfer from azides to isocyanides and the subsequent intramolecular cyclization of functionalized 2-azidobenzoic acids achieving the desired benzooxazinones (**84a–84h**) and quinazolinones (**85a–85h**) in yields up to 94%.

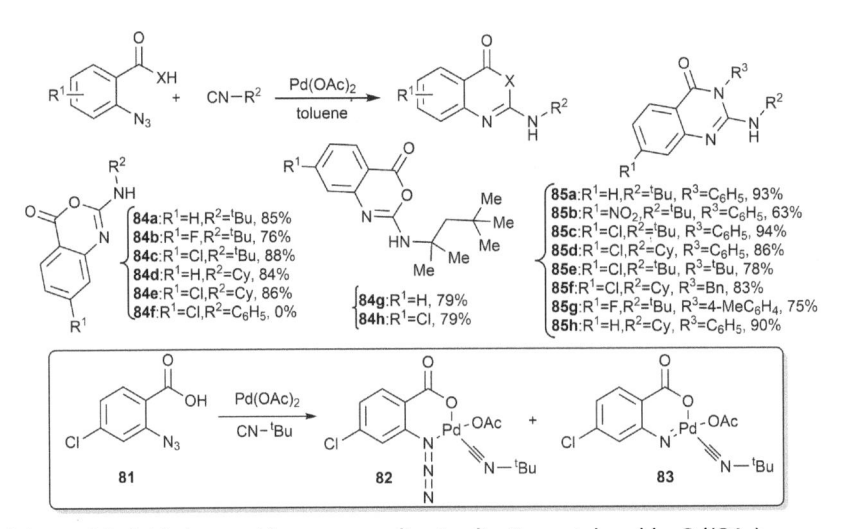

Scheme 33 Azide-isocyanide cross-coupling/cyclization catalyzed by Pd(OAc)$_2$.

A series of experiments were performed to study the reaction mechanism of this intramolecular nitrene transfer. In particular, the model reaction between substrate **81** and *tert*-butyl isocyanide in presence of Pd(OAc)$_2$ afforded the palladium imido complex **83** that was detected in the reaction medium by ESI-MS spectroscopy. These data, together with detailed DFT calculations, encouraged the authors to propose the reaction mechanism reported in Scheme 34 in which the central role is played by the imido intermediate **D**. This latter compound transfers the nitrene fragment to the coordinated isocyanide molecule by a concerted process which affords carbodiimide **E**. At this point, the coordination of acetic acid to palladium (compound **F**) was responsible for the de-coordination of carbodiimide moiety (compound **G**) that was moved closer to the carboxylate of benzoate. The hydrogen bond in intermediate **H** activates the azomethine group allowing the intramolecular cyclization (intermediates **I** and **J**) which forms the desired heterocycle and restores the starting palladium acetate.

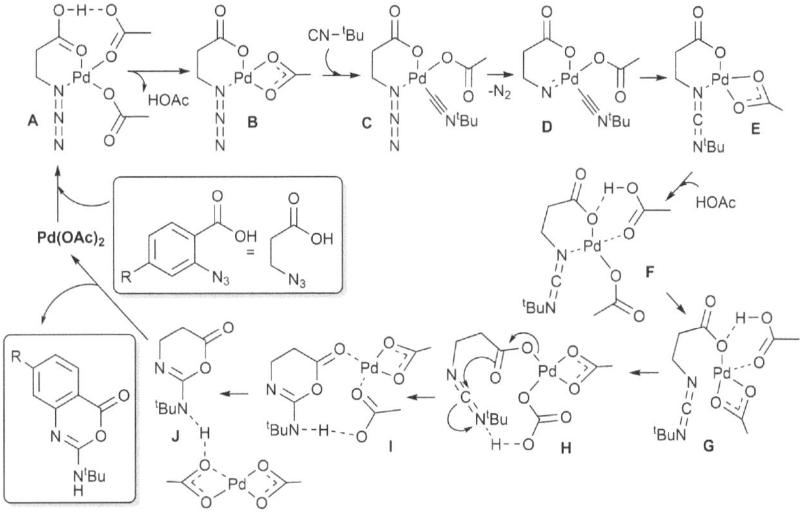

Scheme 34 Reaction mechanism proposed for azide-isocyanide cross-coupling/cyclization catalyzed by Pd(OAc)$_2$.

4. Summary and outlook

Even if imido complexes are usually considered key intermediates of nitrene transfer reactions promoted by transition metal catalysts, the high reactivity and instability of these species have often prevented their full characterization. High efforts have been employed to isolate the putative imido intermediates but in several cases their molecular structures remain only suggested by theoretical calculations. This review is only focused on imido complexes of groups 8–10 showing a good chemical stability/reactivity relationship which allows both their structural characterization and their use as nitrene transfer promoters.

The analysis of literature reported after 2010 has revealed that the strength of the $L_nM=NR$ bond decreases moving from group 8 to group 10 due to a worse orbital overlap between the filled $d\pi$ metal orbitals and the N-valence orbitals of the imido fragment. This causes an increasing instability of corresponding imido intermediates and explains why iron and ruthenium imido complexes have been more easily isolated and fully studied. It should be noted that the high reactivity and instability of cobalt, nickel, and palladium imido derivatives has also been ascribed to their radical character.

The analysis of reported data revealed the importance of the nature of both the ligand and nitrene fragment in determining the dual relationship between stability and reactivity of the resulting imido. For example, the use of bidentate ligands for synthesizing iron and nickel metal complexes have allowed the isolation of active imido species and simultaneously, the capacity of the imido fragment in delocalizing the electron density on the nitrogen imido moiety enhanced the reactivity toward organic substrates.

References

1. Knipe AC, Gras E, Chassaing S. Carbenes and nitrenes. In: Knipe AC, ed. *Organic Reaction Mechanisms 2015*. John Wiley & Sons Ltd; 2019:219–249.
2. Sweeney JB. Aziridines: epoxides' ugly cousins? *Chem Soc Rev*. 2002;31(5):247–258.
3. Chang JWW, Ton TMU, Chan PWH. Transition-metal-catalyzed aminations and aziridinations of C-H and C-C bonds with iminoiodinanes. *Chem Rec*. 2011;11(6):331–357.
4. Minakata S. Utilization of N−X bonds in the synthesis of N-heterocycles. *Acc Chem Res*. 2009;42(8):1172–1182.
5. Huang D, Yan G. Recent advances in reactions of azides. *Adv Synth Catal*. 2017;359(10):1600–1619.
6. Intrieri D, Zardi P, Caselli A, Gallo E. Organic azides: "energetic reagents" for the intermolecular amination of C-H bonds. *Chem Commun*. 2014;50(78):11440–11453.
7. Damiano C, Intrieri D, Gallo E. Aziridination of alkenes promoted by iron or ruthenium complexes. *Inorg Chim Acta*. 2018;470:51–67.
8. Park Y, Kim Y, Chang S. Transition metal-catalyzed C–H amination: scope, mechanism, and applications. *Chem Rev*. 2017;117(13):9247–9301.
9. Hayashi H, Uchida T. Nitrene transfer reactions for asymmetric C–H amination: recent development. *Eur J Org Chem*. 2020;2020(8):909–916.
10. Uchida T, Katsuki T. Asymmetric nitrene transfer reactions: sulfimidation, aziridination and C–H amination using azide compounds as nitrene precursors. *Chem Rec*. 2014;14:117–129.

11. Straub BF. Organotransition metal chemistry. From bonding to catalysis. Edited by John F. Hartwig. *Angew Chem Int Ed*. 2010;49(42):7622.
12. Nugent VWA, Mayer JM. *Metal-Ligand Multiple Bonds*. Chichester: Wiley; 1988.
13. Kuijpers PF, van der Vlugt JI, Schneider S, de Bruin B. Nitrene radical intermediates in catalytic synthesis. *Chem Eur J*. 2017;23(56):13819–13829.
14. Berry JF. Terminal nitrido and imido complexes of the late transition metals. *Comments Inorg Chem*. 2009;30(1–2):28–66.
15. Fantauzzi S, Caselli A, Gallo E. Nitrene transfer reactions mediated by metalloporphyrin complexes. *Dalton Trans*. 2009;28:5434–5443.
16. Driver TG. Recent advances in transition metal-catalyzed N-atom transfer reactions of azides. *Org Biomol Chem*. 2010;8(17):3831–3846.
17. Ray K, Heims F, Pfaff FF. Terminal oxo and imido transition-metal complexes of groups 9–11. *Eur J Inorg Chem*. 2013;2013(22 – 23):3784–3807.
18. Shimbayashi T, Sasakura K, Eguchi A, Okamoto K, Ohe K. Recent progress on cyclic nitrenoid precursors in transition-metal-catalyzed nitrene-transfer reactions. *Chem Eur J*. 2019;25(13):3156–3180.
19. Breslow R, Gellman SH. Tosylamidation of cyclohexane by a cytochrome P-450 model. *J Chem Soc Chem Commun*. 1982;(24):1400–1401.
20. Barton DHR, Hay-Motherwell RS, Motherwell WB. Functionalization of saturated hydrocarbons. Part 1. Some reactions of a ferrous chloride-chloramine-T complex with hydrocarbons. *J Chem Soc Perkin Trans 1*. 1983;445–451.
21. Zhang L, Deng L. C-H bond amination by iron-imido/nitrene species. *Chin Sci Bull*. 2012;57(19):2352–2360.
22. Driver TG. An aminated reaction. *Nat Chem*. 2013;5(9):736–738.
23. Wang P, Deng L. Recent advances in iron-catalyzed C—H bond amination via iron imido intermediate. *Chin J Chem*. 2018;36(12):1222–1240.
24. Cowley RE, Holland PL. C–H activation by a terminal imidoiron(III) complex to form a cyclopentadienyliron(II) product. *Inorg Chim Acta*. 2011;369(1):40–44.
25. Wang L, Hu L, Zhang H, Chen H, Deng L. Three-coordinate iron(IV) bisimido complexes with aminocarbene ligation: synthesis, structure, and reactivity. *J Am Chem Soc*. 2015;137(44):14196–14207.
26. Spasyuk DM, Carpenter SH, Kefalidis CE, Piers WE, Neidig ML, Maron L. Facile hydrogen atom transfer to iron(iii) imido radical complexes supported by a dianionic pentadentate ligand. *Chem Sci*. 2016;7(9):5939–5944.
27. Anderson CM, Aboelenen AM, Jensen MP. Competitive intramolecular amination as a clock for Iron-catalyzed nitrene transfer. *Inorg Chem*. 2019;58(2):1107–1119.
28. King ER, Hennessy ET, Betley TA. Catalytic C-H bond amination from high-spin iron imido complexes. *J Am Chem Soc*. 2011;133(13):4917–4923.
29. Hennessy ET, Betley TA. Complex N-heterocycle synthesis via iron-catalyzed, direct C-H bond amination. *Science*. 2013;340(6132):591–595.
30. Hennessy ET, Liu RY, Iovan DA, Duncan RA, Betley TA. Iron-mediated intermolecular N-group transfer chemistry with olefinic substrates. *Chem Sci*. 2014;5(4):1526–1532.
31. Iovan DA, Betley TA. Characterization of iron-imido species relevant for N-group transfer chemistry. *J Am Chem Soc*. 2016;138(6):1983–1993.
32. Wilding MJT, Iovan DA, Betley TA. High-spin iron imido complexes competent for C–H bond amination. *J Am Chem Soc*. 2017;139(34):12043–12049.
33. Wilding MJT, Iovan DA, Wrobel AT, et al. Direct comparison of C–H bond amination efficacy through manipulation of nitrogen-valence centered redox: imido versus iminyl. *J Am Chem Soc*. 2017;139(41):14757–14766.
34. Leeladee P, Jameson GNL, Siegler MA, Kumar D, de Visser SP, Goldberg DP. Generation of a high-valent iron imido corrolazine complex and NR group transfer reactivity. *Inorg Chem*. 2013;52(8):4668–4682.

35. Hong S, Sutherlin KD, Vardhaman AK, et al. A mononuclear nonheme iron(V)-imido complex. *J Am Chem Soc*. 2017;139(26):8800–8803. 2017/07/05.
36. Hong S, Lu X, Lee Y-M, et al. Achieving one-electron oxidation of a mononuclear nonheme iron(V)-imido complex. *J Am Chem Soc*. 2017;139(41):14372–14375.
37. Lu X, Li X-X, Lee Y-M, et al. Electron-transfer and redox reactivity of high-valent iron imido and oxo complexes with the formal oxidation states of five and six. *J Am Chem Soc*. 2020;142(8):3891–3904.
38. Patra R, Maldivi P. DFT analysis of the electronic structure of Fe(IV) species active in nitrene transfer catalysis: influence of the coordination sphere. *J Mol Model*. 2016; 22(11):278.
39. Mukherjee G, Reinhard FGC, Bagha UK, Sastri CV, de Visser SP. Sluggish reactivity by a nonheme iron(IV)-tosylimido complex as compared to its oxo analogue. *Dalton Trans*. 2020;49(18):5921–5931.
40. Coin G, Patra R, Rana S, et al. Fe-catalyzed aziridination is governed by the electron affinity of the active imido-iron species. *ACS Catal*. 2020;10(17):10010–10020.
41. Bellow JA, Yousif M, Cabelof AC, Lord RL, Groysman S. Reactivity modes of an iron bis(alkoxide) complex with aryl azides: catalytic nitrene coupling vs formation of iron(III) imido dimers. *Organometallics*. 2015;34(12):2917–2923.
42. Yousif M, Wannipurage D, Huizenga CD, et al. Catalytic nitrene homocoupling by an iron(II) bis(alkoxide) complex: bulking up the alkoxide enables a wider range of substrates and provides insight into the reaction mechanism. *Inorg Chem*. 2018;57(15): 9425–9438.
43. Gouré E, Avenier F, Dubourdeaux P, et al. A diiron(III,IV) imido species very active in nitrene-transfer reactions. *Angew Chem Int Ed*. 2014;53(6):1580–1584.
44. Intrieri D, Carminati DM, Gallo E. Recent advances in C–H bond aminations catalyzed by ruthenium porphyrin complexes. *J Porphyrins Phthalocyanines*. 2016;20(01n04): 190–203.
45. Guo Z, Guan X, Huang J-S, Tsui W-M, Lin Z, Che C-M. Bis(sulfonylimide) ruthenium(VI) porphyrins: X-ray crystal structure and mechanism of C–H bond amination by density functional theory calculations. *Chem Eur J*. 2013;19(34):11320–11331.
46. Zardi P, Pozzoli A, Ferretti F, Manca G, Mealli C, Gallo E. A mechanistic investigation of the ruthenium porphyrin catalysed aziridination of olefins by aryl azides. *Dalton Trans*. 2015;44(22):10479–10489.
47. Bizet V, Bolm C. Sulfur imidations by light-induced ruthenium-catalyzed nitrene transfer reactions. *Eur J Org Chem*. 2015;2015(13):2854–2860.
48. Qin J, Zhou Z, Cui T, Hemming M, Meggers E. Enantioselective intramolecular C–H amination of aliphatic azides by dual ruthenium and phosphine catalysis. *Chem Sci*. 2019;10(11):3202–3207.
49. Manca G, Mealli C, Carminati DM, Intrieri D, Gallo E. Comparative study of the catalytic amination of benzylic C–H bonds promoted by Ru(TPP)(py)2 and Ru(TPP) (CO). *Eur J Inorg Chem*. 2015;2015(29):4885–4893.
50. Law S-M, Chen D, Chan SL-F, et al. Ruthenium porphyrins with axial π-conjugated arylamide and arylimide ligands. *Chem Eur J*. 2014;20(35):11035–11047.
51. Intrieri D, Caselli A, Ragaini F, Macchi P, Casati N, Gallo E. Insights into the mechanism of the ruthenium–porphyrin-catalysed allylic amination of olefins by aryl azides. *Eur J Inorg Chem*. 2012;(3):569–580.
52. Zardi P, Gallo E, Solan GA, Hudson AJ. Resonance Raman spectroscopy as an in situ probe for monitoring catalytic events in a Ru-porphyrin mediated amination reaction. *Analyst*. 2016;141(10):3050–3058.
53. Zardi P, Intrieri D, Carminati DM, Ferretti F, Macchi P, Gallo E. Synthesis and catalytic activity of μ-oxo ruthenium(IV) porphyrin species to promote amination reactions. *J Porphyrins Phthalocyanines*. 2016;20(08n11):1156–1165.

54. Zardi P, Savoldelli A, Carminati DM, Caselli A, Ragaini F, Gallo E. Indoles rather than triazoles from the ruthenium porphyrin-catalyzed reaction of alkynes with aryl azides. *ACS Catal.* 2014;4(11):3820–3823.
55. Intrieri D, Carminati DM, Zardi P, et al. Indoles from alkynes and aryl azides: scope and theoretical assessment of ruthenium porphyrin-catalyzed reactions. *Chem Eur J.* 2019;25(72):16591–16605.
56. Gephart RT, Warren TH. Copper-catalyzed sp3 C-H amination. *Organometallics.* 2012;31(22):7728–7752.
57. Ye L-W, Zhu X-Q, Sahani RL, Xu Y, Qian P-C, Liu R-S. Nitrene transfer and carbene transfer in gold catalysis. *Chem Rev.* 2020. https://doi.org/10.1021/acs.chemrev.0c00348.
58. Dong Y, Clarke RM, Porter GJ, Betley TA. Efficient C–H amination catalysis using nickel-dipyrrin complexes. *J Am Chem Soc.* 2020;142(25):10996–11005.
59. Timsina YN, Gupton BF, Ellis KC. Palladium-catalyzed C–H amination of C(sp2) and C(sp3)–H bonds: mechanism and scope for N-based molecule synthesis. *ACS Catal.* 2018;8(7):5732–5776.
60. van Vliet KM, de Bruin B. Dioxazolones: stable substrates for the catalytic transfer of acyl nitrenes. *ACS Catal.* 2020;10(8):4751–4769.
61. Geer AM, Tejel C, López JA, Ciriano MA. Terminal imido rhodium complexes. *Angew Chem Int Ed.* 2014;53(22):5614–5618.
62. Fujita D, Sugimoto H, Morimoto Y, Itoh S. Noninnocent ligand in rhodium(III)-complex-catalyzed C–H bond amination with tosyl azide. *Inorg Chem.* 2018;57(16):9738–9747.
63. Reckziegel A, Pietzonka C, Kraus F, Werncke CG. C – H bond activation by an imido cobalt(III) and the resulting amido cobalt(II) complex. *Angew Chem Int Ed.* 2020;59(22):8527–8531.
64. King ER, Sazama GT, Betley TA. Co(III) imidos exhibiting spin crossover and C–H bond activation. *J Am Chem Soc.* 2012;134(43):17858–17861.
65. Liu Y, Du J, Deng L. Synthesis, structure, and reactivity of low-spin cobalt(II) imido complexes [(Me3P)3Co(NAr)]. *Inorg Chem.* 2017;56(14):8278–8286.
66. Mindiola DJ, Waterman R, Iluc VM, Cundari TR, Hillhouse GL. Carbon–hydrogen bond activation, C–N bond coupling, and cycloaddition reactivity of a three-coordinate nickel complex featuring a terminal imido ligand. *Inorg Chem.* 2014;53(24):13227–13238.
67. Laskowski CA, Miller AJM, Hillhouse GL, Cundari TR. A two-coordinate nickel imido complex that effects C – H amination. *J Am Chem Soc.* 2011;133(4):771–773.
68. Wiese S, McAfee JL, Pahls DR, McMullin CL, Cundari TR, Warren TH. C–H functionalization reactivity of a nickel–imide. *J Am Chem Soc.* 2012;134(24):10114–10121.
69. Kuijpers PF, Tiekink MJ, Breukelaar WB, et al. Cobalt-porphyrin-catalysed intramolecular ring-closing C – H amination of aliphatic azides: a nitrene-radical approach to saturated heterocycles. *Chem Eur J.* 2017;23(33):7945–7952.
70. Goswami M, Rebreyend C, De Bruin B. Porphyrin cobalt(III) "nitrene radical" reactivity; hydrogen atom transfer from ortho-YH substituents to the nitrene moiety of cobalt-bound aryl nitrene intermediates (Y = O, NH). *Molecules.* 2016;21(2):242.
71. Baek Y, Das A, Zheng S-L, Reibenspies JH, Powers DC, Betley TA. C–H amination mediated by cobalt organoazide adducts and the corresponding cobalt nitrenoid intermediates. *J Am Chem Soc.* 2020;142(25):11232–11243.
72. Lorpitthaya R, Xie Z-Z, Sophy KB, Kuo J-L, Liu X-W. Mechanistic insights into the substrate-controlled stereochemistry of glycals in one-pot rhodium-catalyzed aziridination and aziridine ring opening. *Chem Eur J.* 2010;16(2):588–594.
73. Fujita D, Sugimoto H, Shiota Y, Morimoto Y, Yoshizawa K, Itoh S. Catalytic C–H amination driven by intramolecular ligand-to-nitrene one-electron transfer through a rhodium(iii) centre. *Chem Commun.* 2017;53(35):4849–4852.

74. Qi X, Li Y, Bai R, Lan Y. Mechanism of rhodium-catalyzed C–H functionalization: advances in theoretical investigation. *Acc Chem Res*. 2017;50(11):2799–2808.
75. Grünwald A, Munz D. How to tame a palladium terminal imido. *J Organomet Chem*. 2018;864:26–36.
76. Ke Z, Cundari TR. Palladium-catalyzed C – H activation/C – N bond formation reactions: DFT study of reaction mechanisms and reactive intermediates. *Organometallics*. 2010;29(4):821–834.
77. Goswami M, Lyaskovskyy V, Domingos SR, et al. Characterization of porphyrin-Co(III)-'nitrene radical' species relevant in catalytic nitrene transfer reactions. *J Am Chem Soc*. 2015;137(16):5468–5479.
78. Dong Y, Lukens JT, Clarke RM, Zheng S-L, Lancaster KM, Betley TA. Synthesis, characterization and C–H amination reactivity of nickel iminyl complexes. *Chem Sci*. 2020;11(5):1260–1268.
79. Wiese S, Aguila MJB, Kogut E, Warren TH. β-Diketiminato nickel imides in catalytic nitrene transfer to isocyanides. *Organometallics*. 2013;32(8):2300–2308.
80. Ansari AJ, Pathare RS, Maurya AK, et al. Synthesis of diverse nitrogen heterocycles via palladium-catalyzed tandem azide–isocyanide cross-coupling/cyclization: mechanistic insight using experimental and theoretical studies. *Adv Synth Catal*. 2018;360(2): 290–297.

CHAPTER FOUR

Recent progress on group 10 metal complexes of pincer ligands: From synthesis to activities and catalysis

Krishna K. Manar and Peng Ren*
School of Science, Harbin Institute of Technology (Shenzhen), Shenzhen, China
*Corresponding author: e-mail address: renpeng@hit.edu.cn

Contents

1. Introduction	185
2. Well-defined group 10 (nickel, palladium, and platinum) pincer complexes	188
3. General synthetic routes to group 10 pincer complexes	200
3.1 Nickel (Ni) pincer metal compounds	200
3.2 Palladium (Pd) pincer metal compounds	221
3.3 Platinum (Pt) pincer metal compounds	230
4. Catalyzed cross-coupling reactions	237
4.1 Mizoroki–Heck cross-coupling reactions	238
4.2 Suzuki–Miyaura cross-coupling reactions	240
4.3 Sonogashira cross-coupling reactions	241
4.4 Kumada cross-coupling reactions	242
4.5 Hiyama and Negishi coupling reactions	242
5. Miscellaneous reactions	243
6. Conclusions and perspectives	250
Acknowledgments	250
References	250

1. Introduction

Recent advances in modern organometallic chemistry include new catalytic methodology and materials synthesis with improved characteristics. The reactivity of organometallic compounds depends on ligand properties, such as chelating nature, steric behavior, and electronic properties (electron donating/accepting or ambiphilic). Chelation is probably the most important factor in the stability of organometallic and coordination complexes.

Growing interest in pincer ligand chemistry results from their tunable electronic properties and the emerging catalytic behavior of their stable organometallic complexes. Pincer ligands have special electronic properties that result in their complexes showing high thermal stability. Furthermore, unusual/unstable oxidation states of the transition metal center can be stabilized by chelate ring formation.

Shaw and coworkers first reported "pincer" ligands in 1976, and new designs continue to emerge.[1] Early on, pincer ligands were defined by van Koten and Albrecht as tridentate ligands in which the central aryl σ-donating group is flanked by two side groups, such as NR_2, PR_2, or CER_2 (E = N, P).[2] In general, pincer ligands form two five-membered metallocycles through complexation, although some examples of six-membered metallocycles are known. Consequently, the chelating nature of pincer ligands provides thermal and kinetic stability, and wide-ranging catalytic reactivity, which has been extensively explored in homogeneous catalytic transformations, and some other emerging industrial and academic applications. Moderation of the electron donor, acceptor, and steric properties of the pincer ligands allows for multiple chemical changes, which makes the pincer model a powerful synthetic tool. Pincer ligands can be generalized as E^1YE^2, where E is a neutral side arm donor atom that provides two electrons and Y is commonly an anionic carbon or neutral nitrogen donor atom. The central carbon or nitrogen atom forms C–M or N–M covalent bonding with the metal center in complexes (Fig. 1). Pincer ligands have versatile, tunable, and remarkable properties that can be applied to metal centers and their resulting complexes.[3–15] During catalysis, pincer ligands play a crucial role in providing improved chemical, thermal, and kinetic stability, which can prevent the leaching of metal in the catalytic cycle during the catalytic reaction.[2,3,6–15] In general, the metal center has a vacant coordination site that can accept a substrate or activate polar or nonpolar bonds to promote metal-catalyzed reactions. Accordingly, coordinately unsaturated metal complexes can catalyze significant fundamental reactions, such as bond cleavage reactions and oxidative addition (OA) to polar and nonpolar bonds.

Nickel has attracted much attention owing to its high availability, low toxicity, and cost effectiveness compared with its group 10 congeners.[16–18] Palladium catalysts are commonly known as good candidates for catalysis. However, the redox behavior of nickel provides extended reactivity and a broad opportunity for reaction discovery.[19,20] Nickel catalysis has been studied since the 1970s and employed in cross-coupling reactions,[21,22] with single electron transfer (SET) processes usually being proposed to proceed

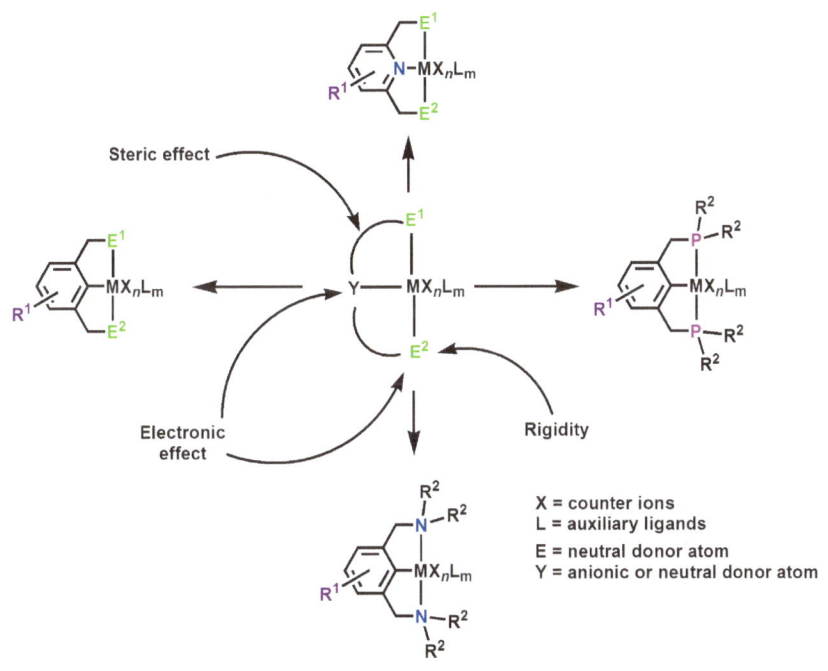

Fig. 1 Steric and electronic control by the pincer donor, and their effect on newly synthesized pincer complexes.

through classical Ni(0)/Ni(II) vs. Ni(I)/Ni(III) pathways.[23–25] High-valent Ni(III) and Ni(IV) key intermediates were recently employed in the C–C and C–heteroatom bond forming reactions, although their isolation and characterization were difficult to achieve.[26–31] Unlike Pt and Pd, the chemistry of organonickel species in the +4 oxidation state is under developed.

Organometallic Pd(II) pincer compounds have received much interest owing to their suitable balance between catalytic activity and complex stability. Furthermore, the high reactivity of Pd-based catalysts toward C–C bond forming reactions is well explored.[32–34] Recently, Morales-Morales and coworkers reviewed recent advances in palladium pincer compound-catalyzed cross-coupling reactions. The authors explored various pincer-ligated palladium complexes with central aryl or pyridine rings and many side arm donor atoms. These palladium complexes have been applied in cross-coupling, allylation, and arylation reactions.[35] Pd(0)/Pd(II) species are the most common oxidation states in palladium catalytic cycles. Furthermore, the air and moisture stability of palladium complexes facilitates their synthesis and storage, resulting in an extended catalyst lifetime and broad substrate scope.

Among the group 10 triad metals, Pt compounds are less reactive catalysts than Ni and Pd compounds. Interestingly, some Pt metal complexes have industrial applications in catalysis, exhibiting some selectivities different to Ni and Pd. The well-known catalytic ability of Pd compounds is greater than that of Pt analogs, making them suitable for many purposes. In contrast, Pt affords a smaller substrate scope and lower yield. Furthermore, unlike its group 10 congeners, Pt(II) is a kinetically inert species, making carboplatination more challenging than carbopalladation. However, non-pincer Pt compounds have been effectively employed in catalysis, including hydroamination,[36] hydrovinylation,[37] and cycloisomerization reactions.[38]

Group 10 organometallic complexes featuring pincer ligands have been synthesized by several research groups, and their coordination chemistry, reactivity, and possible catalytic applications are reviewed herein. This review summarizes some synthetic aspects of more popular group 10 metal pincer complexes, focusing on recently developed promising systems. Particular attention has been paid to the utility and bonding behavior of various synthesized pincer metal complexes employed in the last 3 years.

2. Well-defined group 10 (nickel, palladium, and platinum) pincer complexes

Pincer metal chemistry was first reported in 1976, and has received great interest owing to its increasing catalytic applications. Recently, a broad range of pincer ligand-based metal complexes have been prepared. Therefore, transition metal pincer complex research is flourishing, with their chemistry expanding constantly. These complexes bear central aryl or pyridine-based donor groups and are utilized in coupling reactions and many other organic transformations. Transition metal complexes, especially group 10 metal complexes, have been well explored as catalysts. Furthermore, Ni and Pd metal complexes comprise a broad area of research, exhibiting excellent catalytic activity. Therefore, Ni and Pd are of great synthetic importance in organic synthesis, and have been widely utilized in cross-coupling reactions. Ni and Pd metal complexes with central aryl or pyridine-based pincer ligand donors have been extensively explored. Studies from the last 3 years are summarized in this review to the best of our knowledge, with more than 150 group 10 metal pincer compounds reported in the literature (Table 1).

Table 1 Recently synthesized Ni, Pd, and Pt pincer metal complexes and their utility.

Compound No.	Compounds	Recent advancement	Year	References
1	Chiral bis(phospholane)–PCP–Ni	Asymmetric phosphine alkylation	2018	39
2	[(^{Me2}NNN^Q)–NiCl]	C–H bond alkylation	2016	40
3	[(^{Me2}NNN^Q)–NiBr]	C–H bond alkylation	2016	40
4	[(^{Et2}NNN^Q)–NiCl]	C–H bond alkylation	2016	40
5	[(^{Et2}NNN^Q)–NiBr]	C–H bond alkylation	2016	40
6	[(^{Me2}NNN^Q)–NiOAc]	C–H bond alkylation	2016	40
7	[(^{Me}quinolinyl)NNN–NiCl]	C–H bond alkylation	2020	41
8	[(^{Me}Phenanthridine)N′N′N–NiCl]	C–H bond alkylation	2020	41
9	[(^{tBu}Phenanthridine)N′N′N–NiCl]	C–H bond alkylation	2020	41
10	[(^{CF3}Phenanthridine)N′N′N–NiCl]	C–H bond alkylation	2020	41
11	[(NNN)–NiCl]	Formation of NNN pincer based Ni complex	2020	42
12	[(NNN)–PdCl]	Formation of NNN pincer based Pd complex	2020	42
13	[(NNN)–NiH]	Formation of Ni–H bond	2020	42
14	[(NNN)–NiPh]	Formation of Ni–Ph bond	2020	42
15	[(NNN)–NiCOPh]	CO insertion in to Ni–Ph bond	2020	42
16	[(NNN)–NiN$_3$]	C–H functionalization	2019	43

Continued

Table 1 Recently synthesized Ni, Pd, and Pt pincer metal complexes and their utility.—cont'd

Compound No.	Compounds	Recent advancement	Year	References
17	[(NNN)-Ni-NH$_2$-]	C–H functionalization	2019	43
18	[(^{Tol,Ph}DHPyc)-Ni]	Activation of water	2019	44
19	[(^{Tol,Ph}DHPyc$^•$)-Ni]	Activation of water	2019	44
20	[^{nBu}(CNN)-NiH]	Hydrodehalogenation	2018	45
21	[^{iPr}(CNN)-NiH]	Hydrodehalogenation	2018	45
22	[^{Bn}(CNN)-NiH]	Hydrodehalogenation	2018	45
23	[Me$_4$PNP^{iPr}NiBr]Br	Sterically hindered tetra methylated PNP stabilize Ni(II) center	2019	46
24	[Me$_4$PNP^{iPr}NiCl]Cl	Sterically hindered tetramethylated PNP stabilize Ni(II) center	2019	46
25	[Me$_4$PNP^{tBu}NiBr]Br	Sterically hindered tetramethylated PNP stabilize Ni(II) center	2019	46
26	[Me$_4$PNP^{tBu}NiCl]Cl	Sterically hindered tetramethylated PNP stabilize Ni(II) center	2019	46
23'	[Me$_4$PNP^{iPr}NiBr]B(Ar^F)$_4$	Electrochemical properties	2019	46
24'	[Me$_4$PNP^{iPr}NiCl]B(Ar^F)$_4$	Electrochemical properties	2019	46
25'	[Me$_4$PNP^{tBu}NiBr]B(Ar^F)$_4$	Electrochemical properties	2019	46
26'	[Me$_4$PNP^{tBu}NiCl]B(Ar^F)$_4$	Electrochemical properties	2019	46

27	[Me$_4$PNP^{iPr}NiBr]	PNP pincer stabilize Ni(I) complex	2019	46
28	[Me$_4$PNP^{iPr}NiCl]	PNP pincer stabilize Ni(I) complex	2019	46
29	[Me$_4$PNP^{tBu}NiBr]	PNP pincer stabilize Ni(I) complex	2019	46
30	[Me$_4$PNP^{tBu}NiCl]	PNP pincer stabilize Ni(I) complex	2019	46
31	[Me$_4$PNP^{iPr}NiBr]BPh$_4$	Sterically hindered tetramethylated PNP stabilize Ni(II) center	2019	47
32	[Me$_4$PNP^{tBu}NiBr]BPh$_4$	Sterically hindered tetramethylated PNP stabilize Ni(II) center	2019	47
33	[Me$_4$PNP^{iPr}NiH]B(Ar^F)$_4$	Formation of Ni-H bond	2019	47
34	[Me$_4$PNP^{iPr}NiH]BPh$_4$	Formation of Ni-H bond	2019	47
35	[Me$_4$PNP^{tBu}NiH]B(Ar^F)$_4$	Formation of Ni-H bond	2019	47
36	[Me$_4$PNP^{tBu}NiH]BPh$_4$	Formation of Ni-H bond	2019	47
37	[Me$_4$PNP^{iPr}NiMe]B(Ar^F)$_4$	Formation of Ni-Me bond	2019	47
38	[Me$_4$PNP^{iPr}NiMe]BPh$_4$	Formation of Ni-Me bond	2019	47
39	[Me$_4$PNP^{tBu}NiMe]B(Ar^F)$_4$	Formation of Ni-Me bond	2019	47
40	[Me$_4$PNP^{tBu}NiMe]BPh$_4$	Formation of Ni-Me bond	2019	47
41	[Me$_4$PNP^{iPr}NiH]	Dearomatization of pyridine ring	2019	47
42	[Me$_4$PNP^{tBu}NiH]	Dearomatization of pyridine ring	2019	47
43	[Me$_4$PNP^{iPr}NiH]	Dearomatization of pyridine ring	2019	47

Continued

Table 1 Recently synthesized Ni, Pd, and Pt pincer metal complexes and their utility.—cont'd

Compound No.	Compounds	Recent advancement	Year	References
44	[Me$_4$PNP^{iPr}NiMe]$_2$	C–C bond formation and dearomatization of pyridine ring	2019	47
45	[^{Ph2}PNP^{Ph2}Ni-MeCN]$_2$(BF$_4$)$_2$	C–C bond formation	2020	48
46	[^{Ph2}PNP^{Ph2}Ni-MeCN]$_2$	C–C bond formation and dearomatized acridine ring	2020	48
47	[^{iPr}PNP^{iPr}Ni-Cl]	PNP pincer base Ni complex	2017	49
48	[^{iPr}PNP^{iPr}Ni]	Homolytic cleavage of H–H, N–N and C–C	2017	49
49	[^{iPr}PNP^{iPr}Ni-Cl]Cl	[2+2+2] cycloaddition reaction	2017	50
50	[^{tBu}PNP^{tBu}Ni-Cl]Cl	[2+2+2] cycloaddition reaction	2017	50
51	[^{iPr}PNP^{iPr}Ni-Cl]BPh$_4$	[2+2+2] cycloaddition reaction	2017	50
52	[^{tBu}PNP^{tBu}Ni-Cl]BPh$_4$	[2+2+2] cycloaddition reaction	2017	50
53	[^{Ph}PNP^{Ph}Ni-Br]NiBr$_4$	[2+2+2] cycloaddition reaction	2017	50
54	[^{Ph}PNP^{Ph}Ni-Cl]OTf	[2+2+2] cycloaddition reaction	2017	50
55	[PCN–NiCl]	Kharasch addition	2018	51
56	[PCN–NiCl$_3$]	Formation of PCN pincer based Ni complex	2018	51
57	[PCN–NiBr]	Kharasch addition	2018	51
58	[PCN–Ni(III)Cl$_2$]	PCN pincer ligand stabilize Ni(III) center	2018	51
59	[PCN–Ni (III)Br$_2$]	PCN pincer ligand stabilize Ni(III) center	2018	51

60	[PCN-Ni-ONO$_2$]	Formation of PCN pincer based Ni complex	2018	51
61	[PCN-Ni-TFA]	Formation of PCN pincer based Ni complex	2018	51
62	[PCN-Ni-OH]	Formation of Ni-hydroxy complex	2018	51
63	[PCN-Ni-OCO$_2$H]	Carboxylation reaction	2018	51
64	{[PCN]Ni}$_2$(μ-CO$_2$)	Formation of PCN pincer based CO$_2$ bridged Ni complex	2018	51
65	[PCN-Ni-NH$_2$]	Formation of Ni-NH$_2$ bond	2018	51
66	[PCN-Ni-OCONH$_2$]	CO$_2$ insertion in to Ni-NH$_2$ bond	2018	51
67	[PCN-Ni-Me]	Formation of Ni-Me bond	2018	51
68	[PCN-Ni-Ph]	Formation of Ni-Ph bond	2018	51
69	[PCN-Ni-p-tolylacetylide]	Formation of the Ni-acetylide complex	2018	51
70	[PCN-NiOAc]	Formation of PCN pincer based Ni complex	2018	51
71	[PCN-NiBr]	Formation of PCN pincer based Ni complex	2020	52
72	[PCN-NiBr$_3$]	Formation of PCN pincer based Ni complex	2020	52
73	[PCN-NiCH$_3$]	Formation of the Ni-Me complex	2020	52
74	[PCN-NiOCOCH$_3$]	Carboxylation of the Ni-Me complex	2020	52
75	[$^{i-Pr}$PONNP-NiBr$_2$]	Formation of PONNP pincer based Ni complex	2020	53
76	[PSCOP-NiCl]	Cross coupling reactions	2020	54
77	[PSCOP-NiCl]	Cross coupling reactions	2020	54
78	[PSCOP-NiCl]	Cross coupling reactions	2020	54

Continued

Table 1 Recently synthesised Ni, Pd, and Pt pincer metal complexes and their utility.—cont'd

Compound No.	Compounds	Recent advancement	Year	References
79	[^{tBu}NCP-NiCl]	Ethylene oligomerization	2020	55
80	[^{tBu}NCP-NiBr]	Ethylene oligomerization	2020	55
81	[^{tBu}NCP-NiCl$_2$]	PCN pincer stabilize Ni(III) center	2020	55
82	[^{tBu}NCP-NiBr$_2$]	PCN pincer stabilize Ni(III) center	2020	55
83	[^{tBu}PCN-Ni(BH$_4$)]	Formation of Ni-borohydride complex	2020	55
84	{[NHC(CH$_2$Py)$_2$]NiCl}(PF$_6$)	Kumada-Tamao-Corriu coupling	2020	56
85	{[NHC(CH$_2$Py)$_2$]NiCl}(BF$_4$)	Kumada-Tamao-Corriu coupling	2020	56
86	[^{Ph}POCSP^{Ph}]Ni–Cl	C–S cross-coupling reactions	2018	57
87	[^{iPr}PSCSP^{iPr}–NiCl]	Kumada coupling	2018	58
88	[^{iPr}POCSP^{iPr}–NiCl]	Kumada coupling	2018	58
89	[(^{iPr}POCOP^{iPr})NiCl]	Electrocatalytic proton reduction	2019	59
90	[(^{iPr}POCSP^{iPr})NiCl]	Electrocatalytic proton reduction	2019	59
91	[(^{iPr}PSCSP^{iPr})NiCl]	Electrocatalytic proton reduction	2019	59
92	[(^{iPr}POCOP^{iPr})NiSC$_6$H$_4$CH$_3$]	Formation of POCOP pincer based Ni complex	2019	59
93	[(^{iPr}POCSP^{iPr})NiSC$_6$H$_4$CH$_3$]	Formation of POCSP pincer based Ni complex	2019	59
94	[(^{iPr}PSCSP^{iPr})NiSC$_6$H$_4$CH$_3$]	Formation of PSCSP pincer based Ni complex	2019	59
95	[ONN–NiCl]	Kumada-Tamao-Corriu coupling	2018	60
96	[ONP–NiCl]	Kumada-Tamao-Corriu coupling	2018	60

97	[NNN–NiCl]	Kumada–Tamao–Corriu coupling	2018	60
98	[NNP–NiCl]	Cross-coupling reaction	2019	61
99	[NCN–NiCl$_2$NCCH$_3$]	N-Alkylation and dehydrogenative coupling	2020	62
100	[CNN–NiBr]	Transfer hydrogenation of ketones	2019	63
101	[CNN–NiBr]	Transfer hydrogenation of ketones	2019	63
102	[CNN–NiBr]	Kumada coupling reactions	2014	64
103	[CNN–NiBr]	Kumada coupling reactions	2014	64
104	[^{Ph}POCOP^{Ph}–PdCl]	Suzuki–Miyaura couplings	2018	65
105	[^{iPr}POCOP^{iPr}–PdCl]	Suzuki–Miyaura couplings	2018	65
106	[^{tBu}POCOP^{tBu}–PdCl]	Suzuki–Miyaura couplings	2018	65
107	[SCS–PdCl]	Suzuki coupling reactions	2009	66
108	[SeCSe–PdCl]	Heck coupling reaction	2004	67
109	[SeCSe–PdOAc]	Allylation of aldehydes	2006	68
110	[OCO–PdCl]	Heck coupling reactions	2009	69
111	[OCO–PdOTf]	Heck coupling reactions	2009	69
112	[NNN–PdCl]	Cross dehydrogenative coupling	2020	70
113	[CNN–PdL] (L = 2-phenylimidazo[1,2-*a*]pyridine)	Cross dehydrogenative coupling	2020	70
114	[CNN–PdL] (L = benzothiazole)	Cross dehydrogenative coupling	2020	70

Continued

Table 1 Recently synthesized Ni, Pd, and Pt pincer metal complexes and their utility.—cont'd

Compound No.	Compounds	Recent advancement	Year	References
115	[CNN-PdI] (L=1-methyl-1H-imidazole)	Cross dehydrogenative coupling	2020	70
116	[NNN-PdNCCH$_3$] (H, OMe substitute NNN pincer backbone)	Suzuki-Miyaura cross-coupling	2018	71
117	[NNN-PdNCCH$_3$] (OMe substitute NNN pincer backbone)	Suzuki-Miyaura cross-coupling	2018	71
118	[NNN-PdNCCH$_3$] (Cl substitute NNN pincer backbone)	Suzuki-Miyaura cross-coupling	2018	71
119	[(^{Tol,Cy}DIPy)PdCl]	Ligand protonation activity	2018	72
120	[(^{Tol,Cy}DIPyH)PdCl$_2$]	Formation of diimino-Pyrrole based Pd Complex	2018	72
121	[(^{Tol,Cy}DIPy)Pd(PMe$_3$)$_2$Cl]	Redox behavior	2018	72
122	[(^{Tol,Cy}DIPy)PdL] (L=THF or PMe$_3$)	Formation of diimino-Pyrrole based Pd Complex	2018	72
123	[C$_6$H$_4$-1,2-(NCH$_2$P^{t}Bu$_2$)$_2$B-PdCl]	Suzuki-Miyaura cross-coupling	2019	73
124	[C$_6$H$_3$-2,6-(OP^{t}Bu$_2$)$_2$-PdCl]	Suzuki-Miyaura cross-coupling	2019	73
125	[Ni(bC^N^bC)Cl]OTf	Electrocatalytic CO$_2$ reduction	2018	74
126	[Pd(bC^N^bC)Cl]OTf	Electrocatalytic CO$_2$ reduction	2018	74
127	[Pt(bC^N^bC)Br]OTf	Electrocatalytic CO$_2$ reduction	2018	74
128	[Ni(bC^N^bC)CH$_3$CN](OTf)$_2$	Electrocatalytic CO$_2$ reduction	2018	74
129	[Pd(bC^N^bC)CH$_3$CN](OTf)$_2$	Electrocatalytic CO$_2$ reduction	2018	74
130	[Pt(bC^N^bC)CH$_3$CN](OTf)$_2$	Electrocatalytic CO$_2$ reduction	2018	74

131	[PNO-PdCl]Cl	Suzuki-Miyaura couplings	2020	75
132	[PNO-PdCl]Cl	Suzuki-Miyaura couplings	2020	75
133	[CNN-PdCl] (OMe substituted)	Arylation of azoles and Mizoroki-Heck reaction	2020	76
134	[CNN-PdCl] (NMe$_2$ substituted)	Arylation of azoles and Mizoroki-Heck reaction	2020	76
135	[CNN-PdCl] (NEt$_2$ substituted)	Direct Csp^2–H arylation	2020	76
136	[CNN-PdCl] (morpholine substituted)	Direct Csp^2–Harylation	2020	76
137	[N$_{py}$N$_{im}$O$_{ph}$PdCl]	Cross coupling reaction and allylation of aldehydes	2019	77
138	[N$_{py}$N$_{im}$O$_{ph}$PdCl]	Cross coupling reaction and allylation of aldehydes	2019	77
139	[N$_{py}$N$_{im}$O$_{ph}$PdCl]	Cross coupling reaction and allylation of aldehydes	2019	77
140	[N$_{py}$N$_{im}$O$_{ph}$PdCl]	Cross coupling reaction and allylation of aldehydes	2019	77
141	[SCS-Pd] {Phenylene-Bridged Bis(thione)}	Transfer hydrogenation of quinolines	2020	78
142	[SCS-Pd] {Phenylene-Bridged Bis(thione)}	Transfer hydrogenation of quinolines	2020	78
143	[SCS-Pd] {Phenylene-Bridged Bis(thione)}	Transfer hydrogenation of quinolines	2020	78
144	[CNN-PdCl]	Heck coupling reactions	2016	79
145	[CNN-PdCl]	Heck coupling reactions	2016	79
146	[PCN-PdCl]	Heck coupling reactions	2012	80

Continued

Table 1 Recently synthesized Ni, Pd, and Pt pincer metal complexes and their utility.—cont'd

Compound No.	Compounds	Recent advancement	Year	References
147	[PCP-PdCl]	Sonogashira cross-coupling reactions	2009	81
148	[NCN-PdBr]	Sonogashira cross-coupling reactions	2015	82
149	[NCN-PdBr]	Sonogashira cross-coupling reactions	2015	82
150	[NCN-PdBr]	Sonogashira cross-coupling reactions	2015	82
151	[PCN-PdCl]	Suzuki, Sonogashira, and Hiyama Couplings	2008	83
152	[SNS-PdCl]	Negishi coupling reactions	2009	84
153	[SNS-PdCl]	Negishi coupling reactions	2009	84
154	[PCP-PdNCCH$_3$]	Intermolecular hydroamination	2020	85
155	[PCP-PdNCCH$_3$]	Intermolecular hydroamination	2020	85
156	[PCP-PdNCCH$_3$]	Intermolecular hydroamination	2020	85
157	[PCP-PdNCCH$_3$]	Intermolecular hydroamination	2020	85
158	[PCP-PdCl]	Carbonylative Sonogashira and Suzuki–Miyaura Cross-Coupling	2019	86
159	[TBA][Pt(CNC)TzQn]	Photoredox catalyst	2019	87
160	[TBA][Pt(CNC)TzH]	Photoredox catalyst	2019	87
161	[TBA][Pt(CNC)TzMe]	Photoredox catalyst	2019	87
162	[TBA][Pt(CNC)TzBr]	Photoredox catalyst	2019	87
163	[^{Me,Quin}NNN^{Quin,Me}–PtCl]	Luminescent platinum(II) complexes	2019	88

164	[^{Me,Phen}NNN^{Quin,Me}-PtCl]	Luminescent platinum(II) complexes	2019	88
165	[^{Me,Phen}NNN^{Phen,Me}-PtCl]	Luminescent platinum(II) complexes	2019	88
166	[(^{Me}BIMCA-Pt-CNCy)](BPh$_4$)	Luminescent complex	2020	89
167	[(^{Me}BIMCA-Pt-CNMe$_4$Bu)](BPh$_4$)	Luminescent complex	2020	89
168	[(^{Me}BIMCA-Pt-CNAd)](BPh$_4$)	Luminescent complex	2020	89
169	[(^{Me}BIMCA-Pt-CNBzy)](BPh$_4$)	Luminescent complex	2020	89
170	[(^{Me}BIMCA-Pt-CNMe$_2$Ph)](BPh$_4$)	Luminescent complex	2020	89
171	BIMCAPt phenyl acetylides(CH$_3$ substituted)	Phosphorescent platinum(II) alkynyls	2020	90
172	BIMCAPt phenyl acetylides(OMe substituted)	Phosphorescent platinum(II) alkynyls	2020	90
173	BIMCA-Pt phenyl acetylides(NMe$_2$ substituted)	Phosphorescent platinum(II) alkynyls	2020	90
174	BIMCA-Pt phenyl acetylides(NPh$_2$ substituted)	Phosphorescent platinum(II) alkynyls	2020	90
175	[Pt(C^N^C)(C^C≡C$_6$H$_5$)]$^+$	Luminescent and photoinduced dehydrogenation catalysis	2018	91
176	[Pt(C^N^C)(C≡N)]$^+$	Luminescent and photoinduced dehydrogenation catalysis	2018	91
177	[CNNPt^{IV}Me(Cl)]	Nucleophilic selectivity	2020	92
178	[CNNPt^{IV}Me(I)]	Nucleophilic selectivity	2020	92
179	[CNNPt^{IV}Me(OCH$_2$CF$_3$)]	Nucleophilic selectivity	2020	92
180	[CNC-Pt-DMSO]	Luminescence and cytotoxic evaluation	2020	93

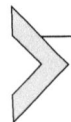

3. General synthetic routes to group 10 pincer complexes

Numerous synthetic methodologies have been established to provide various novel pincer metal compounds efficiently, with metal and ligand combination playing a significant role. In many cases, the catalysts preparation method and yield should be considered carefully since they may occupy the most part of the cost for the catalysis reaction. The application of these synthetic strategies depends on both the transition metal and pincer ligands. Therefore, this section mainly focuses on recent synthetic advances in group 10 pincer metal compounds, including well-known neutral and anionic donor pincer ligands, and some other significant pincer ligands.

3.1 Nickel (Ni) pincer metal compounds

PCP-containing pincer metal complexes are the most common Ni pincer complexes. 1,3-(But$_2$PCH$_2$)$_2$C$_6$H$_4$, synthesized in 1976, was the first reported PCP pincer ligand, with the foremost Ni pincer metal compound, {2,6-bis[(di-*tert*-butylphosphino)methyl]-phenyl}chloronickel, being prepared using the reported PCP pincer ligand.[1] This successful result triggered the exploration in the field of pincer ligand coordination chemistry. POCOP–Ni pincer compounds are also similar to the reported PCP–Ni pincer compounds.[94–99] 4-Dimethylaminopyridine and trimethylamine have been used as bases to improve the yield of the synthesized Ni pincer compounds.[100] In this section, we have summarized recently synthesized Ni pincer complexes (Fig. 2) and their efficacy.

Rheingold and coworkers have reported a chiral bis(phospholane) PCP–Ni pincer complex (Scheme 1). Interestingly, the enantioselective alkylation of phosphines was achieved using this chiral PCP–Ni pincer complex as catalyst.[39]

The asymmetric catalytic activity of the chiral PCP–Ni pincer complex promoted the alkylation of secondary phosphines, which had previously been described using expensive chiral Pt, Pd, or Ru metal complexes. The secondary phosphines were alkylated with the corresponding alkyl halide in the presence of base and the chiral PCP–Ni catalyst (10 mol% loading) at room temperature in tetrahydrofuran to afford stereogenic tertiary phosphines with the usual selectivity (Scheme 2).[39] Furthermore, the catalyst was recovered and showed recyclability after the catalytic reaction, affording identical enantioselectivity. This chiral catalyst is also widely used in asymmetric catalysis, with development of this research ongoing.

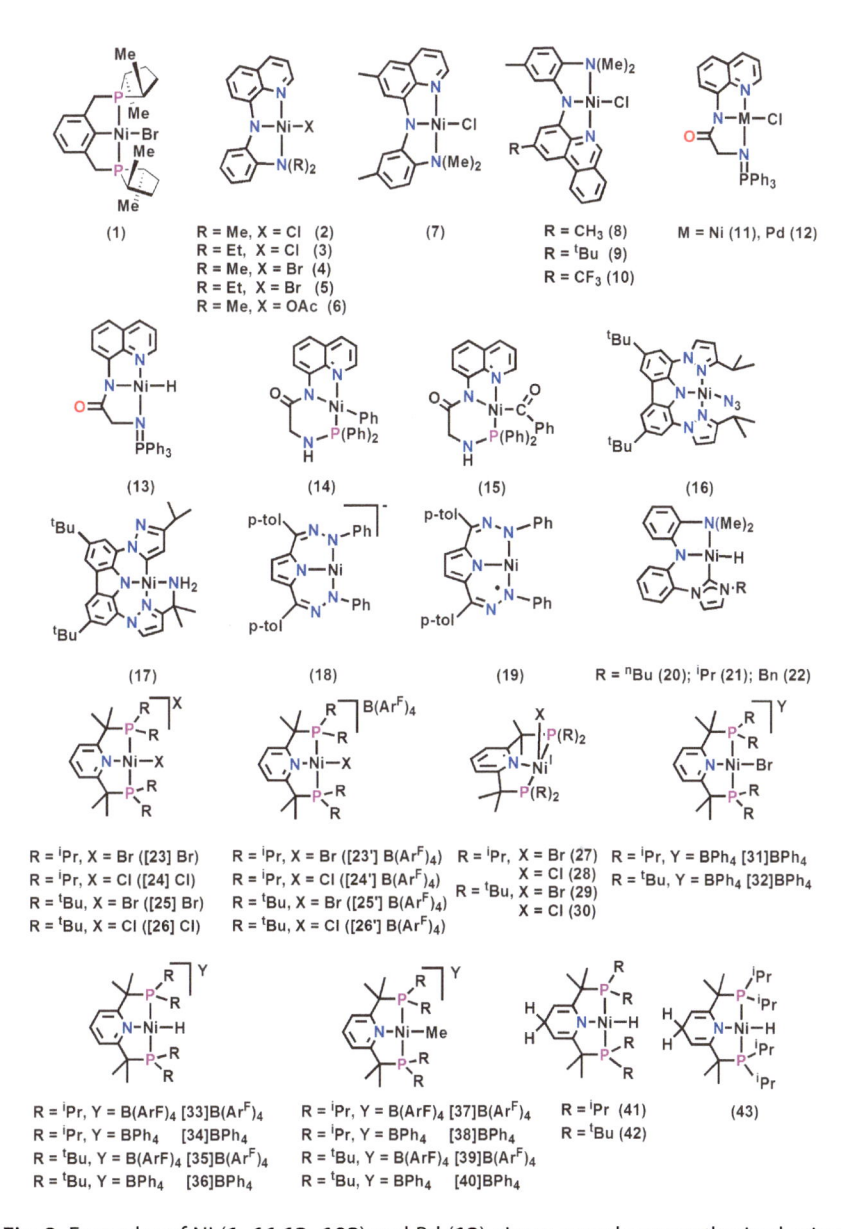

Fig. 2 Examples of Ni (**1–11,13–103**) and Pd (**12**) pincer complexes synthesized using different types of pincer ligand.

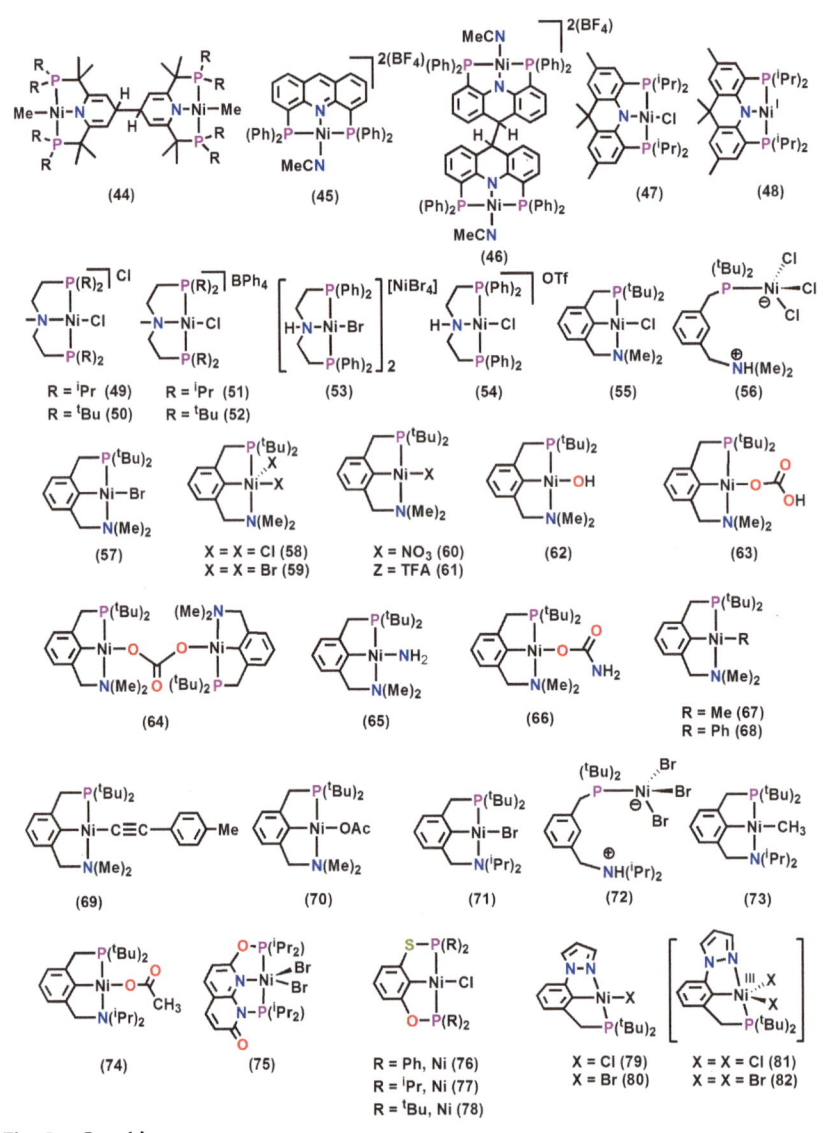

Fig. 2—Cont'd

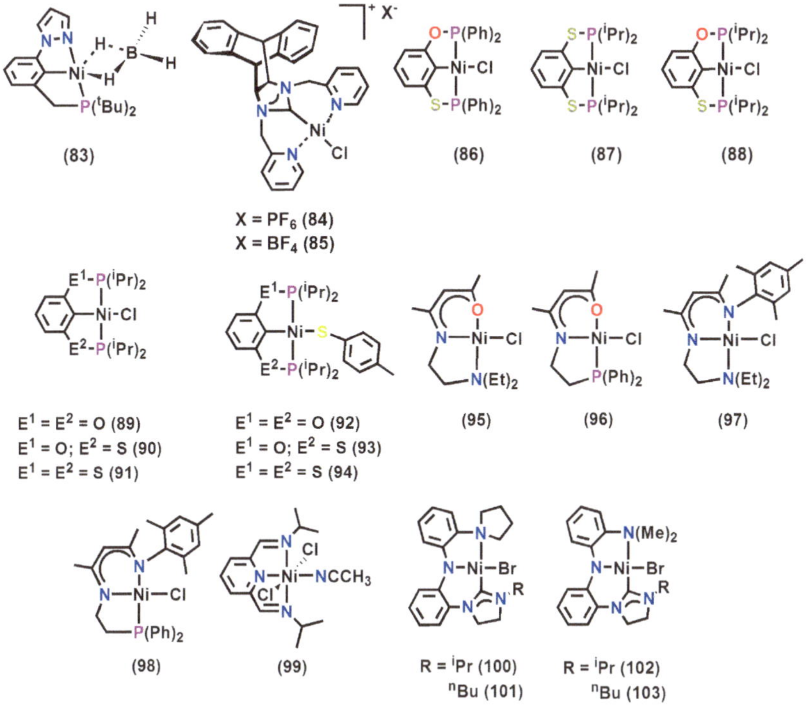

Fig. 2—Cont'd

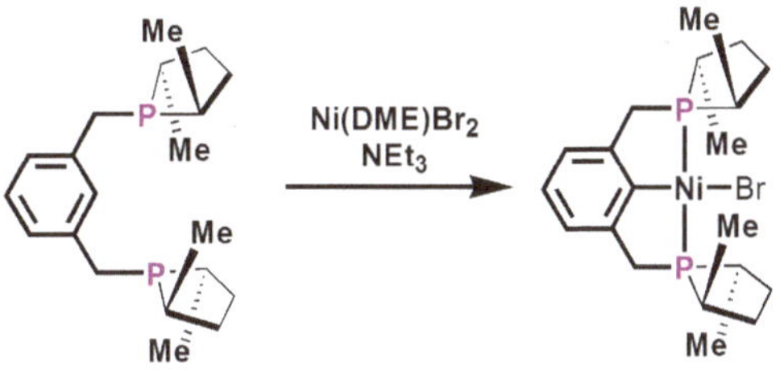

(1)

Scheme 1 General synthetic route for chiral bis(phospholane) Ni metal complex.

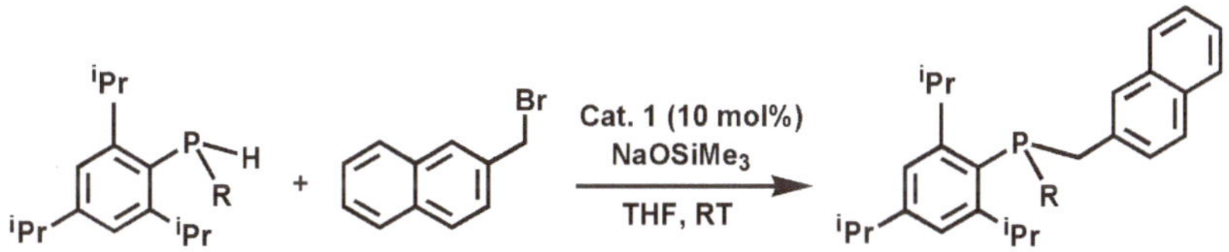

enantioenriched

R = Me 55:45 er
R = Ph 66:34 er

Scheme 2 Enantioselective alkylation of secondary phosphines by chiral Ni pincer metal complex **1**.

In general, high temperatures and harsh conditions are essential for C(sp³)–H bond activation, which is key to the synthesis of many target complexes. The PCP–Ni pincer compounds act as catalysts for many organic transformations. The first Ni pincer compounds with NCN-based pincer ligands have been prepared following the protocol employed with the previously reported PCP compounds.[101] The chemistry of NCN pincer metal complexes has developed rapidly owing to their emerging reactivity. NCN pincer ligands have rigid donor atoms that can stabilize Ni(III) oxidation states.[101,102] NCN–Ni pincer metal compounds are mostly thermally stable and can sublime, can have different substituents on the ligands, with nitrogen atoms able to have both cyclic and linear substituents. However, more commonly studied pincer ligands contain nitrogen-containing heterocycles, such as imidazole, pyrazole, oxazoline, or indazole.

Punji and coworkers have reported NNN–Ni pincer metal compounds, prepared by treating substituted quinoline-based pincer ligands with the Ni precursor in the presence of triethylamine (Scheme 3). Synthesized compound **2** was highly stable and used as a catalyst, showing activity for C–H alkylation with alkyl halides bearing β-hydrogen atoms (Scheme 4).

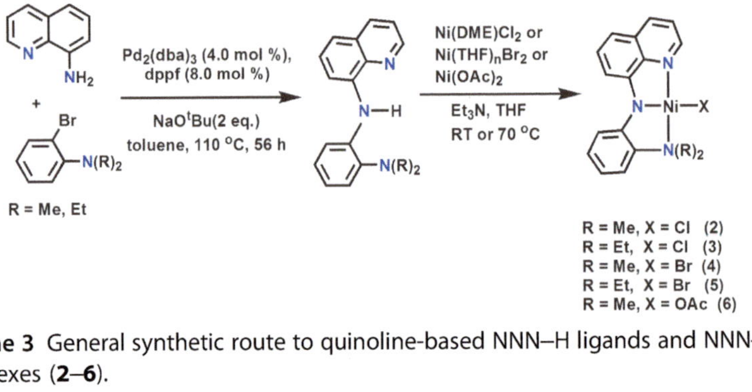

Scheme 3 General synthetic route to quinoline-based NNN–H ligands and NNN–Ni–X complexes (**2–6**).

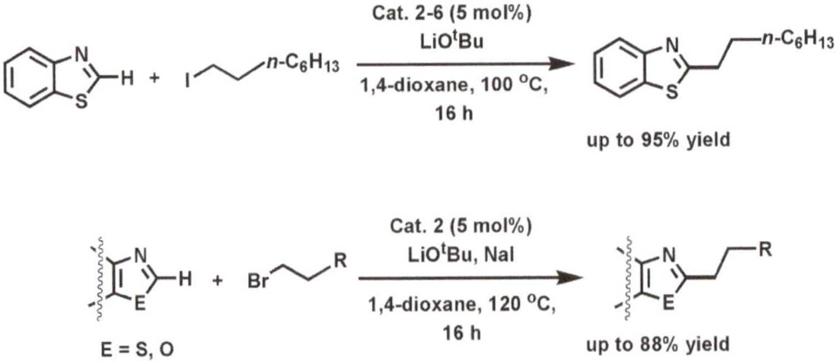

Scheme 4 Alkylations of benzothiazole and azole using alkyl halides catalyzed by a Ni pincer complex.

Furthermore, this catalyst was recycled in five catalytic alkylation reaction runs without any loss of activity. The catalytic reactivity of compound **2** showed that the alkylation reaction proceeded via a radical mechanism.[40]

Punji and coworkers also performed a detailed mechanistic study of the alkylation of azoles catalyzed by this quinolinyl-based Ni(II) pincer complex **2** (Scheme 5). Compound (**2**) was also utilized as a catalyst for the alkylation of azoles through an iodine atom transfer mechanism. In the mechanistic study, Ni(II)/Ni(III) was proposed as the active catalytic species. Radical species were identified in the catalytic cycle and their role was confirmed using deuterium labeling experiments. Furthermore, active intermediate [(^{Q}NNN^{Me2})Ni(benzothiazolyl)] was isolated and its crystal structure was studied using single-crystal X-ray diffraction (SCXRD) analysis.[103]

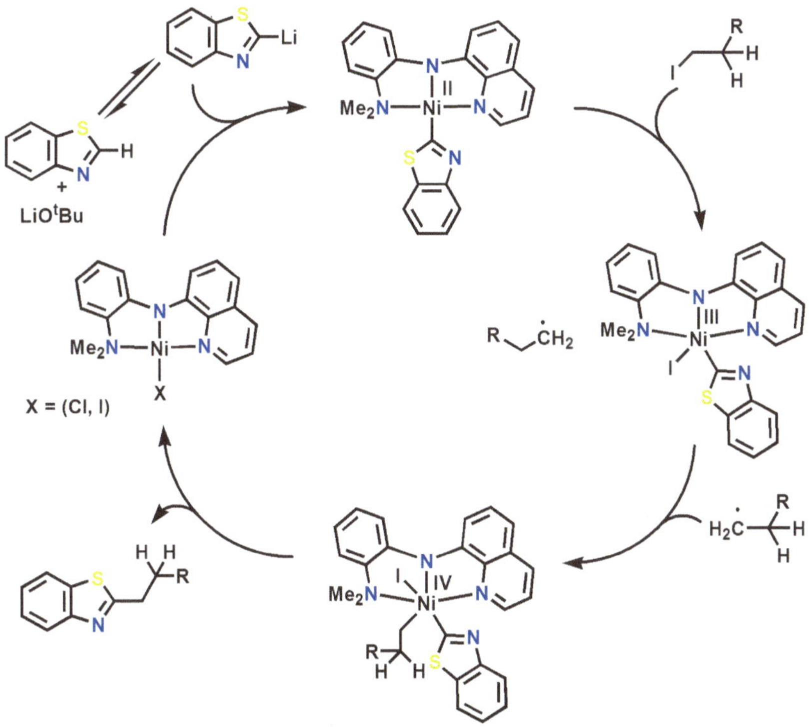

Scheme 5 Proposed catalytic mechanistic cycle for the alkylation of azoles by quinolinyl-based Ni complex (^{Q}NNN^{Me2})NiCl.

Herbert and coworkers synthesized Ni(II) compounds by reacting NNN diarylamido-based pincer-type ligands with Ni salt (Schemes 6 and 7). The synthesized NNN–Ni pincer compounds were used as catalysts in C–C bond forming reactions, while direct C–H activation of azoles with R–X

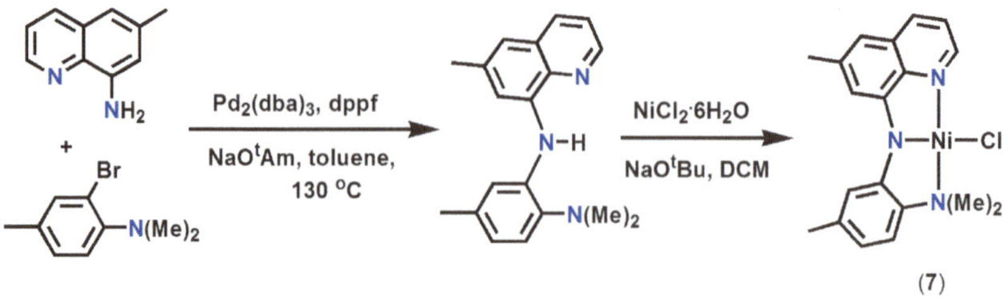

Scheme 6 Synthetic route to methyl-substituted quinolinyl-based Ni pincer metal complex **7**.

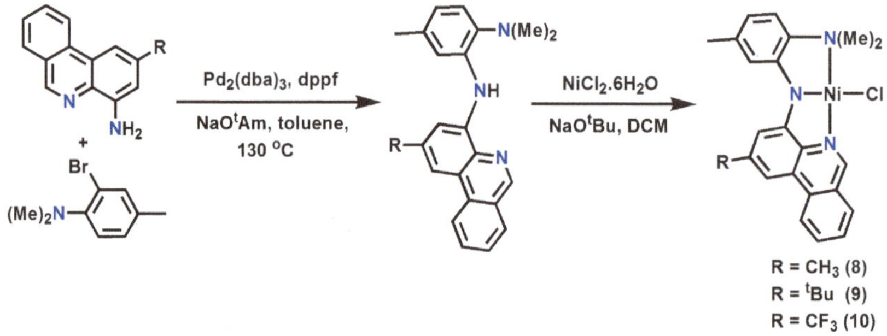

Scheme 7 Synthetic route to NNN–Ni pincer metal complexes **8–10**.

was extensively explored. The NNN–Ni pincer complexes were also characterized by SCXRD analysis and cyclic voltammetry.[41]

Auffrant and coworkers synthesized a tridentate NNN ligand (L^{R}H, R = Ph, Cy) by combining an amidoquinoline and iminophosphorane (Scheme 8). Square planar Ni(II) and Pd(II) compounds were obtained by treating the respective metal salts with NNN pincer ligand. The most exciting results were obtained by treating [Ni(COD)$_2$] with L^{Ph}H, which furnished a Ni(II) phenyl complex (**14**) bearing a tridentate amidoquinoline–aminophosphine ligand. This rearrangement might be due to proton transfer supported by the Ni(0) species through stabilization by a phenyl substituent on the phosphorus atom. These findings were corroborated by density functional theory (DFT) calculations and related experiments. Furthermore, the reactivity of the Ni phenyl complex was studied in the presence of CO, which afforded the corresponding benzoyl complex (**15**).[42]

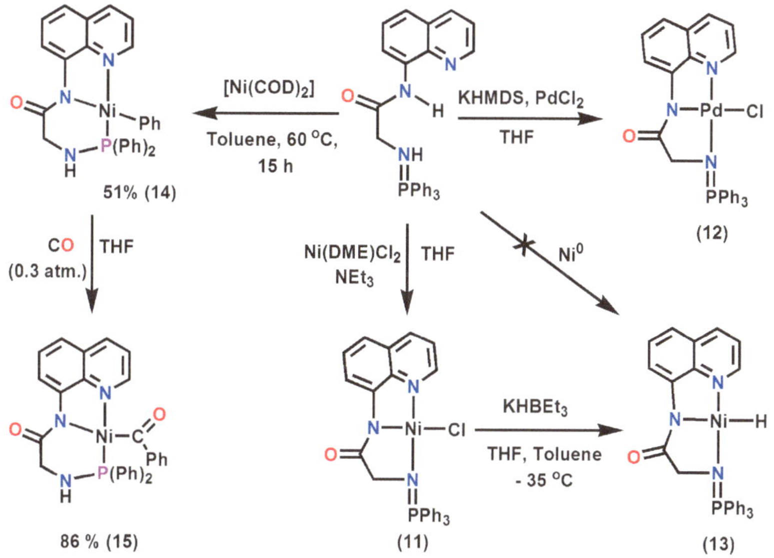

Scheme 8 Synthesis of tridentate NNN pincer metal (Ni, Pd) complexes and carbonylation reaction.

Lee and coworkers disclosed the preparation of Ni bismetallocyclic complex **17** via a Ni–nitridyl radical transition species. The authors photochemically irradiated a Ni(II) azido complex, affording unique Ni complex **17** in >90% isolated yield. Interestingly, complex **17** was obtained by intramolecular C–H activation of a Ni nitridyl intermediate via [2σ + 2π] addition (Scheme 9). Double intramolecular C–H functionalization of the transient moiety was demonstrated by SCXRD analysis. Complex **17** was also obtained by treating a Ni(I) precursor with TMS–N$_3$. DFT calculations corroborated the formation of complex **17** via a transient species with radical character after N$_2$ release.[43]

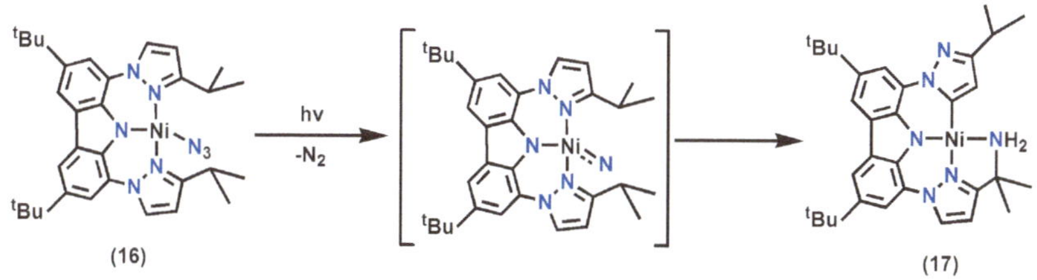

Scheme 9 Synthesis of NNC–Ni complex **17** via NNN–Ni-nitride intermediate.

Anderson and coworkers synthesized T-shaped [^{Tol,Ph}DHPy]Ni complexes, which were also fully characterized using spectroscopic techniques (Scheme 10). Theoretical calculations and EPR studies confirmed that the Ni(II) center was high-spin and antiferromagnetically coupled with a radical-containing ligand. The resulting compound activated small molecules, such as water at room temperature to give the hydroxyl bridged dimeric compound.[44] This dimeric complex was fully analyzed using SCXRD and spectroscopic techniques, and experimentally characterized using deuterium labeling. Water activation was the most notable feature of this report, and was corroborated by DFT calculations.[44]

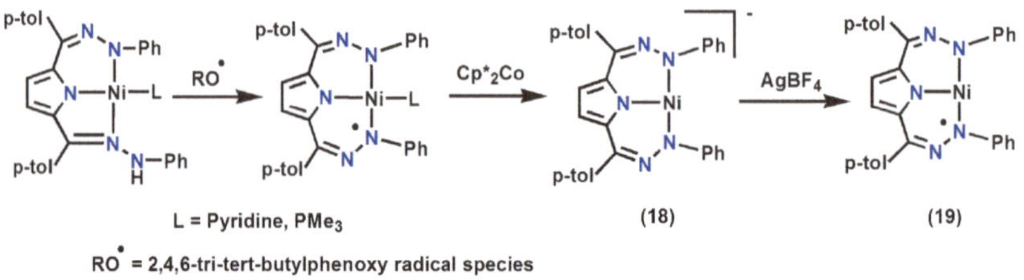

Scheme 10 General synthetic route to [^{Tol,Ph}DHPy]-based Ni pincer compounds.

Fenske and coworkers synthesized CNN–Ni(II) hydride compounds **20–22** by treating the corresponding Ni(II) bromide pincer complexes with hydride sources such as silane or NaBH$_4$ (Scheme 11).[45] The resulting hydride complexes were subjected to full spectroscopic analysis and their molecular structures were elucidated by SCXRD analysis. CNN–Ni(II) hydride complexes are used as catalysts for hydrodehalogenation reactions.[45]

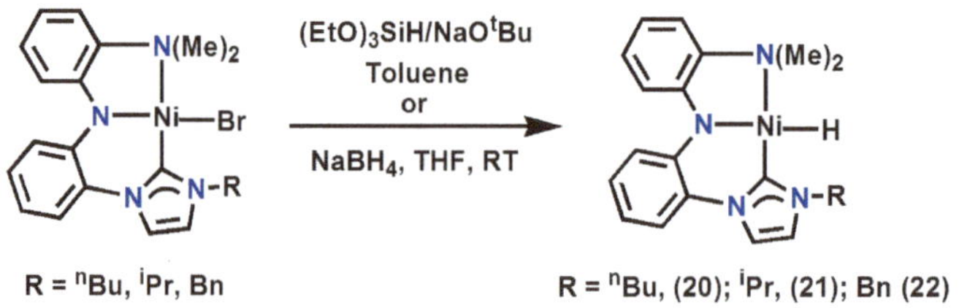

Scheme 11 General synthetic route to NNC–Ni(II) hydride pincer complexes **20–22**.

The tunability of the electron donor/acceptor character and steric behavior of tetramethylated PNP-based pincer ligands was discovered by Khusnutdinova and coworkers (Scheme 12). The methylation of PNP-based pincer ligands played a crucial role in avoiding dearomatization of the pincer backbone and tuned their steric and electronic behavior. Furthermore, the authors synthesized highly reactive Ni(I)–PNP pincer metal complexes through bulk electroreduction and chemical reduction (Scheme 13).[46] Complexes **27–30** were structurally characterized by SCXRD analysis. Furthermore, DFT calculations for the above Ni(I) complexes showed that the spin density mainly occurred at the Ni center.[46]

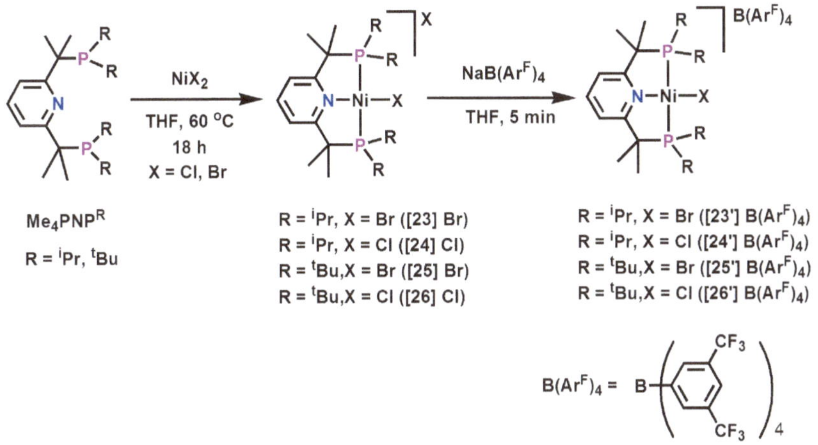

Scheme 12 Synthetic route to PNP–Ni(II) cationic complexes **23–26**.

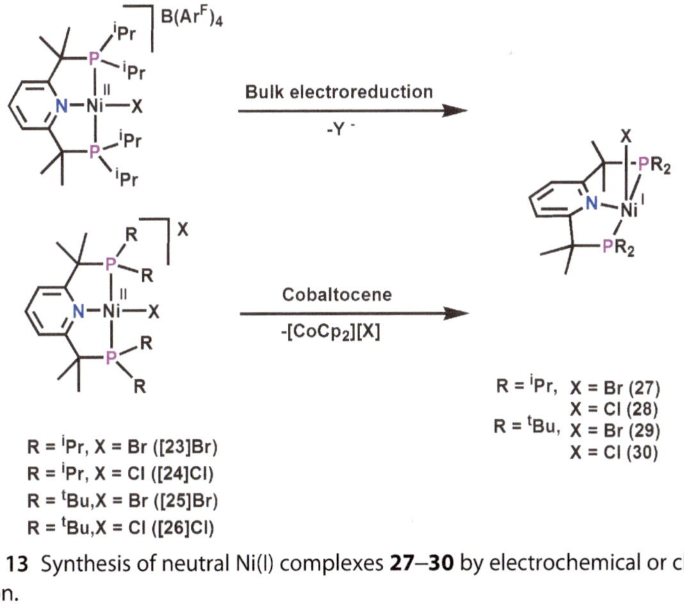

Scheme 13 Synthesis of neutral Ni(I) complexes **27–30** by electrochemical or chemical reduction.

The same group also reported the activity of tetramethylated PNP pincer complexes (Scheme 14), which showed the uncommon reactivity of pyridine ring dearomatization at the *para*-position (Scheme 15). Furthermore, the dimerization of two tetramethylated PNP-Ni-Me complexes through C–C bond formation at the *para*-position was achieved by reduction with KC$_8$ at room temperature (Scheme 16).[47]

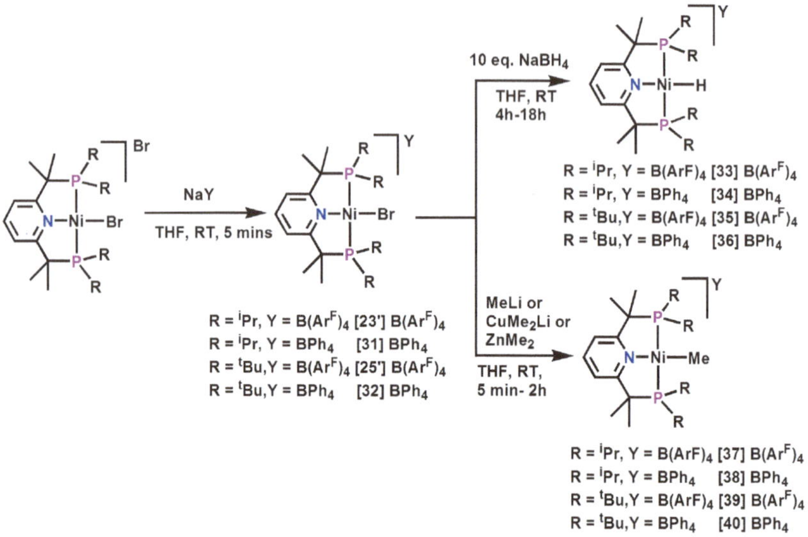

Scheme 14 Synthesis of Ni(II) cationic complexes **31–40**.

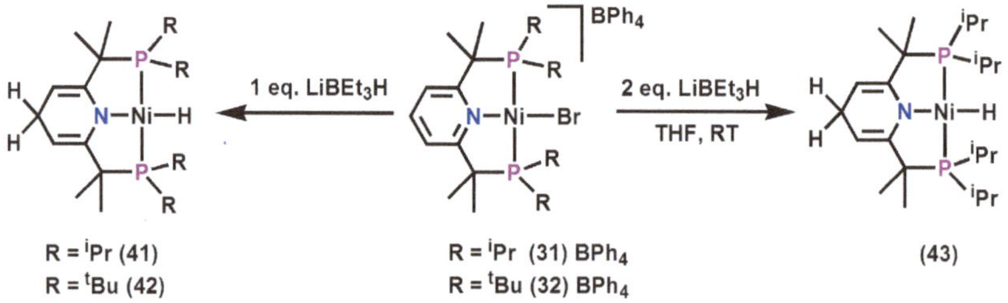

Scheme 15 Synthesis of PNP-based Ni complexes **41–43**.

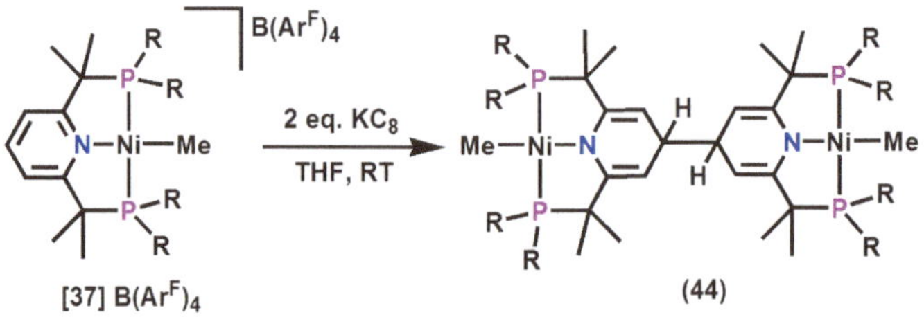

Scheme 16 Reduction of PNP-based Ni(II) complex **37** to form complex **44**.

Milstein and coworkers reported acridine-based noninnocent redox active PNP pincer complexes (metals: Ni, Co, Fe, and Mn) with C–C bond formation at the electrophilic site of the aromatic ring (Scheme 17). Dication Ni(II) complex **45** was synthesized by treating the PNP ligand with Ni metal salt and then further reduced by Na/Hg to obtain the Ni(I) dimeric complex **46**. In the dimeric complex, C–C bond formation occurred between the acridine ligand at the dearomatized central acridine ring, where one-electron transfer at the electron deficient C9 carbon supported complex dimerization (Scheme 17).[48]

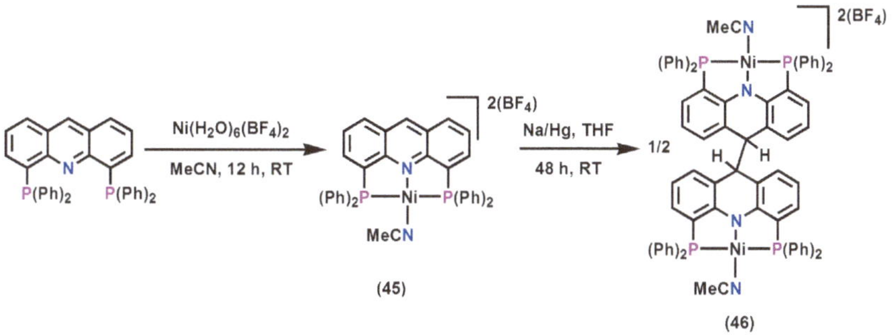

Scheme 17 Synthesis of acridine-based PNP pincer metal complexes.

Recently, Yoo and Lee reported a T-shaped metalloradical Ni(I) compound bearing a rigid acridane-based pincer ligand (Scheme 18).[49] The Ni(I) acridane complex **48** was prepared by reducing corresponding Ni(II) complex **47** with NaC$_{10}$H$_8$. Complex **48** induced the activation of a number of substrates, such as carbon dioxide and ethylene molecules to form CO$_2$ and ethane-bridged bimetallic complexes, respectively, owing to the sterically exposed half-filled d_{x-y}^{22} orbital (Fig. 3). Furthermore, this complex activated a range of molecules for homolytic bond cleavage, including dihydrogen, methanol, methyl iodide, phenol, diphenyl disulfide, acetonitrile, and hydrazine, among others.[49]

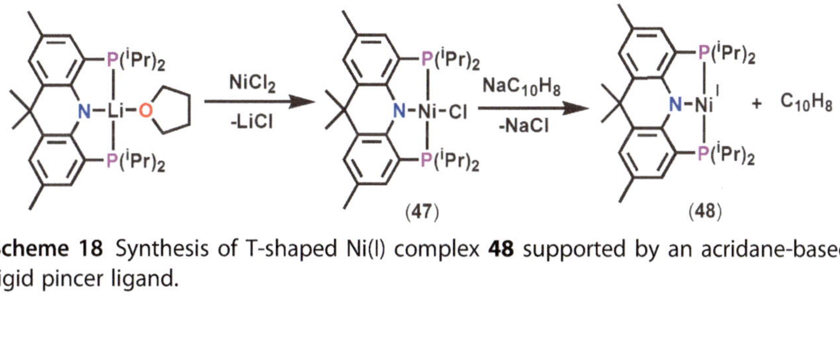

Scheme 18 Synthesis of T-shaped Ni(I) complex **48** supported by an acridane-based rigid pincer ligand.

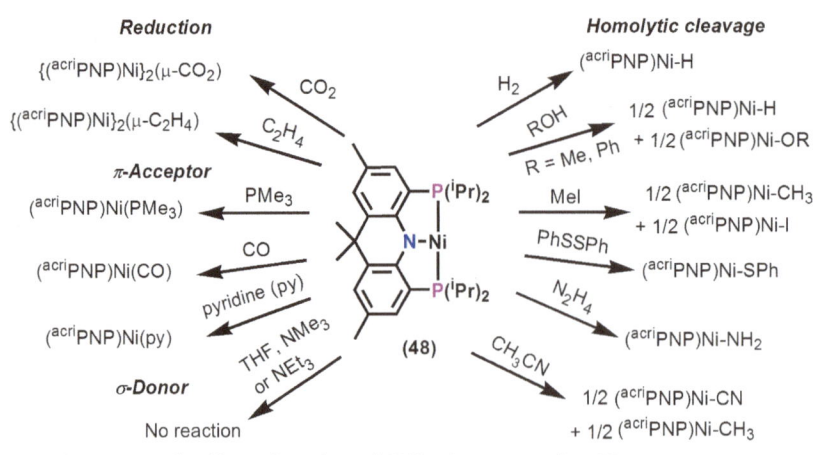

Fig. 3 Reactivity of Ni(I) acridane-based PNP pincer complex **48**.

Sivasankar and coworkers synthesized and fully characterized novel PNP–Ni(II) pincer metal complexes (Scheme 19).[50] The authors reported cycloaddition [2 + 2 + 2] reactions of alkynes with high regioselectivity. Complexes **49** and **50** were synthesized by treating the PNP-based ligand with NiCl$_2$6H$_2$O. Furthermore, complexes **51** and **52** were synthesized by treating **49** and **50**, respectively, with NaBPh$_4$ in methanol. Furthermore, complex **53** was synthesized by treating the PNP ligand with NiBr$_2$ (Scheme 19). Complex **53** was treated with silver triflate in dichloromethane solution to afford complex **54** through C–Cl bond cleavage of the dichloromethane solvent.[50]

Scheme 19 Synthesis of PNP-based Ni(II) pincer complexes **49–54**.

Pincer metal complexes with unsymmetrical PCN ligands have also been explored. PCN-based pincer ligands have unique properties owing to the combined ambiphilic (σ-donor and π-acceptor) electronic properties of ligands containing amine and phosphine moieties. The ambiphilic property of PCN ligands may promote direct metalation via C–H activation in the presence of Et$_3$N.

Unsymmetrical diamagnetic PCN–Ni(II) pincer compounds **55–57** were reported by Wendt and coworkers in 2018 (Scheme 20).[51] Furthermore, these complexes were oxidized by anhydrous CuX$_2$ (X=Cl, Br) to successfully obtain paramagnetic PCN–Ni(III) complexes **58** and **59** (Scheme 21). Complexes **58** and **59** showed no NMR peaks (^{1}H and ^{31}P) owing to their paramagnetic behavior. Interestingly, the Kharasch addition reaction was catalyzed by PCN–Ni(II) halide complexes **55** and **57**, which promoted the addition of carbon tetrachloride to olefins.[51]

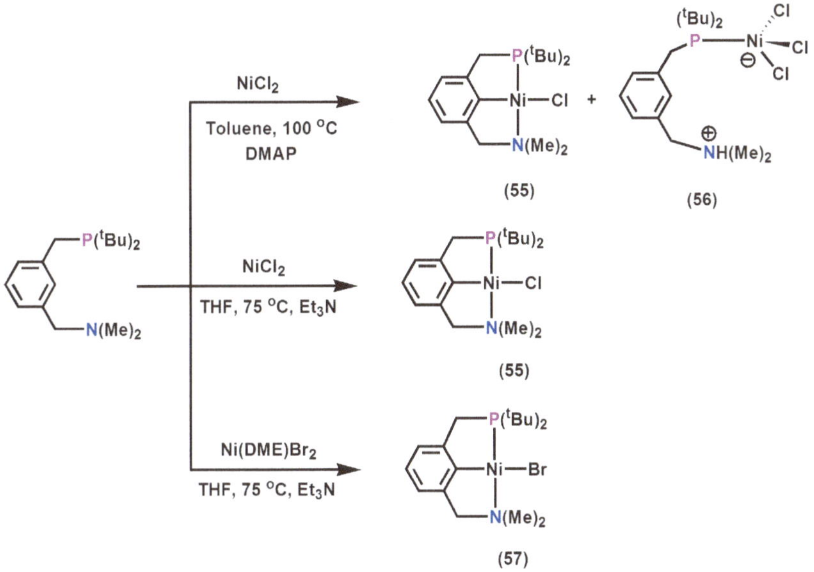

Scheme 20 Synthesis of PCN–Ni pincer complexes **55–57**.

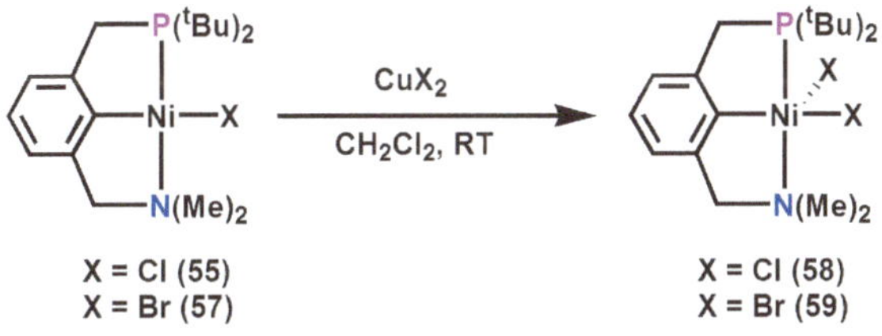

Scheme 21 Synthesis of (PCN)Ni-X$_2$ pincer complexes **58** and **59**.

The synthesis of [PCN–Ni–OH] complex **62** by reacting **57** with a silver salt gave intermediate complexes [PCN–Ni–NO₃] (**60**) and [PCN–Ni–TFA] (**61**). Complexes **60** and **61** were reacted with KOH/NaOH to obtain desired complex **62** (Scheme 22).[51]

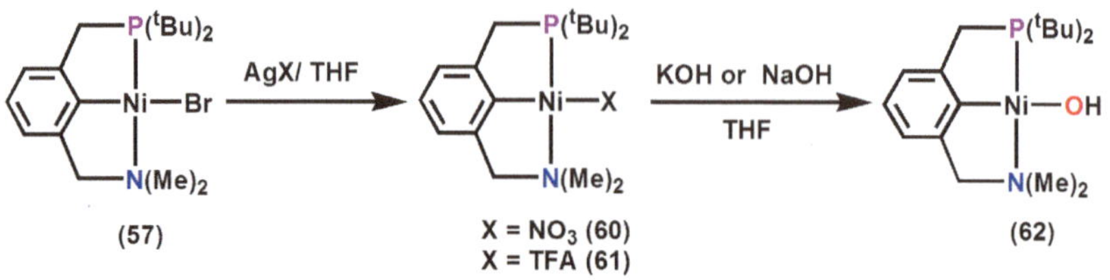

Scheme 22 Synthesis of (PCN)Ni–OH pincer complex **62**.

Complex **62** was treated with CO_2 (4 atm) to instantly form new pincer complex [PCN–Ni–OCO₂H] (**63**), which was detected by ^1H and ^{31}P NMR spectroscopy (Scheme 23).[51] The removal of volatiles from compound **63**, followed by crystallization, afforded dimeric carbonate compound **64**, exhibiting reversible CO_2 insertion (Scheme 23).[51]

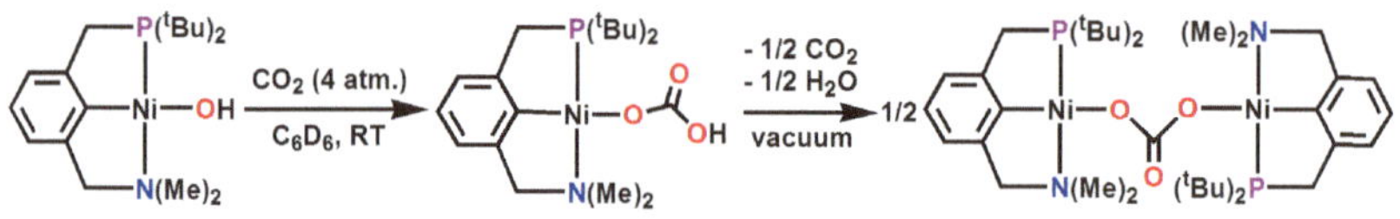

Scheme 23 Reaction of complex **62** with CO_2.

PCN–Ni–NH₂ complex **65** was obtained by the salt metathesis reaction of complex **57** and NaNH₂ (Scheme 24).[51] Complex **65** was air and moisture-sensitive. When **65** was treated with CO_2 (4 atm.) in deuterated benzene at room temperature, carbamate compound **66** was obtained (Scheme 24).[51]

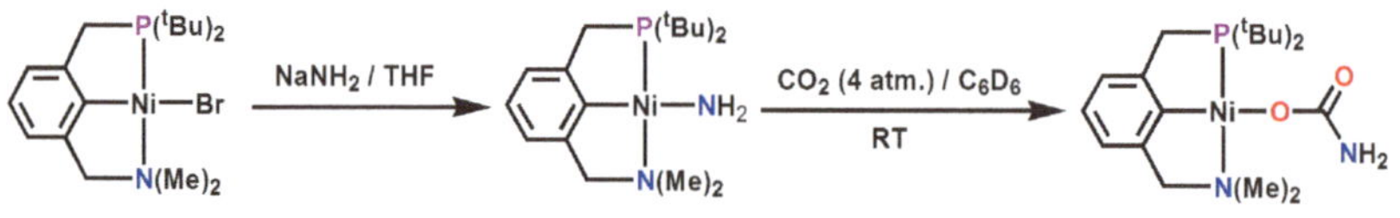

Scheme 24 Carbon dioxide insertion into Ni–NH₂ bond.

The reaction of complexes **55** and **57** with electrophiles afforded PCN–Ni hydrocarbyl complexes **67** and **68**. Treating complex **55** with methylmagnesium chloride in THF generated complex **67**, with phenyl complex **68** obtained in a similar manner (Scheme 25).[51]

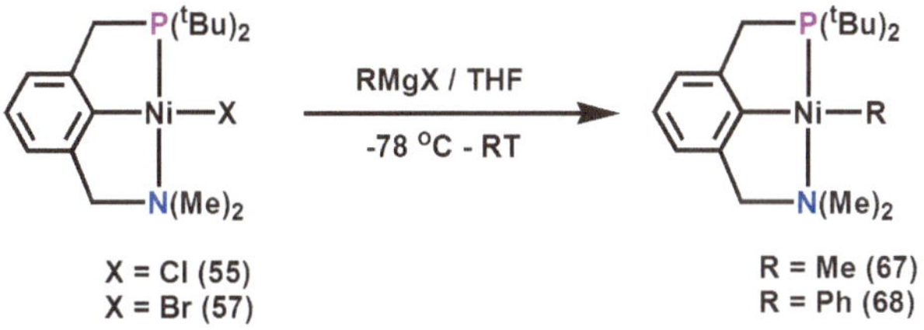

Scheme 25 Synthesis of PCN–Ni–R (R=Me, pH) from PCN–Ni–X (X=Cl, Br).

Complex **69** was prepared by treating PCN–Ni–NO$_3$ with p-tolylacetylene and K$_2$CO$_3$ as base (Scheme 26).[51] The resulting complex was expected to be more stable than similar methyl and phenyl complexes.

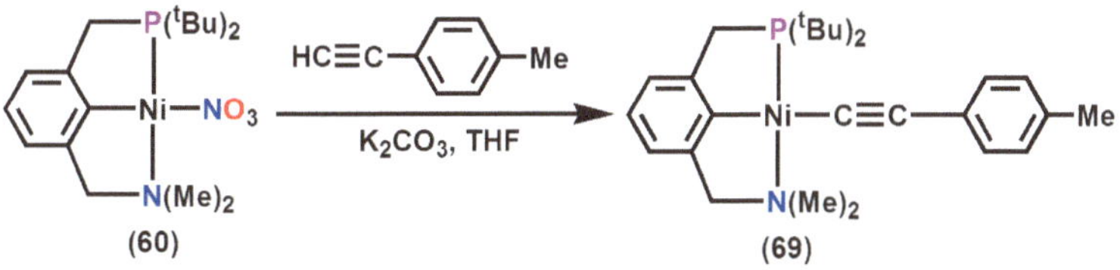

Scheme 26 General synthesis of (PCN^{Me})Ni acetylide complex from (PCN^{Me})Ni–NO$_3$ complex.

PCN–Ni–Me complex **67** was also treated with phenylbromide in deuterated benzene (Scheme 27).[51] The reaction did not proceed at low temperatures, but afforded the cross-coupling product and complex **57** when the temperature was increased to 120 °C. Furthermore, the PCN–Ni–Me complex was reacted with CO$_2$ (4 atm.) to give carboxylated product **70**. The carboxylated compound was also produced by treating a related chloride complex with silver acetate (Scheme 27).[51]

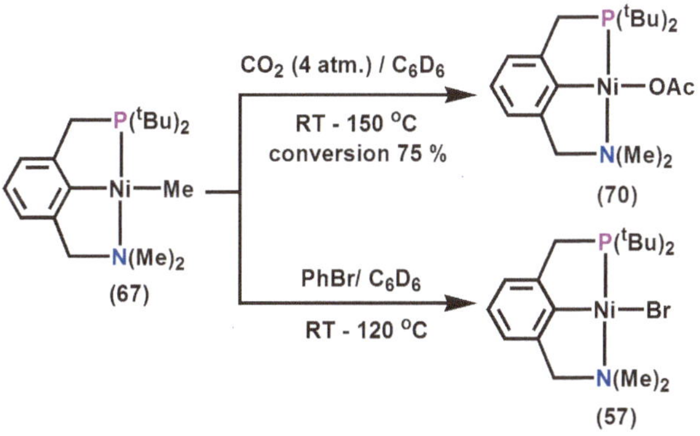

Scheme 27 Reactivity of complex **67** with electrophiles (CO$_2$ and phenylbromide).

Very recently, Wendt and coworkers reported bulky PCN–Ni pincer complexes obtained by direct cyclometalation. A $^{t\text{-}Bu}$PCN$^{i\text{-}Pr}$–Ni complex was treated with methylmagnesium chloride to give the corresponding transmetalated PCN–Ni–Me pincer compound (Scheme 28).[52] This PCN–Ni–Me complex showed high reactivity for carboxylation of the Ni–carbon bond by CO_2 compared with previously reported Ni pincer complexes, and similar reactivity to the PCP–Pd–Me compound (Scheme 28).[52]

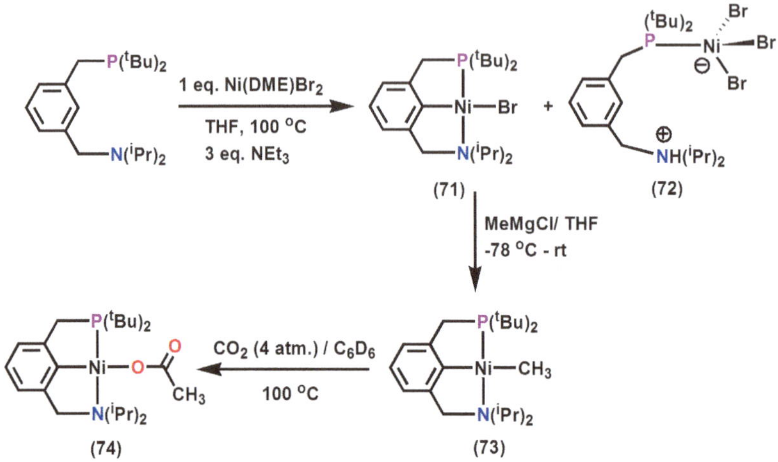

Scheme 28 General preparation of PCN^{Me}–Ni complexes and their reactivity toward CO_2.

Recently, Broere and coworkers reported a naphthyridine-based PONNOP pincer ligand and its coordination behavior (Scheme 29).[53] The Ni complex **75** was obtained by treating pincer ligand $^{i\text{-}Pr}$PONNOP with NiBr$_2$ in tetrahydrofuran or dichloromethane solvent. The unexpected rearrangement product **75** was obtained, and its molecular structure was established by SCXRD analysis, showing the rearrangement of the $^{i\text{-}Pr}$PONNOP ligand at the Ni center.[53]

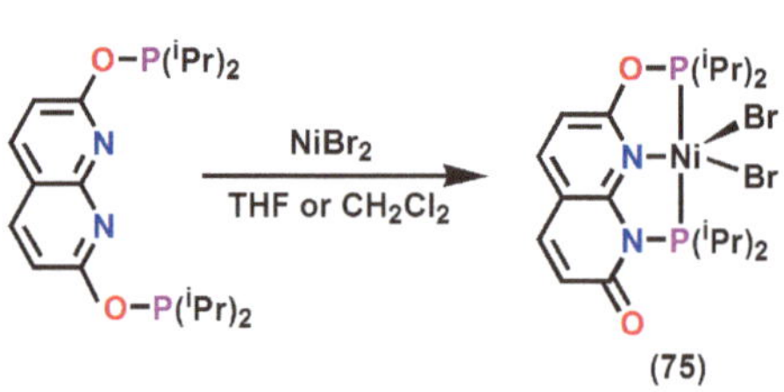

Scheme 29 Synthesis of Ni(II) pincer complex **75** with $^{i\text{-}Pr}$PONNOP ligand.

Morales-Morales and coworkers synthesized Ni (**76–78**) complexes using an unsymmetrical PSCOP-based pincer ligand (Scheme 30).[54] Interestingly,

C–S and C–Se coupling reactions were conducted using these PSCOP–Ni pincer complexes as catalysts. Owing to the ᵗBu substituent increasing electron density on the metal center, the Ni (**78**) catalyst showed improved catalytic activity in compared with their ⁱPr- and Ph-substituted analogs.[54]

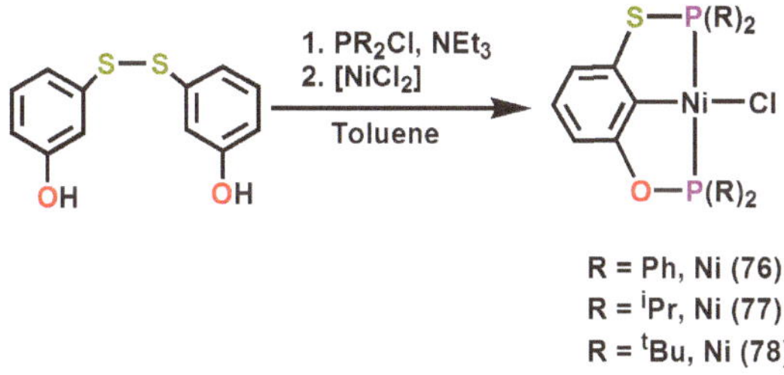

R = Ph, Ni (**76**)
R = ⁱPr, Ni (**77**)
R = ᵗBu, Ni (**78**)

Scheme 30 General synthesis of Ni compounds with an unsymmetrical PSCOP base pincer ligand.

Yakhvarov and coworkers reported the catalytic reactivity of PCN–Ni(II) compounds. Furthermore, oxidizing agents, such as Cu(II) halides, and hydride sources, such as sodium borohydride and lithium aluminum hydride, were utilized to form PCN–Ni(III) complexes (Schemes 31 and 32).[55] The reaction of PCN–NiCl with hydride donor sodium borohydride afforded the corresponding complex (ᵗBuPCN)Ni(BH₄). These Ni(II) halide compounds were used as catalysts for ethylene oligomerization.[55]

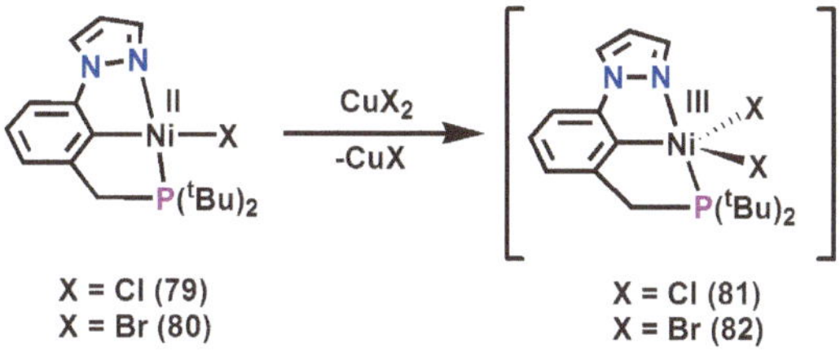

X = Cl (**79**)
X = Br (**80**)

X = Cl (**81**)
X = Br (**82**)

Scheme 31 Reactivity of PCN–Ni(II) compounds with copper(II) halides.

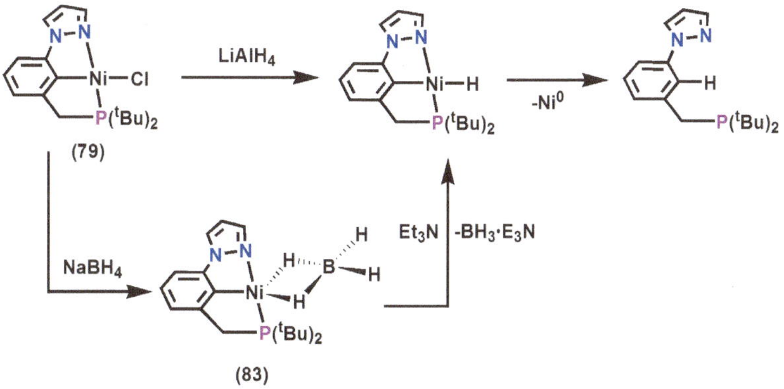

(**83**)

Scheme 32 Reactivity of Ni(II) complex **79** with NaBH₄ and LiAlH₄.

Ishizuka reported the synthesis of stable Ag and Ni compounds bearing NHCs with a bicyclic framework as NCN pincer ligands, and explored the catalytic activity of the synthesized Ni complexes (Scheme 33).[56] The Ni complexes were synthesized by transmetalation of the corresponding silver complex with NiCl$_2$(PPh$_3$) through successive anion exchange to afford the respective cationic complexes (Scheme 33). These NCN–Ni complexes showed catalytic reactivity in the Kumada–Tamao–Corriu (KTC) reaction.[56]

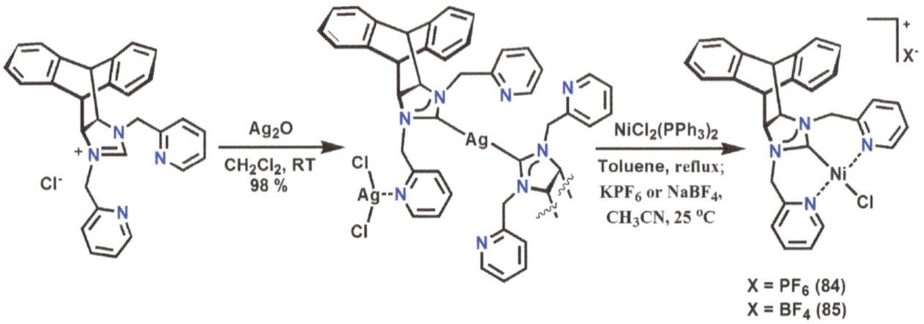

Scheme 33 Synthesis of NCN pincer NHC–Ni complexes **84** and **85**.

Morales-Morales and coworkers reported an unsymmetrical POCSP-Ni(II) pincer compound (Scheme 34).[57] The crystal structure of this complex was elucidated by SCXRD analysis, showing distorted square-planar geometry around the metal center. The synthesized POCSP–Ni(II) pincer complex was used to catalyze C–S coupling reactions, showing good catalytic activity and selectivity.[57]

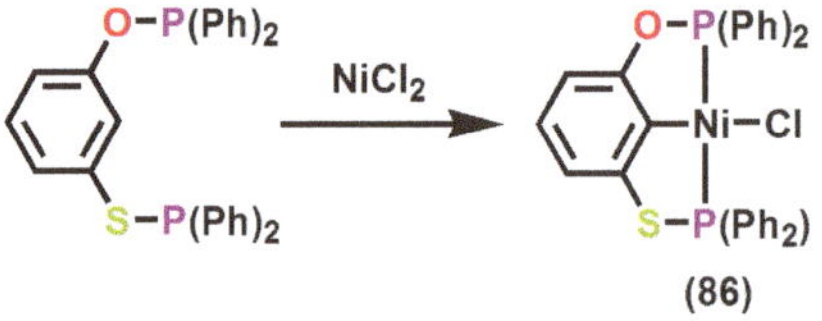

Scheme 34 Synthesis of unsymmetrical Ni(II) pincer complex **86**.

Beweries and coworkers synthesized symmetrical PSCSP and unsymmetrical POCSP-Ni(II) pincer complexes (Scheme 35).[58] These complexes were fully characterized, including SCXRD analysis, and showed activity as catalysts for Kumada coupling reactions.[58]

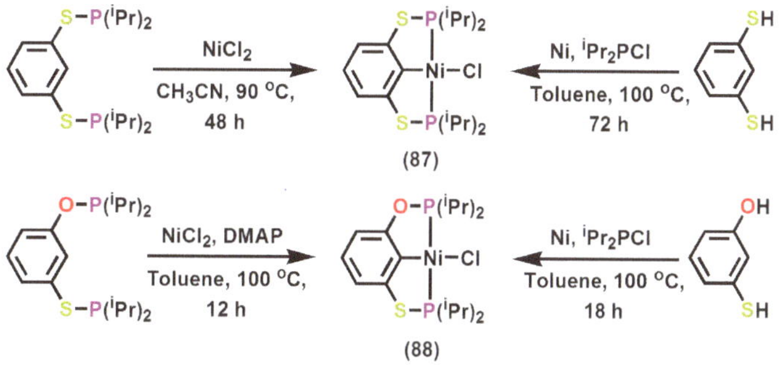

Scheme 35 General preparation of symmetrical and unsymmetrical Ni(II) pincer compounds **87** and **88**.

Furthermore, Beweries and coworkers reported PECEP-Ni(II) pincer complexes, which were explored as electrocatalysts for H$_2$ evolution in the presence of CH$_3$COOH and TFA (Scheme 36).[59]

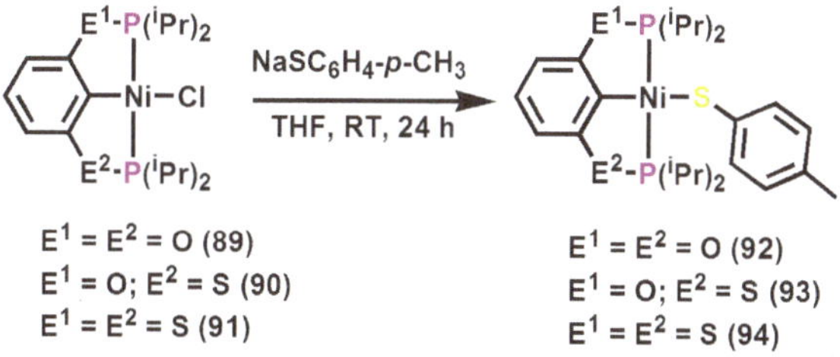

Scheme 36 Synthesis of Ni(II)–thiolate complexes from Ni–Cl precursor compounds.

However, there are few reports on SCS ligand metal complexes compared with complexes containing similar ligands.[104–106] The foremost SCS–Ni pincer complex was reported in 2008.[104] The crystallographic and electrochemical behavior of SCS pincer complexes shows that SCS ligands possess higher electron-donor ability compared with analogous SNS ligands.[105] Ni pincer metal complexes with a pyridine backbone instead of an aryl moiety have been widely exploited in catalytic alcohol dehydrogenation.[107]

Yamaguchi and coworkers reported acetylacetonato-based Ni(II) pincer-type complexes (Scheme 37).[60] These complexes were used as catalysts for Kumada–Tamao–Corriu reactions. Compound **96** was a particularly efficient catalyst among these complexes.[60]

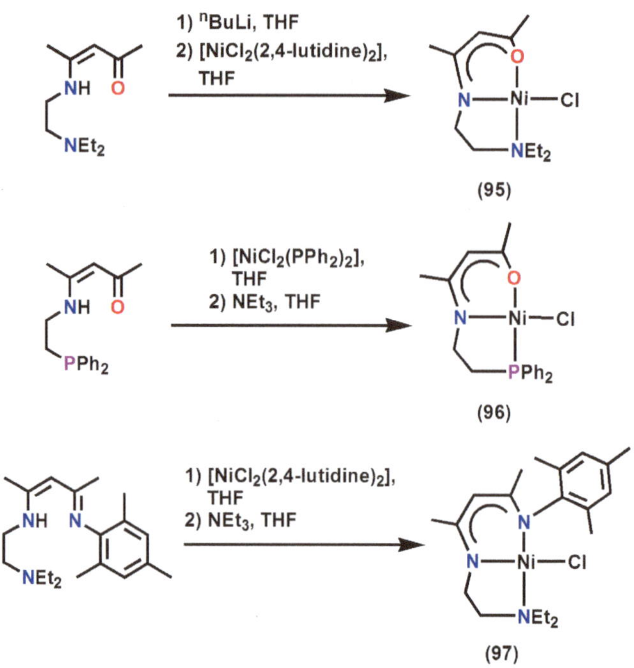

Scheme 37 Synthesis of ONN, ONP, and NNN Ni(II) pincer metal complexes **95–97**.

Yamaguchi and coworkers also reported β-diketiminato-based NNP–Ni(II) pincer-type complex **98** (Scheme 38)[61] and its reactivity in coupling reactions. SCXRD showed a molecular structure of the complex with a distorted square-planar geometry.[61]

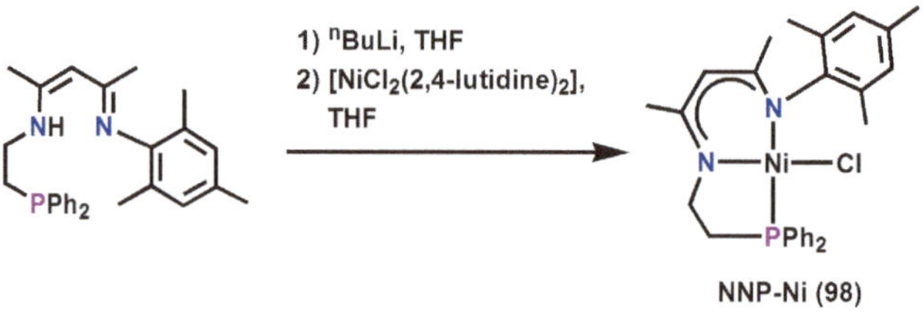

Scheme 38 Synthetic route to Ni(II) pincer metal complex **98**.

Kumar and coworkers reported the (^{iPr2}NNN)NiCl$_2$(CH$_3$CN)–Ni pincer complex **99** (Scheme 39).[62] The authors demonstrated the efficacy of this compound in the catalytic alkylation of amines by alcohols. Interestingly, yields of up to 98% were obtained under solvent-free conditions in alkylations of 2-aminopyridine with naphthyl-1-methanol and 4-methoxybenzyl alcohol.[62]

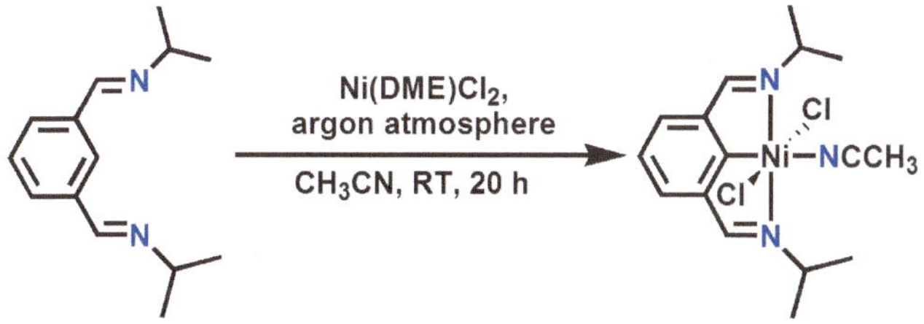

Scheme 39 Synthesis of complex (^{iPr2}NNN)NiCl$_2$(CH$_3$CN).

3.2 Palladium (Pd) pincer metal compounds

Traditionally, Pd catalysts such as bis(triphenylphosphine)palladium chloride, tetrakis(triphenylphosphine)palladium(0), PdCl$_2$(RCN)$_2$, Pd$_2$(dba)$_3$, (η^3-allyl-PdCl)$_2$, have been widely used in cross-coupling reactions and other organic transformations. These catalysts have some limitations, such as storage problems, sensitivity, and a low activity that requires high catalyst loading. As a recent development, Pd pincer complexes (Fig. 4) have been used in various catalytic reactions and developed into acceptable alternatives to conventional catalysts. Pd pincer metal complexes are widely used in various catalytic reactions, as summarized in this section.

Symmetric PCP pincer metal complexes are more easily obtained compared with asymmetric PCP complexes.[108,109] Heinekey reported noteworthy features of symmetric PCP–Pd complexes, producing the first Pd dihydrogen complex that exhibited σ-bond (H–H and C–H) interactions between the noninnocent PCP ligand and the Pd center under strong acidic conditions.[109] Furthermore, POCOP–Pd complexes have shown thermal stability and robustness that are similar to their phosphine analogs, but exhibit superior catalytic activity, achieving higher conversions.[109–114]

POCOP–Pd pincer complexes have also been synthesized by substituting resorcinol with suitable ClPR$_2$ and Pd precursors in the presence of base. Recently, Morales-Morales synthesized unsymmetrical ketone-functionalized POCOP–Pd(II) complexes **104–106** (Scheme 40).[65] These POCOP pincer-based Pd(II) compounds were applicable as catalysts in Suzuki–Miyaura reactions.[65]

Different donor atoms and steric behavior can be used to tune the properties of pincer ligands and induce potential ligand participation in pincer ligand metal chemistry. The synthesis of asymmetric pincer compounds has resulted in interesting reactivity, in which the ligand plays a significant role. Therefore, growing interest over the last decade has seen unsymmetric

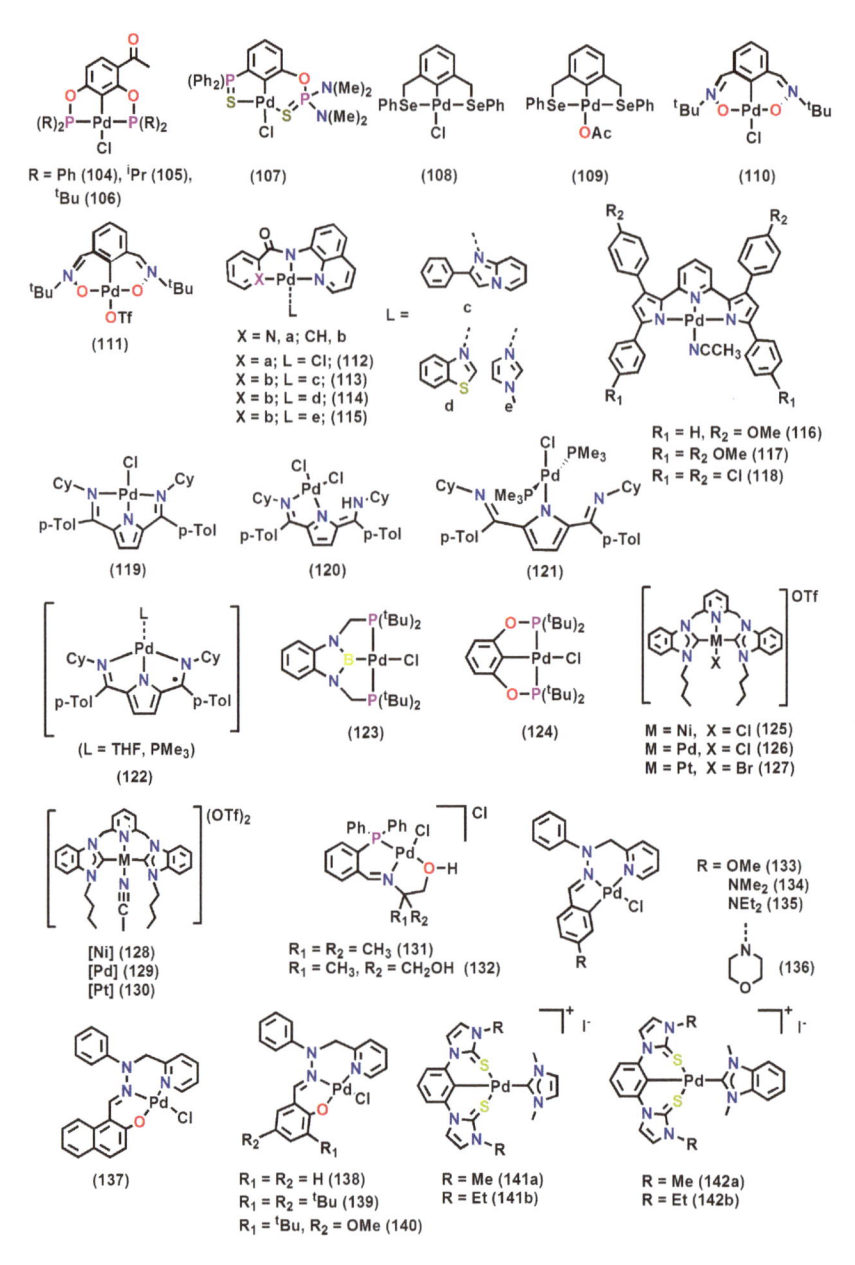

Fig. 4 Examples of Pd pincer complexes synthesized using various types of pincer ligand.

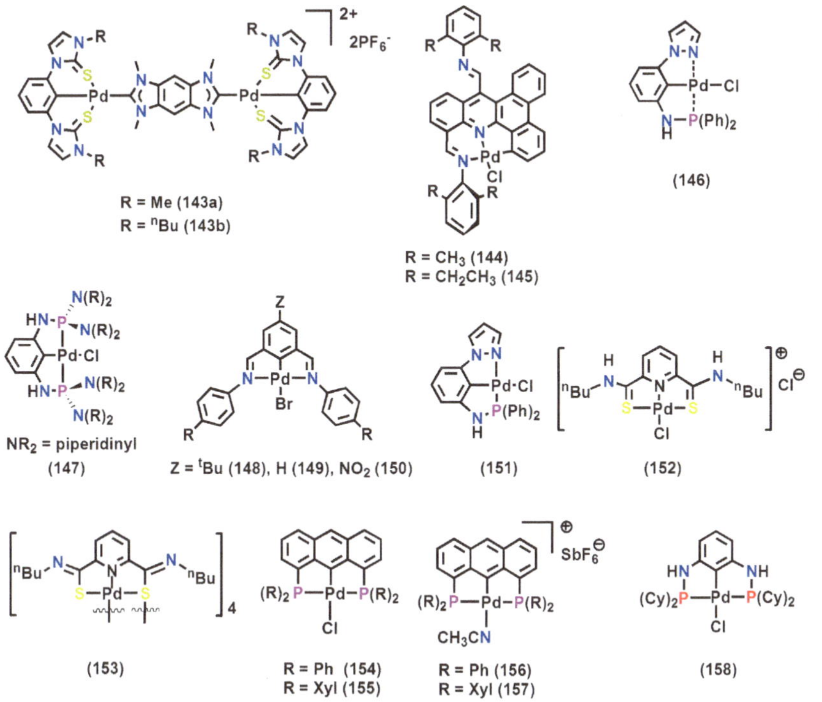

Fig. 4—Cont'd

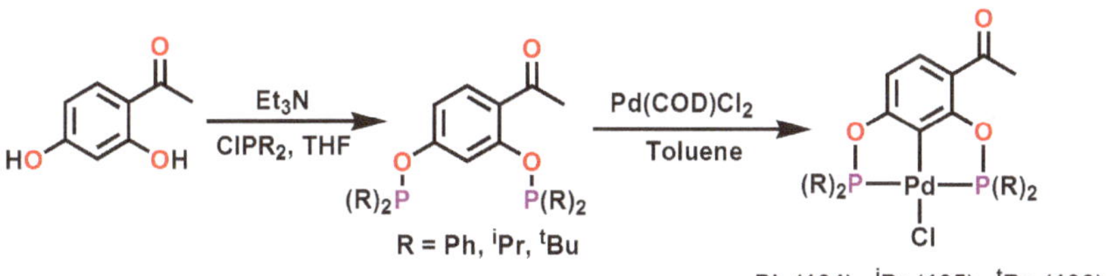

Scheme 40 General synthetic route to POCOP–Pd(II) compounds **104–106**.

pincer complexes explored extensively. However, these complexes have recently received less attention than symmetric pincer complexes owing to their multistep synthetic procedures, which afford low yields and make their preparation difficult.[112,113,115–117]

PCP–Pd(II) pincer metal complexes have gained further interest, while NCN–Pd(II) pincer metal complexes have shown exciting properties and catalytic reactivities in many organic transformations. In contrast to PCP–Pd pincer complexes, most NCN–Pd complexes are stable. Furthermore, NCN pincer metal complexes requireless synthetic efforts compared with other donor atoms, allowing easy tuning of the electronic and steric effects of NCN ligands by varying specific substituents.[118–121] The synthesis of chiral pincer ligands and their corresponding metal complexes result in efficient asymmetric catalysis. The common synthetic strategy used to prepare several PCP-based metal complexes involves treatment of the metal salts with PCP ligands in a reaction involving C-H bond activation at the coordination carbon sites due to strong electron donation of phosphine groups and chelation effects, while current synthetic methodology for NCN complexes has not been successful when employing similar NC(H)N ligand backbones.[122,123] More active ligands, such as N(C–X)N (X = Cl, Br, I) might be required.[122–124] Furthermore, the synthesis of NCN–Pd compounds is generally difficult owing to ligand decomposition, which occurs at normal reaction conditions. This type of activity has been observed for several imine ligands that are unstable and inappropriate for metalation reactions.[125]

Pd pincer complexes were reported in 1980 by Shaw and coworkers using dithioether-based pincer ligands.[126] Numerous symmetrical and unsymmetrical SCS-based pincer metal compounds have been explored using lateral Group 16 element donors, such as thioamides,[127] thioethers,[128,129] and phosphine sulfides.[130] Kozlov and coworkers disclosed unsymmetrical SCS-based pincer ligand and the corresponding unsymmetrical SCS–Pd pincer complex **107** (Scheme 41) prepared by an easy and unique method.[66] These complexes showed good catalytic activity for the Suzuki coupling reactions.[66] Although SCS pincer metal complexes have not been well explored compared with NCN and PCP pincer metal complexes, they have been used to catalyze many cross-coupling reactions,[129] and the borylation of allylic alcohols.[131] Furthermore, the foremost SeCSe–Pd pincer metal compound was reported in 2004 by Yao and coworkers. This complex **108** (Scheme 42) showed high catalytic activity toward Mizoroki–Heck cross-coupling reactions.[67] The catalytic reactivity of SeCSe–Pd pincer metal complex is similar to the analogs with phosphorus and sulfur donor atoms. Selenium-containing pincer metal complexes are easy to synthesize and show better reactivity, while selenium-ligated metal complexes (Scheme 42) are more stable toward air oxidation.[68] The synthesized SeCSe based Pd pincer complex (**109**) showed catalytic activity toward allylation of aldehydes with allyltributyltin.[68] The OCO pincer -molybdenum complex have somewhat limited applications owing to the weak donor ability of oxygen atoms and the donor ability of anionic oxygen atoms.[132] In contrast, Li et al. reported a modified OCO ligand with a nitrone-based donor group in which the dipolar nature of the nitrone caused the O atom to exhibit improved electron donor behavior, resulting in an OCO–Pd pincer compounds **110–111** (Scheme 43). This OCO–Pd pincer compound showed high catalytic activity in Kumada and Mizoroki–Heck reactions.[69]

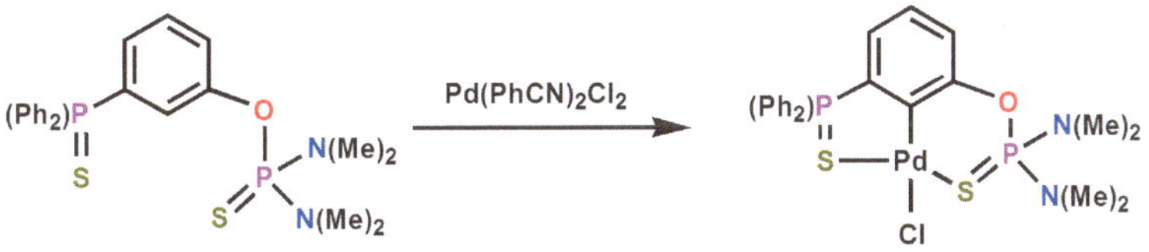

Scheme 41 General synthetic route of unsymmetrical SCS based Pd pincer complex.

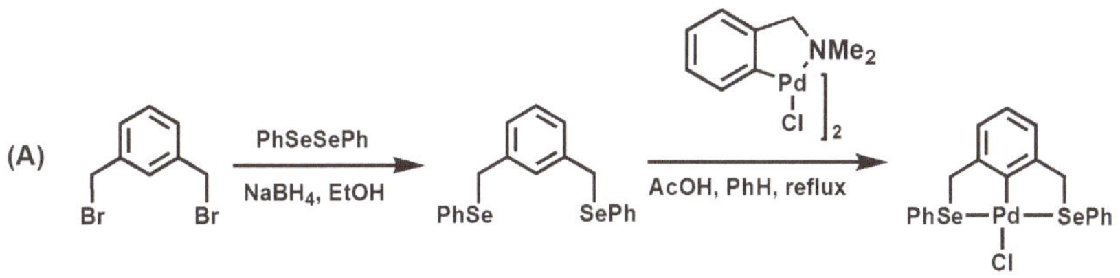

Scheme 42 General synthetic route of SeCSe based Pd pincer complex.

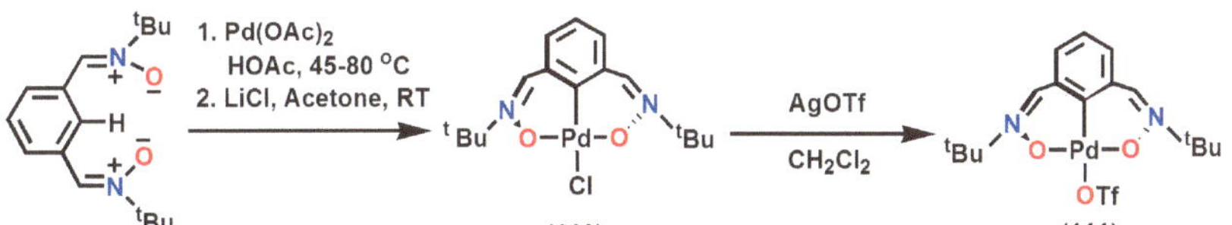

Scheme 43 General synthetic route of OCO based Pd pincer complex.

Joshi and coworkers reported Pd(II) pincer metal complexes with NNN and CNN coordination (Scheme 44),[70] their structures being elucidated by SCXRD analysis. These complexes are stable and applicable as catalysts to cross-dehydrogenative coupling (CDC) reactions.

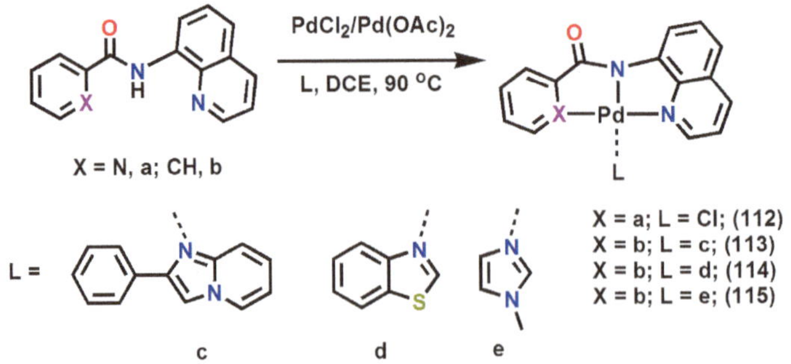

Scheme 44 Synthesis of Pd pincer complexes **112–115**.

Dash and coworkers have reported phosphine-free pincer-based Pd complexes (Scheme 45).[71] Their molecular structures showed distorted square-planar geometry. The synthesized NNN–Pd complexes were employed as catalysts in the Suzuki–Miyaura reaction, showing good catalytic activity.[71]

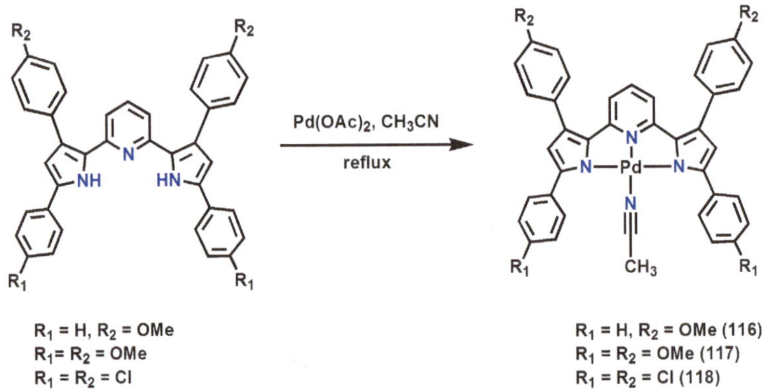

Scheme 45 General synthesis of NNN–Pd pincer compounds **116–118**.

Anderson and coworkers have reported the variable coordination modes of Pd complexes containing diamino–pyrrole ligands, and their redox activity and ligand protonation behavior (Schemes 46 and 47).[72] These pyrrole-based pincer ligands display interesting redox activity, protonation reactions and coordination abilities.

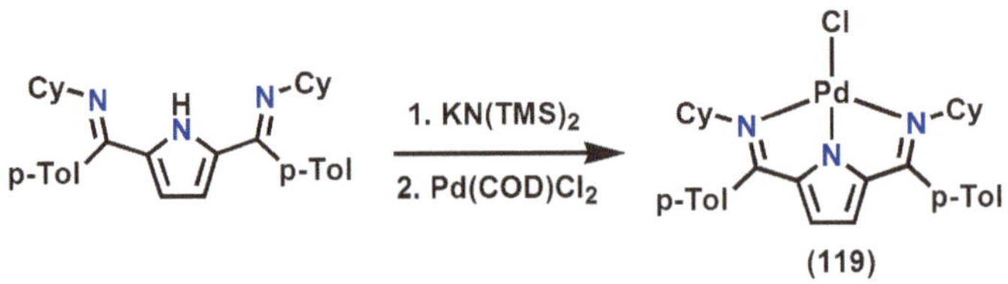

Scheme 46 General synthetic method of NNN–Pd compound **119** using a diamino–pyrrole ligand.

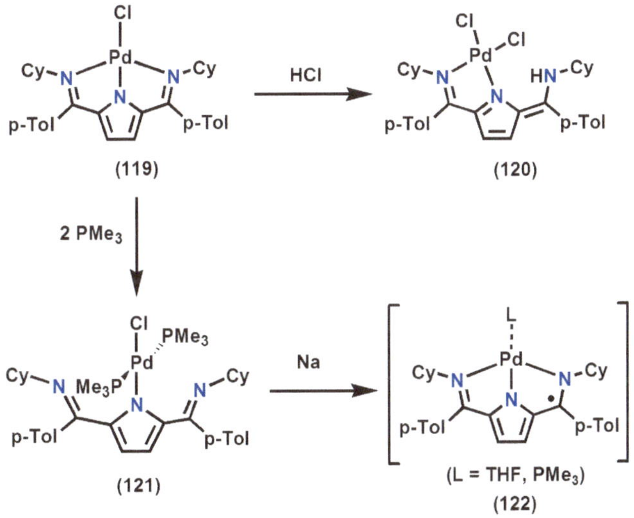

Scheme 47 Reactivity of complex **119** to give complexes **120–122**.

Chen and coworkers have disclosed the synthesis of Pd(II)–chloride complex **123** using a PBP-based pincer ligand (Scheme 48).[73] The same group also synthesized a POCOP–Pd complex using the method reported by Guan et al. using Pd(COD)Cl$_2$ instead of PdCl$_2$ as the Pd source.[133] The authors observed that PBP–Pd complex **123** was more electron-rich than POCOP–Pd complex **124**, and more prone to oxidize. Furthermore, complex **123** showed superior catalytic activity in the Suzuki–Miyaura coupling reaction compared with the POCOP–Pd pincer complex.[73]

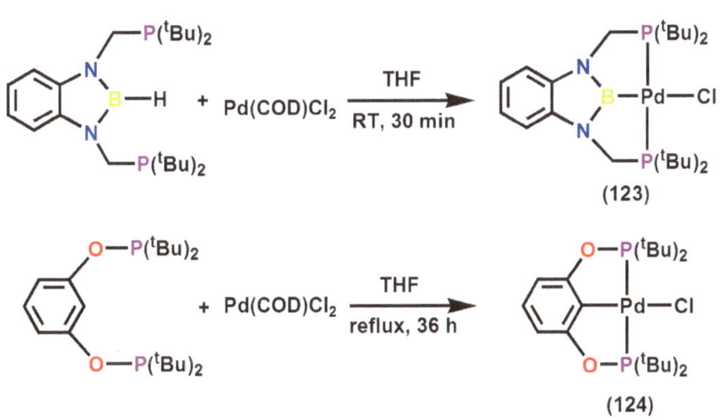

Scheme 48 Synthesis of PBP–Pd and POCOP–Pd complexes **123–124**.

Patrick and coworkers reported bis-NHC-based CNC metal (M=Ni, Pd, and Pt) complexes **125–130** (Scheme 49).[74] All complexes showed electrocatalytic reactivity toward carbon dioxide reduction, with DFT calculations also performed. The CNC–Pd complex showed better reactivity toward CO_2 reduction into CO as compared with the corresponding [Ni] and [Pt] complexes, the order of reactivity being [Pd] >> [Pt] > [Ni].[74]

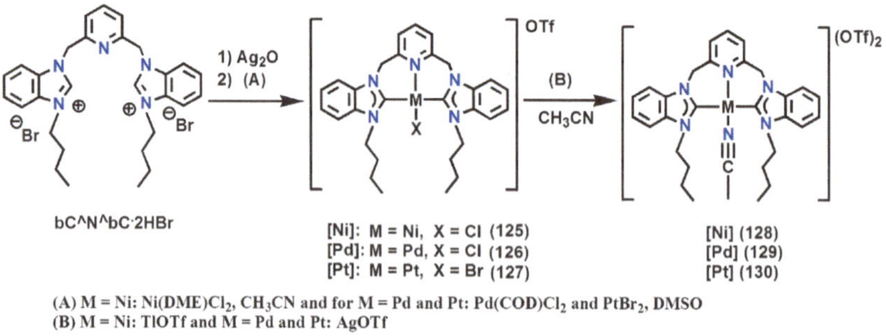

Scheme 49 General synthetic route to group 10 metal complexes supported by CNC pincer-type ligands.

Morales-Morales and coworkers reported the reactivity of Pd(II) pincer complexes bearing P–N–OH ligands (Scheme 50).[75] These ligands exhibited a dual nature through either bidentate P–N coordination or tridentate P–N–OH coordination to the metal. Their molecular structures were elucidated by SCXRD analysis. Interestingly, complexes **131** and **132** showed reactivity in microwave-assisted Suzuki–Miyaura reactions, with similar activities and catalytic conversions of up to 86%.[75]

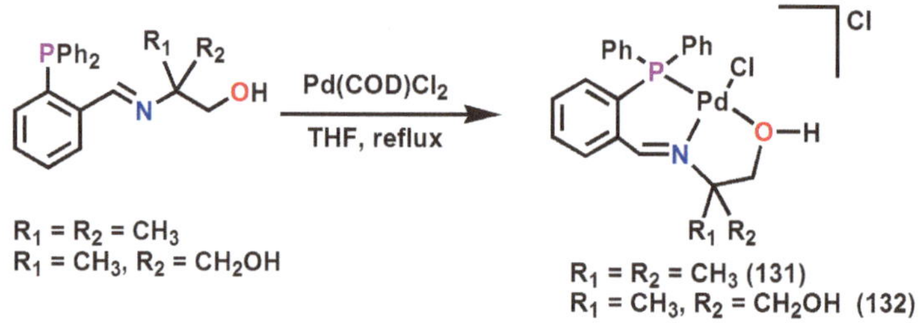

Scheme 50 Synthesis of Pd(II) complexes **131–132** bearing P–N–OH ligands.

Recently, Ghosh and coworkers synthesized Pd complexes using unsymmetrical XYC pincer-based ligands (Scheme 51).[76] The crystal structures of these compounds were elucidated by SCXRD analysis. These Pd pincer complexes were utilized as catalysts in the Mizoroki–Heck coupling reaction. Furthermore, Pd(0) nanoparticles were generated in situ during catalysis

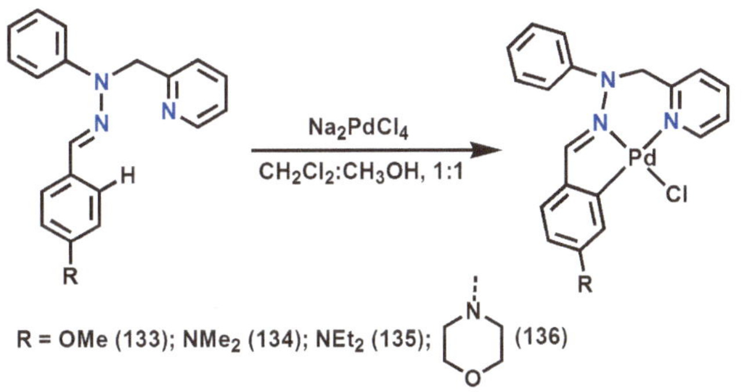

Scheme 51 General synthesis of palladacycles **133–136**.

and characterized by scanning electron microscopy (SEM), transmission electron microscopy (TEM), and X-ray photoelectron spectroscopy (XPS).[76]

Ghosh and coworkers also recently synthesized sterically hindered and unsymmetrical $N_{py}N_{im}O_{ph}$–Pd pincer complexes (Scheme 52).[77] The crystal structures of complexes **137** and **139** were determined by SCXRD analysis. Synthesized Pd pincer metal complexes **137–140** were applied to Suzuki–Miyaura and aldehyde allylation reactions.[77] Furthermore, the mechanism of the Suzuki–Miyaura reaction using these catalysts was investigated theoretically and experimentally. The allylation of aldehydes was also corroborated by theoretical calculations.

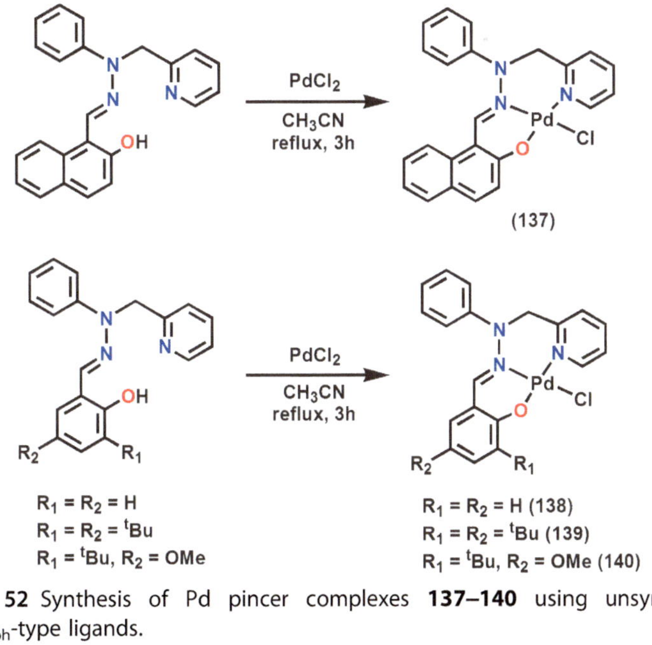

Scheme 52 Synthesis of Pd pincer complexes **137–140** using unsymmetrical $N_{py}N_{im}O_{ph}$-type ligands.

Recently, Han and coworkers synthesized and structurally characterized SCS–Pd(II) mononuclear and binuclear complexes by reacting SCS-based pincer ligands with N-heterocyclic carbenes (NHCs) (Scheme 53).[78] The regioselective reduction of quinolines by ammonia–borane was achieved using these SCS–Pd complexes as catalysts. Notably, the catalytic reactivity of mononuclear Pd complexes was lower than that of binuclear Pd complexes under similar reaction conditions.[78]

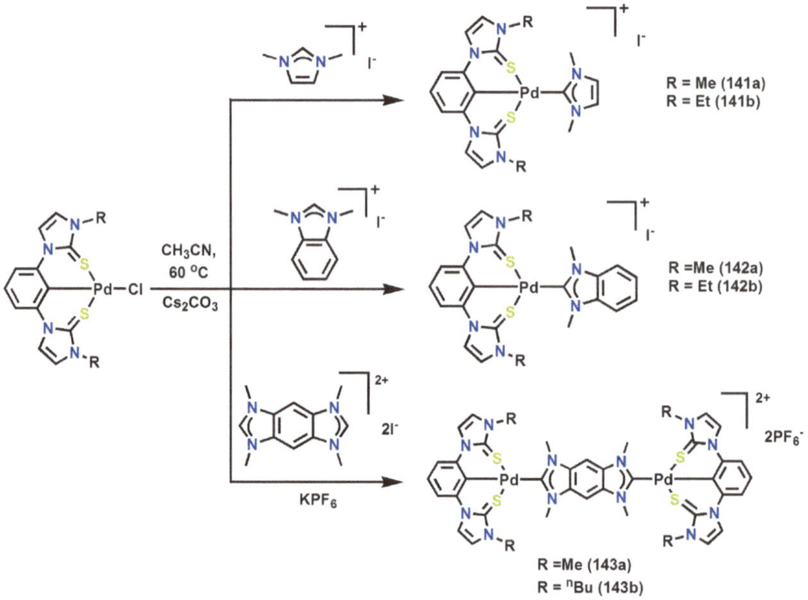

Scheme 53 Synthesis of SCS–Pd pincer metal complexes **141–143**.

3.3 Platinum (Pt) pincer metal compounds

Furthermore, Pt pincer metal complexes have been reported in which the donor atom used by the pincer ligand is altered, from CNN through NCN to CNC among others. Cyclometalated Pt complexes have been widely used as photoluminescent emitters. However, the reactivity of cyclometalated Pt compounds has been little explored. A noteworthy characteristic of Pd compounds in catalysis is their vast exploitation compared with Pt compounds. Pt(II) complexes have been less utilized in catalysis owing to the frequently observed less activity. Although Pt(II) is the most kinetically inert metal atom in the group, carboplatination occurs more readily than carbopalladation. Pt metal complexes partially activate substrates such as allenenes,[134] alkylidene cyclopropanes,[135] and 1,3-dienes.[136] In addition Pt frequently shows the opposite activity to Pd,[137] with non-pincer Pt

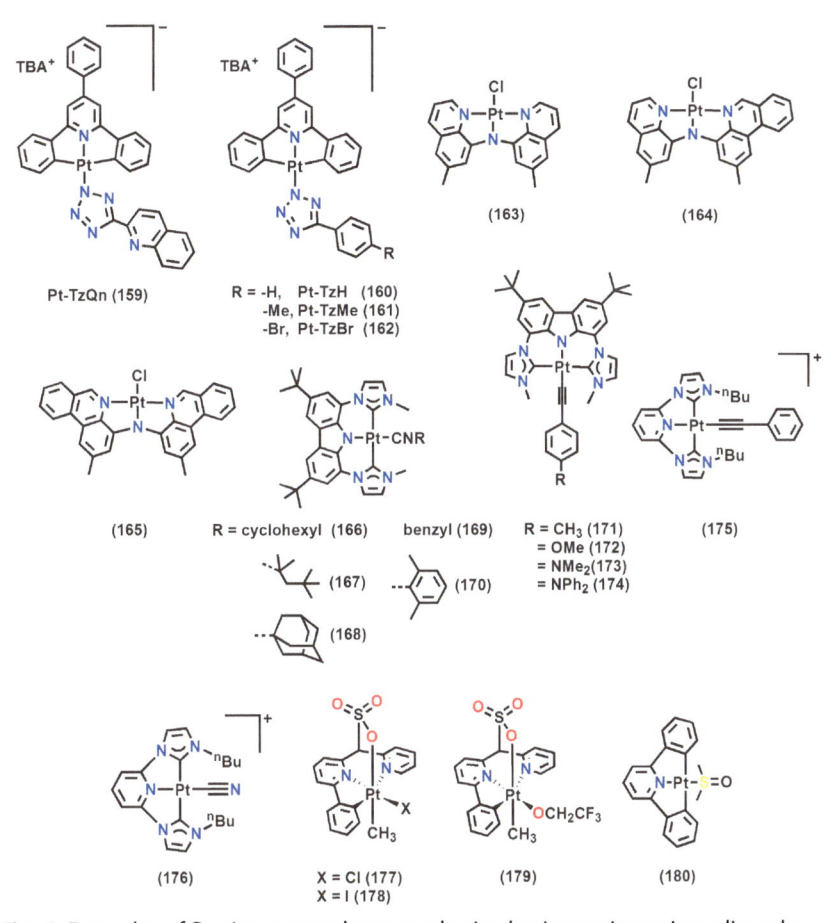

Fig. 5 Examples of Pt pincer complexes synthesized using various pincer ligands.

complexes effectively employed in hydroamination,[138,139] cycloisomerization,[38,140] and hydrovinylation[37,141] reactions. Recent developments in Pt pincer metal complexes (Fig. 5) are discussed in this section.

Massi and coworkers have disclosed the synthesis of Pt(II) CNC-based pincer metal complexes. These Pt complexes were synthesized by replacing dimethyl sulfoxide (DMSO) in the complex [Pt(CNC)(DMSO)] with the related tetrazolato moiety (Scheme 54). Furthermore, the photophysical properties of these complexes have been explored in solution and solid states. Pt(II) complexes in the solid state showed an emission band due to aggregation of the molecules, while no emission band was observed in the solution state. Therefore, the obtained Pt(II) compounds were used as catalysts in various photocatalyzed reactions, such as C–H functionalization reactions, atom transfer radical addition (ATRA) chemistry, and hydrodeiodination reactions.

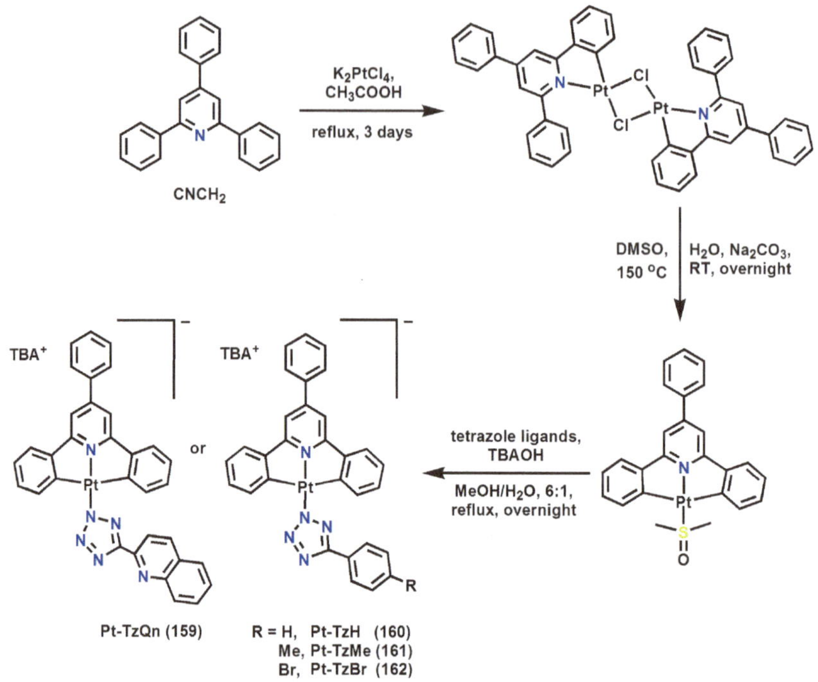

Scheme 54 Synthesis of anionic CNC–Pt(II) tetrazolato complexes **159–162**.

Anionic and neutral Pt–DMSO complexes have shown reactivity in photocatalyzed Povarov-type reactions. Furthermore, Pt(II) complexes have been employed as catalysts for photoredox-catalyzed radical addition reactions. These complexes were also applied as photocatalysts in hydrodeiodination reactions. The catalytic activity was also examined in the ATRA reaction of 1-octene and bromotrichloromethane.[87]

Herbert and coworkers synthesized luminescent Pt(II) complexes with anionic NNN amido-based ligands and N-donor heterocyclic groups (Scheme 55).[88] Phosphorescent NNN–PtCl complexes bearing symmetrical and unsymmetrical proligands were obtained. The emission behavior of the more π-extended (phenanthridinyl)amido complexes showed a blue shift in the emission spectra compared with less π-conjugated complexes. This phenomenon contrasted with conventional assumptions regarding π-extension. Furthermore, the unusual luminescent behavior of the complexes was corroborated using DFT and time-dependent (TD)-DFT calculations.[88] The NNN based Pt(II) pincer complexes **163–165** emitted red phosphorescence from monomolecular excited states. The synthesized phosphorescent Pt(II) compounds were the first reported redshifted complexes in the literature. This demonstrated π-extension of the benzannulated group by NNN amido-based ligands and their Pt(II) complexes **163–165**.[88]

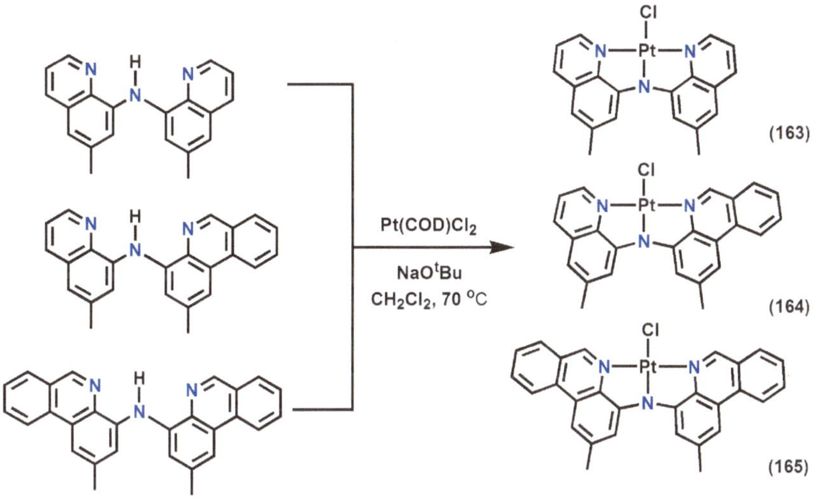

Scheme 55 General synthesis of Pt(II) complexes **163–165** using NNN amido pincer ligands.

The Gray group also reported Pt(II) complexes bearing CNC pincer ligands and various isonitrile ligands (Scheme 56).[89] SCXRD analysis showed square-planar geometry with little distortion around Pt. Interestingly, the synthesized Pt(II) complexes showed in the solid state biexponential phosphorescence decay with a lifetime of up to 25 ms. In contrast, these complexes showed less emissive photophysical behavior and nanosecond-scale lifetimes in the solution state. Furthermore, DFT calculations showed that the highest occupied and lowest unoccupied molecular orbitals (HOMO and LUMO, respectively) were located on the NNN pincer backbone.[89]

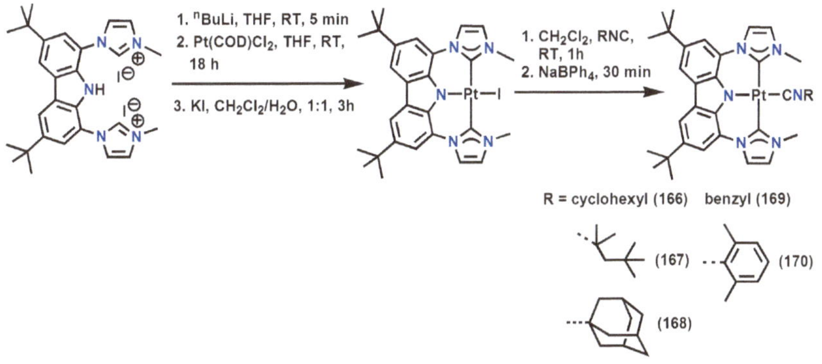

Scheme 56 General synthetic route to isonitrile–Pt(II) complexes **166–170** using CNC pincer ligands.

The same group also reported four-coordinate room-temperature phosphorescent Pt(II) complexes with CNC-based (BIMCA) pincer ligands. The synthesized Pt compounds were efficient photoemitting materials with a long lifetime (Scheme 57).[90] The authors synthesized four bisimidazolylcarbazolide (BIMCA)–Pt phenyl acetylide complexes that showed green emission (λ_{max} = 507–540 nm) and millisecond-scale lifetime decays. Theoretical calculations showed that the CNC pincer (BIMCA) ligands dominate frontier orbitals with the first Franck-Condon singlet and triplet excited states.[90]

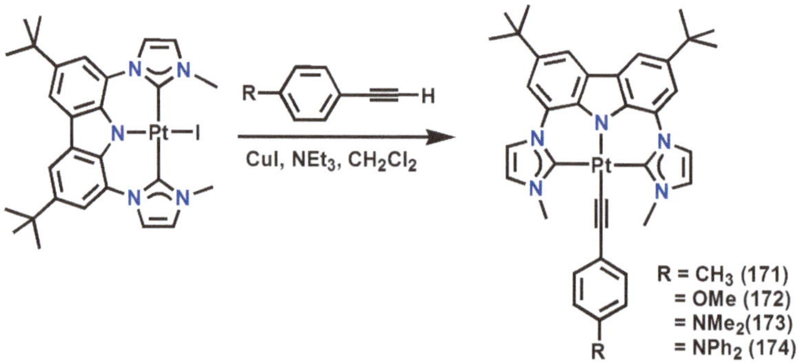

Scheme 57 Synthesis of phosphorescent bisimidazolylcarbazolide (BIMCA) ligand-based Pt(II) complexes **171–174**.

Che reported the incorporation of metal organic framework (MOF) materials into pincer metal complexes (M=Pt(II)), which afforded phosphorescent composite Pt(II)@MOF materials.[91] The different loading of Pt(II) complexes (**175, 176**) with MOFs led to composites such as 175@MOF and 176@MOF. These composite materials were used as catalysts for dehydrogenation reactions through photocatalysis. They have shown better achievements in various catalytic reactions, such as C–C bond formation, and dehydrogenation and dehydrogenative cyclizations, compared with related Pt(II) compounds in solution (Scheme 58).[91] The incorporation of MOFs into organometallic compounds might provide a new research area in which synthesized heterogeneous materials can be used as catalysts for various photocatalytic reactions and promising phosphorescent materials.[91]

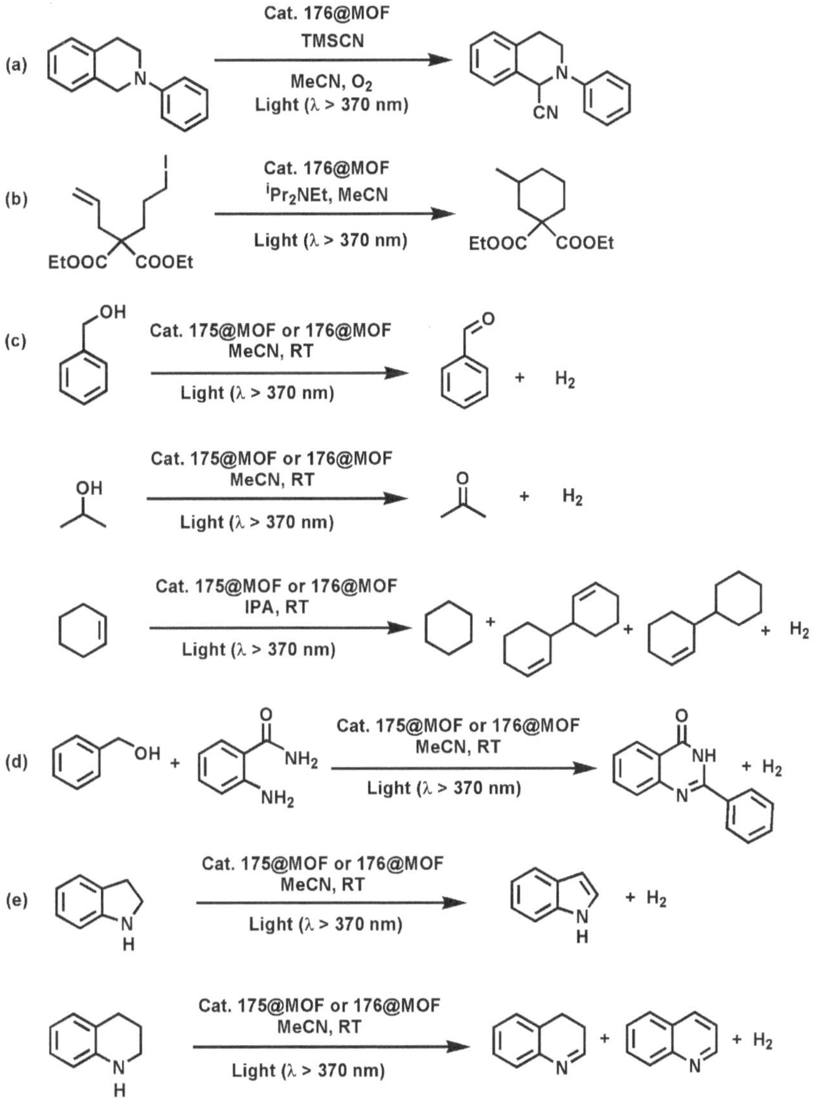

Scheme 58 (A) Amine α-cyanation, (B) cyclization reactions photocatalyzed by the composite material **176@MOF**; (C) dehydrogenation reactions of alcohols and cyclohexene photocatalyzed by **175@MOF**, **176@MOF**; (D) coupling (dehydrogenative) reactions between benzyl alcohol and o-aminobenzamide catalyzed by **175@MOF** and **176@MOF**; and (E) dehydrogenation reaction photocatalyzed by **175@MOF** and **176@MOF**.

Vedernikov and coworkers reported the synthesis and characterization of Pt(IV) complexes using sulfonated CNN-based pincer ligands (Scheme 59).[92] The synthesized complexes showed reactivity toward various nucleophiles, including H_2O, $CF_3CO_2^-$, Me_2SO, and $PhNMe_2$, relevant for the development of related C–X coupled compounds (Scheme 60). The observed reactivities of the complexes in S_N2-type reactions were in the order **177 > 178 > 179**.[92]

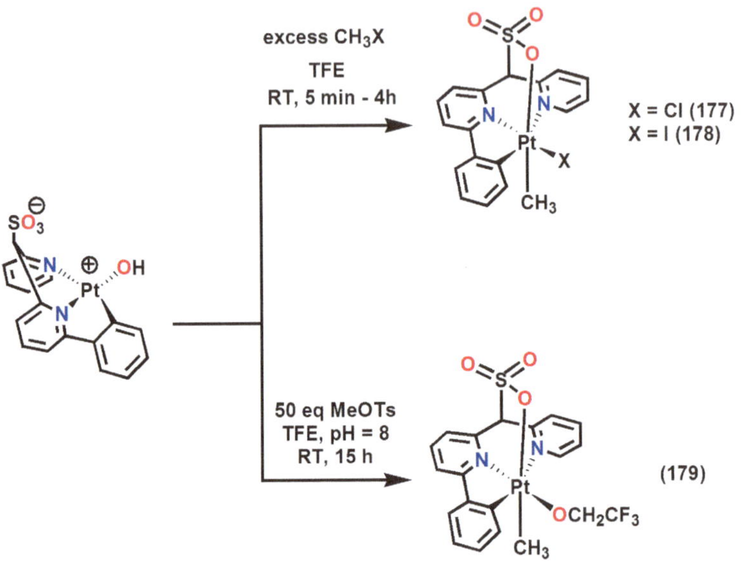

Scheme 59 General synthesis of complexes **177–179** using sulfonated CNN pincer ligands.

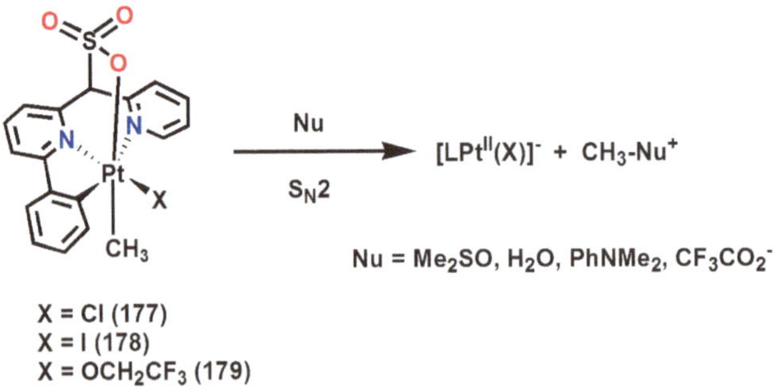

Scheme 60 Reactivity of aryl MePt(IV) complexes toward C–X reductive elimination.

Klein and coworkers synthesized cyclometalated Pt(II) compounds using CNC-based pincer ligands. The synthesized Pt(II) pincer compounds were fully analyzed by spectroscopic methods and crystal structures were elucidated by SCXRD. The authors reported the luminescent behavior, electrochemical study, and antiproliferative properties of these compounds. The luminescent properties were also corroborated by theoretical calculations (Scheme 61).[93]

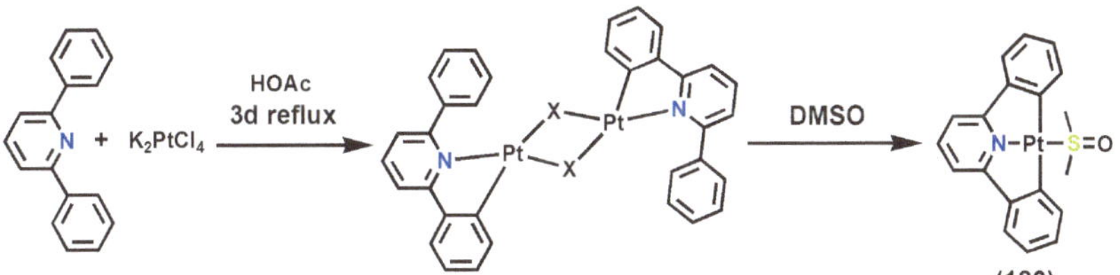

Scheme 61 General synthetic route to cyclometalated Pt(II) pincer complexes.

4. Catalyzed cross-coupling reactions

C–C bond formation is the most prominent application of cross-coupling reactions in organic synthesis.[16–18,21,23,35,54,57,60,76,96] Mostly C–C bond formations have been accomplished using an organic electrophile and nucleophile in the presence of organometallic catalysts. Group 9 and 10 organometallic compounds have been widely used as catalysts.[2,116,142] The extraordinary stability of pincer compounds has resulted in their use as catalysts in wide-ranging organic transformations and industrial applications. For example, Suzuki–Miyaura, Heck, Sonogashira, Negishi, Stille, and Hiyama cross-coupling reactions have been well explored in the literature.[21,22,79,81,143–149]

The three common steps in the catalytic cycles of coupling reactions are oxidative addition (OA), reductive elimination (RE), and transmetalation. Commonly, Pd-catalyzed reactions are performed using Pd(0) species resulting from complex decomposition. However, several group 10 pincer metal complexes have been employed directly as catalysts for reactions that have been studied in considerable detail.[2,116,150,151] The redox activity of Pd complexes for Pd(0) formation has been found to result from Pd–C bond cleavage or simple compound decomposition. Furthermore, the oxidation of Pd(IV) requires a strong oxidant because it is a thermodynamically disfavored process. Despite the extensive use of coupling reactions, mechanistic insight has been difficult to obtain using only experimental data. Therefore, detailed mechanistic behavior has also been proved using DFT modeling.[6,152–154]

4.1 Mizoroki–Heck cross-coupling reactions

The Mizoroki–Heck reaction[67,79,118,119,123,137,155,156] is a significant application of group 10 organometallic catalysts. Many research groups have been dedicated to expanding the scope of coupling reactions for C–C bond formation using numerous catalytic methods. A common Mizoroki–Heck coupling reaction involves an alkyl halide derivative (R–X, X=Cl, Br, or I) and an α-olefin. The most notable C–C bond formation is observed when an electron-withdrawing group is attached to the olefin.

Most Heck reactions are catalyzed by Pd(0) or Pd(II) species in the presence of a supporting ligand, where the highly reactive species or reaction intermediate formed during catalysis is sensitive and unstable. To overcome this drawback, robust and effective catalysts have been developed by Milstein and coworkers.[155] The synthesized PCP–Pd complexes acted as catalysts for the Heck cross-coupling reaction, achieving a high turnover number (TON) without any catalyst degradation.[155] The pincer ligand-based metal complexes showed high stability under harsh reaction conditions and afforded good turnover numbers, which inspired the expansion of reactions catalyzed by pincer metal compounds. Furthermore, Sun and coworkers have been reported bisimino-functionalized dibenzo [a,c]acridines pincer (NNC) based palladium complexes. These complexes showed high catalytic activity in Heck cross coupling reactions (Scheme 62) with high turnover number (174000).[79]

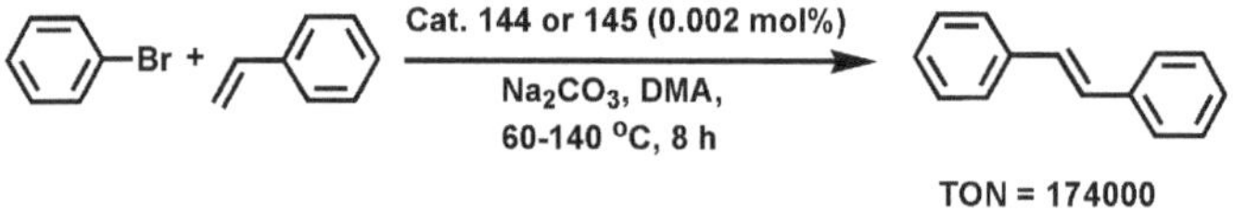

Scheme 62 Heck reaction catalyzed by palladium(II) pincer complexes.

Singh and coworkers reported the synthesis and catalytic reactivity of SeNSe–Pd complexes with a high turnover number in 2009.[156] Another notable study by Ahn and coworkers[118] reported a fused six-membered metallacycle NCN-Pd-type pincer complex.[118] This NCN–Pd pincer complex showed good catalytic reactivity in the Heck reaction, affording a high TON, turnover frequency (TOF), and yield. Furthermore, the

NCP–Pd pincer complex was employed in Heck cross-coupling reactions, exhibiting good to moderate activities in both water (13%) and DMF (69%) as solvents (Scheme 63).[80]

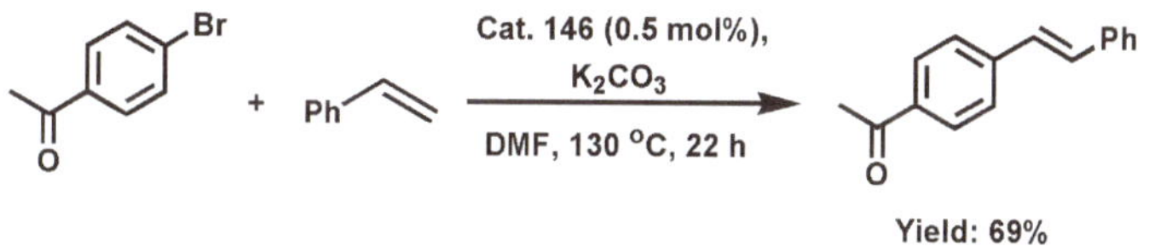

Scheme 63 Heck coupling reaction catalyzed by NCP based Pd pincer complex.

Recently, Ghosh and coworkers reported palladacycles **133–136**. These newly synthesized Pd pincer compounds were used as catalysts to explore Heck coupling reactions of aryl halides with acrylate substituents. A probable reaction pathway and mechanism were proposed (Scheme 64).[76] In the catalytic reactions, Pd(0) species were observed in the catalytic cycle, as analyzed by XPS, SEM, and TEM. Interestingly, the gram-scale synthesis of octinoxate was catalyzed by complex **134** (Scheme 65). Octinoxate is used as a UV-B sunscreen representative.[76]

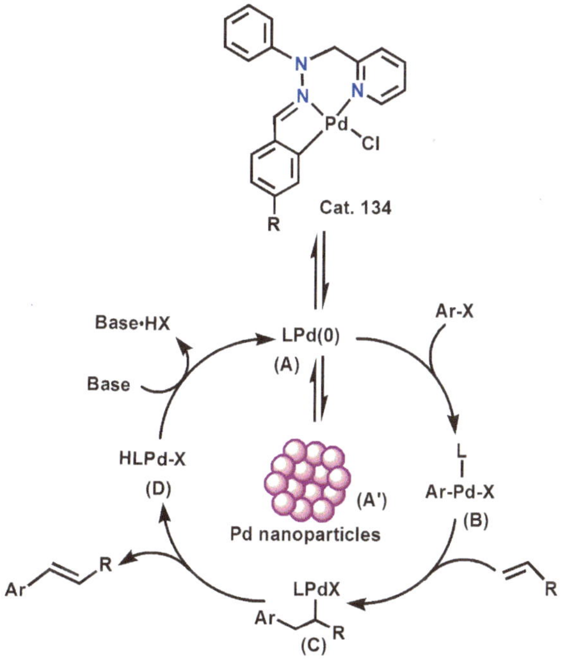

Scheme 64 General mechanism of Mizoroki–Heck reactions.

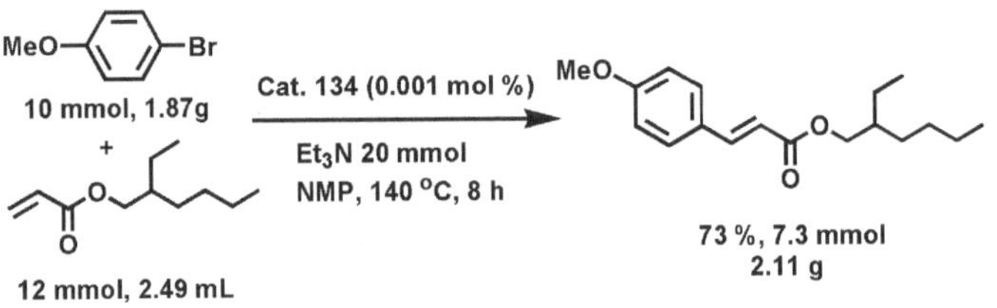

Scheme 65 Gram-scale synthesis of octinoxate catalyzed by a Pd pincer complex.

4.2 Suzuki–Miyaura cross-coupling reactions

Suzuki–Miyaura cross-coupling reactions have been widely used in catalysis to produce various useful substrates from organoboron compounds and aryl halides in many fields, such as pharmaceuticals, agriculture, and natural product synthesis. The formation of a conjugated diene in the presence of tetrakis(triphenylphosphine)palladium(0) as catalyst and base was first reported by Suzuki and Miyaura.[157]

Suzuki–Miyaura (SM) cross-coupling reactions are the most important multifunctional synthetic approach to the synthesis of biaryl compounds. The SM reaction occurs between aryl/alkyl halides or triflates and boronic acids in the presence of base. Pd(II) pincer complexes have been widely used in the SM cross-coupling reaction. Useful compounds have been synthesized by the effective utilization of pincer-containing ligands, including PCP, DCD, hybrid DCD' (D=donor group), and pyridine-based DND' ligands, in SM reactions. PCP pincer metal complexes are the most prominent pincer compounds used in this catalytic reaction.

Complexes **123** and **124** (see Fig. 4) have been utilized as catalysts in SM reactions. The catalytic activity was optimized for SM reactions (Scheme 66) using a low catalyst loading (0.1 mol%). Reactions between aryl halides (X=Cl, Br, or I) and boronic acids in the presence of base (K_3PO_4) were tested. Complex **123** was found to be a much more active catalyst than **124**. Complex **123** achieved a TOF of 582 h^{-1} compared with a TOF of 71 h^{-1} for complex **124** under the same reaction conditions. Furthermore, a much lower catalyst loading (0.002 mol%) was needed for complex **123** in SM reactions, confirming that compound **123** was the better option for SM reactions.[73]

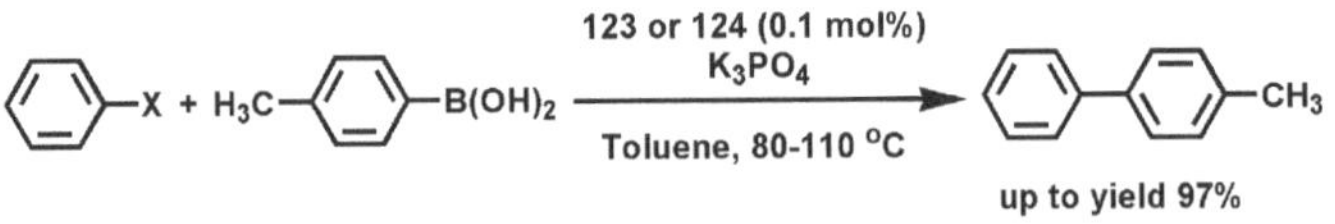

Scheme 66 Suzuki–Miyaura coupling reactions catalyzed by Pd complexes **123** and **124**.

4.3 Sonogashira cross-coupling reactions

Sonogashira reactions have been used extensively for C–C bond formation between arylalkynes and conjugated enynes.[80–82,158] The most common procedure for Sonogashira cross-coupling reactions involves reacting terminal acetylenes and aryl or alkenyl halides in the presence of Pd(0) and copper(I) catalysts with base. This reaction has been widely used to produce natural products, medicines, heteroarenes, and electronics in polymer chemistry or nanochemistry.[158] Mechanistic studies have shown that the main role of the Cu(I) salt in the catalytic cycle is to transfer the alkynyl moiety to the Pd complex via the Cu-mediated acetylide moiety generated in situ, which is known as transmetalation.[158] Owing to the synthetic importance of this reaction, much research has focused on the development of Pd–PCP pincer complexes and their chemistry. Frech[81] investigated aminophosphine-based pincer metal complexes as potential catalysts for Sonogashira reactions (Scheme 67).[81] Interestingly, in this report the Sonogashira coupling reactions were carried without co-catalysts such as CuI, FeCl$_3$ and ZnCl$_2$, amines and other common additives such as TBAB (tetrabutyl ammonium bromide). The model catalytic reaction was performed by using ethylene glycol as solvent and K$_3$PO$_4$ as a base. When copper was not used as cocatalyst, the reaction afforded quantitative yields with a TON of 2×10^6. Recently, Wang and coworkers[82] reported imine-based pincer metal complexes as suitable catalysts for Sonogashira cross-coupling reactions (Scheme 68).[82] In the catalytic reaction the electron-withdrawing nitro-substituted palladium complex showed better catalytic activity in comparison with the ^{t}Bu or H substituted complexes.

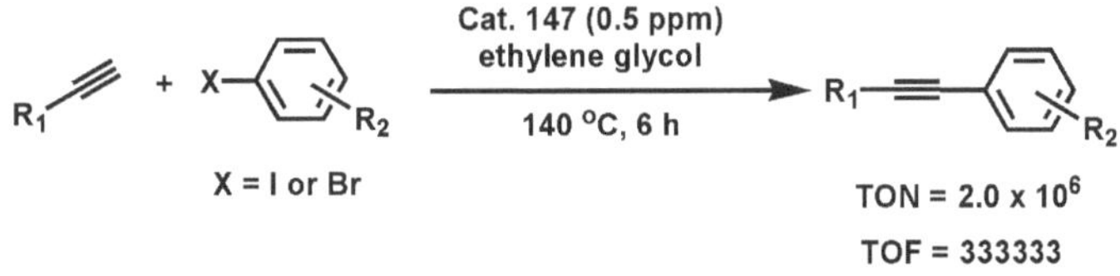

Scheme 67 Sonogashira cross coupling reactions catalyzed by PCP-based palladium pincer complexes.

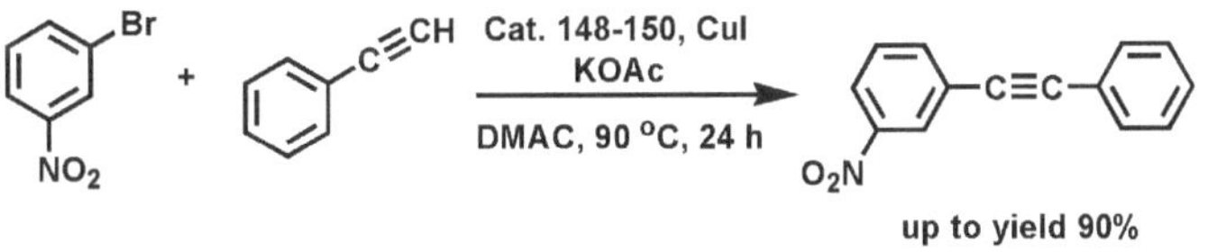

Scheme 68 Sonogashira cross coupling reactions catalyzed by NCN-based palladium pincer complexes.

4.4 Kumada cross-coupling reactions

The Kumada coupling reaction was first disclosed in 1972 by Makoto Kumada[159] and Robert Corriu.[160] These catalytic reactions involve organozinc or organomagnesium reagents, and alkyl or aryl halides.[23,161–165]

As Pd complexes promote various catalytic reactions, Ni metal complexes were also explored as catalysts.[24,166,167] The importance of Ni-catalyzed reactions results from the low toxicity, low cost, and earth abundance of Ni metal compared with Pd.[168] However, in contrast to Pd pincer complexes, Ni-based pincer catalysts have rarely been used in cross-coupling reactions.[166,169–171]

Regarding pincer-containing catalyst with group 10 metals, Beweries and coworkers examined the activity of synthesized Ni compounds **87** and **88** as catalysts in Kumada reactions between haloarenes and p-tolyl magnesium bromide under mild reaction conditions (Scheme 69).[58]

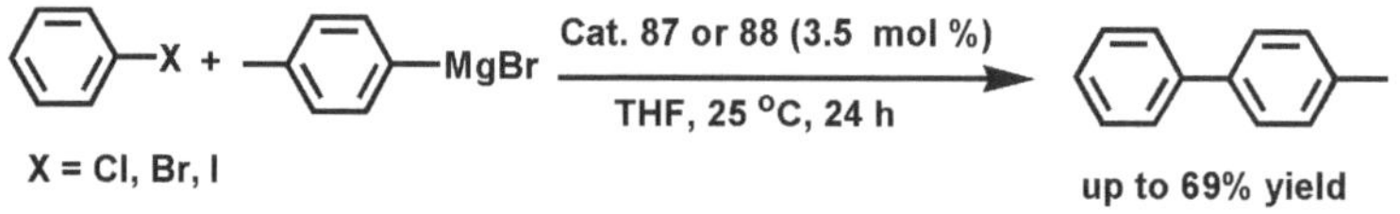

$$\text{Ph-X} + \text{p-Tol-MgBr} \xrightarrow[\text{THF, 25 °C, 24 h}]{\text{Cat. 87 or 88 (3.5 mol \%)}} \text{Ph-Tol}$$

X = Cl, Br, I

up to 69% yield

Scheme 69 Cross-coupling reactions catalyzed by complexes **87** and **88**.

4.5 Hiyama and Negishi coupling reactions

The Hiyama and Negishi cross-coupling reactions are emerging synthetic transformations [2,21,22,83,149,164,165,169] compared with the Mizoroki–Heck, Sonogashira, and Suzuki–Miyaura reactions. Pd pincer catalysts have been utilized as catalysts in Hiyama and Negishi coupling reactions. Arriortua et al. reported the most prominent Hiyama cross-coupling reactions in 2008, using NCP-Pd pincer compounds as catalysts (Scheme 70).[83] The NCP-Pd pincer complex was utilized as catalyst for the Hiyama, Suzuki and Sonogashira cross coupling reactions. The Hiyama coupling reaction was performed by reacting bromobenzene and silane compounds in the presence of NaOH as base in water to give the corresponding products in high yields.

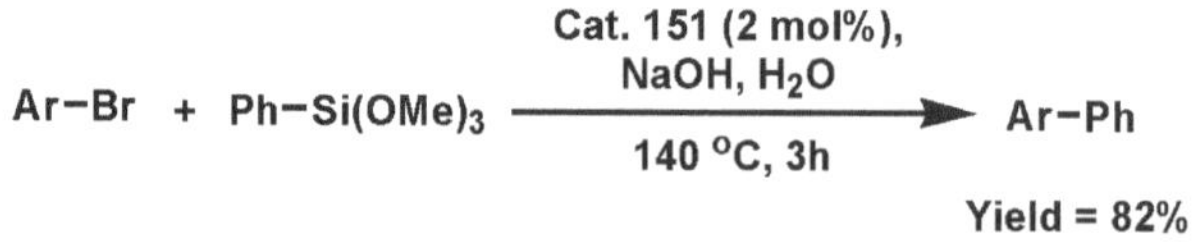

$$\text{Ar-Br} + \text{Ph-Si(OMe)}_3 \xrightarrow[\text{140 °C, 3h}]{\text{Cat. 151 (2 mol\%), NaOH, H}_2\text{O}} \text{Ar-Ph}$$

Yield = 82%

Scheme 70 Hiyama cross coupling reactions catalyzed by NCP based palladium pincer complex.

Lei and coworkers reported Negishi cross-coupling reactions of primary and secondary alkylzinc compounds bearing β-hydrogen atoms catalyzed by Pd–SNS pincer compounds (Scheme 71).[84] The SNS pincer-based Pd complexes showed good catalytic activity at a low catalyst loading with a high TON. This catalytic reaction is also suitable for industrial use owing to its good yield and selectivity. A large scale (19.35 g) reaction was performed using very low catalyst loading and showed high TON (6100000). Mechanistic investigations of this coupling reaction showed that the catalyst had good stability.

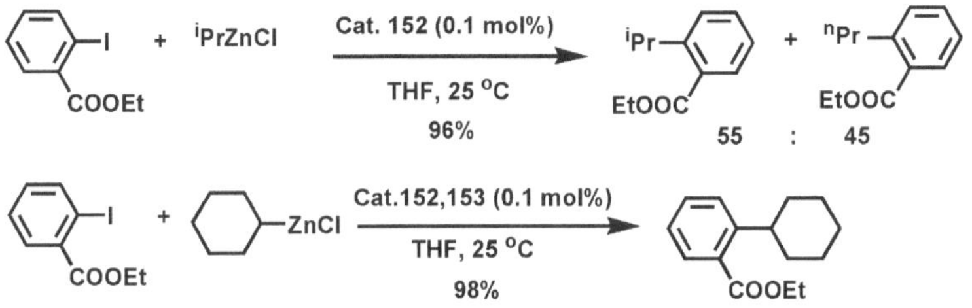

Scheme 71 Negishi cross-coupling reactions catalyzed by SNS-Pd complexes.

5. Miscellaneous reactions

Costly metals (Rh, Ir, Pt, Pd, Ru) are widely used in hydrogenation reactions owing to their selectivity, activity, robustness, and easy handling. In the last decade, noble metal catalysts have been replaced by more earth-abundant 3d metal atoms (Mn, Fe, Co, Ni), resulting in increased focus on the development of less costly pincer metal complexes as catalysts for hydrogenation. Recently, Schneider and coworkers published a good review of hydrogenation/dehydrogenation catalysis using 3d metal complexes with functionalized pincer ligands.[172] As an example of nickel-based catalysts, Sun and coworkers reported using unsymmetrical CNN–Ni pincer complexes to catalyze the transfer hydrogenation of ketones (Scheme 72).[63,64] The activity of complexes (**100**) and (**102**) was better than that of (**101**) and (**103**). The reaction conditions were optimized using NaO^{t}Bu in ^{i}PrOH at 80 °C with a 2% catalyst loading of (**100**) and (**102**), affording 77–98% yields of aromatic alcohols.[63,64]

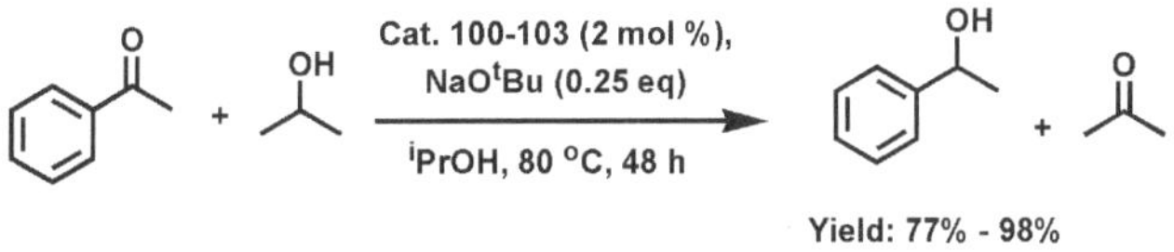

Scheme 72 Transfer hydrogenation reactions catalyzed by unsymmetrical CNN–Ni pincer complexes.

Recently, Wendt et al.,[52] reported a ($^{t\text{-}Bu}$PCN$^{i\text{-}Pr}$)NiMe complex with a sterically demanding unsymmetrical pincer ligand. The synthesized bulky pincer Ni–Me complex showed higher reactivity for CO_2 utilization in the carboxylation of a Ni–C bond than previously described Ni pincer complexes, and similar reactivity to the PCP–Pd–Me complex. The synthesized Ni pincer complexes catalyzed the carboxylation of C–C bond forming reactions.

The carboxylation mechanism was corroborated by DFT calculations, with two feasible routes identified, as follows: (a) Attack of CO_2 at the M–C bond and (b) 1,2-insertion (Scheme 73). These results were comparable with those of similar complexes PCP–NiMe and PCP–PdMe. Route (a) was determined to be lower in energy, with the calculated $\Delta H^\ddagger$ and $\Delta S^\ddagger$ values agreeing with those obtained experimentally.[52]

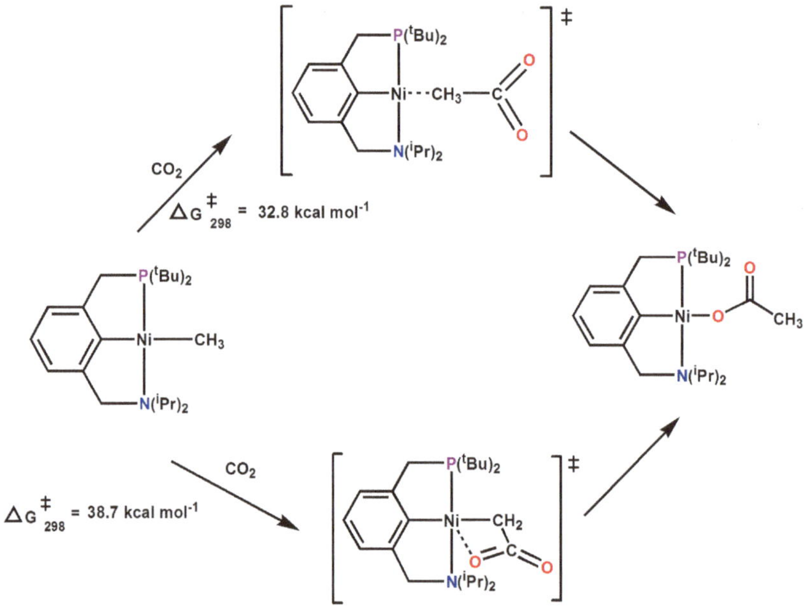

Scheme 73 Probable mechanisms for carboxylation reaction: (A) CO_2 attack and (B) 1,2-insertion.

Other important applications of group 10 pincer metal complexes include intermolecular hydroamination, hydrosilylation, and reduction reactions of some other functional groups. Recently, Leitner and coworkers reported the catalytic exploitation of cationic Pd(II) anthraphos PCP-pincer compounds in the intermolecular hydroamination reaction of aromatic alkynes with aromatic amines (Scheme 74). This catalytic reaction proceeded with a low catalyst loading under neat reaction conditions. Alkynes with a broad substrate scope were selectively reacted with primary aromatic amines to afford valuable imines through Markovnikov product formation.[85]

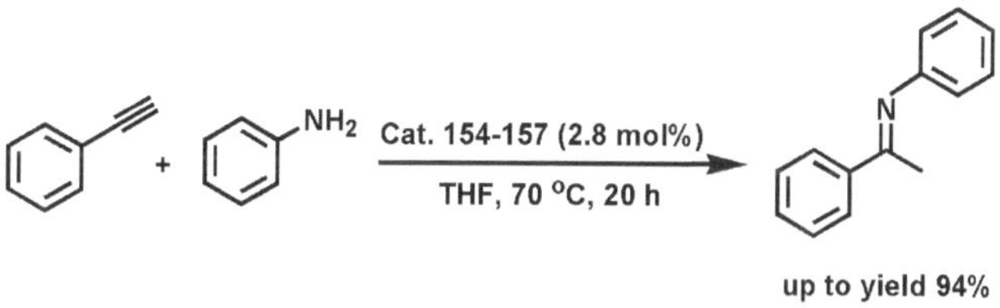

Scheme 74 Hydroamination of phenylacetylene catalyzed by PCP–Pd pincer complexes.

Wendt and coworkers reported PCN–Ni complexes and their various reactivities. Interestingly, the synthesized Ni(III) complexes were used as catalysts for the Kharasch addition reaction, which is an important transformation owing to new C–C bond formation and functionalization occurring simultaneously. This reaction was screened using complexes **55** and **57** as catalysts (Scheme 75). The [(PCN)–NiBr] complex (**57**) was a proficient catalyst compared with the analogous [(PCN)–NiCl] complex (**55**).[51]

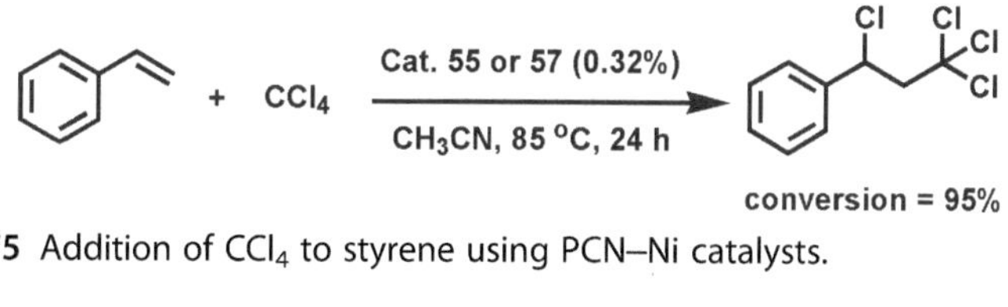

Scheme 75 Addition of CCl$_4$ to styrene using PCN–Ni catalysts.

Herbert and coworkers observed the promising reactivity of Ni pincer complex **10** in the alkylation of azoles with alkyl halides through C–H bond activation (Scheme 76). Azole groups were successfully coupled with alkyl halides bearing carbazole moieties.[41]

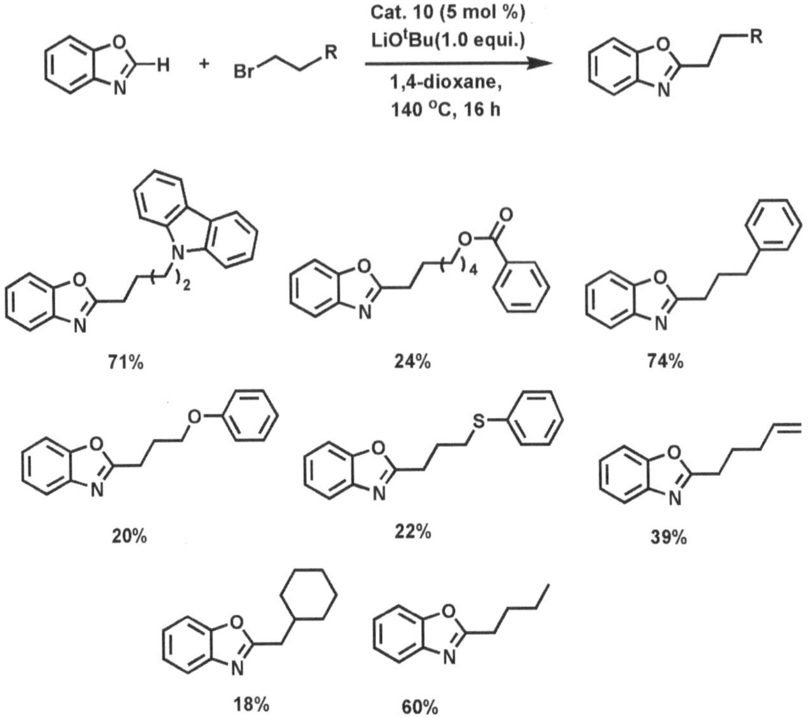

Scheme 76 Substrate scope for alkylation of benzoxazole with alkylbromides catalyzed by Ni pincer metal complex **10**.

Fenske and coworkers reported the high catalytic reactivity of complex **21** in hydrodehalogenation reactions (Scheme 77). Interestingly, this reduction of haloalkanes to alkanes was catalyzed by Ni pincer metal complexes, with complex **21** showing higher reactivity. This reaction was also proposed to occur via a radical mechanism, as somewhat confirmed by experimental data.[45]

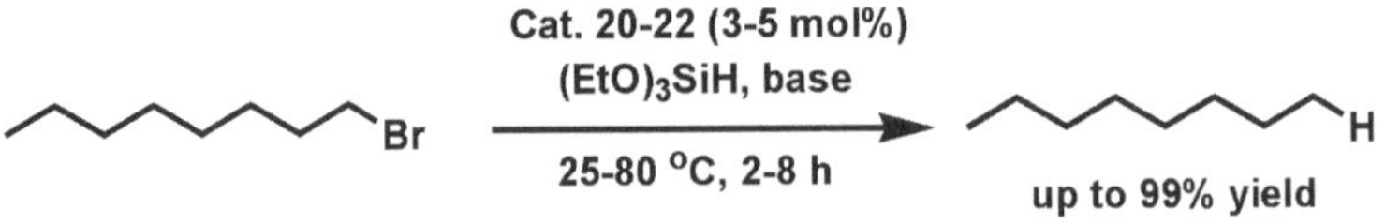

Scheme 77 Catalytic hydrodehalogenation reactions catalyzed by Ni pincer complexes **20–22**.

Sivasankar and coworkers have reported interesting [2+2+2] cycloadditions of alkynes using Ni(II) catalysts **49–54** (Scheme 78).[50] PNP-based Ni compounds catalyzed the [2+2+2] cycloaddition reaction of phenyl

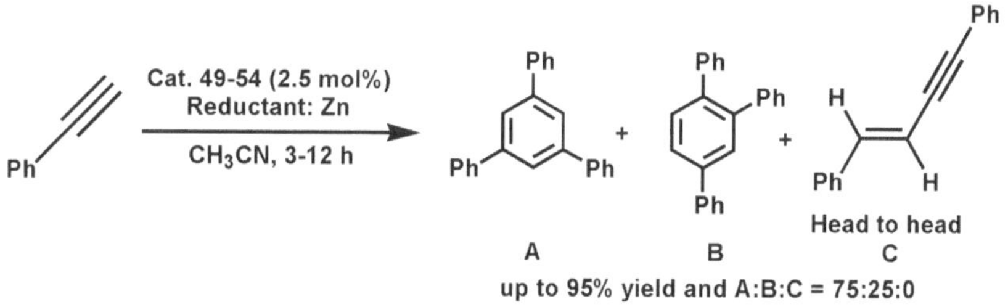

Scheme 78 Cycloaddition reactions of alkynes using Ni(II) catalysts.

acetylene as reactant.[50] The authors observed regioselectivity that was in good agreement with the electronic behavior of alkynes. Complex **50** has shown better catalytic activity when compared with other complexes.

Morales-Morales and coworkers reported the coupling reaction of phenyl iodide with C–S and C–Se reaction partners using Ni(II) complexes **76–78** as catalysts. Furthermore, the authors observed the best catalytic performance and conversion using the ᵗBu derivative of the Ni complex (**78**) (Scheme 79).[54] In this coupling reaction, zinc metal was used as a reducing agent and DMF was used as solvent.

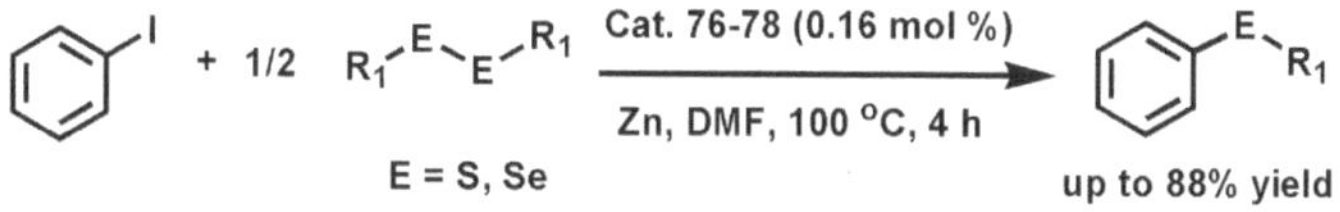

Scheme 79 C–S and C–Se coupling reactions catalyzed by Ni(II)–PSCOP pincer complexes.

Morales-Morales and coworkers reported C–S coupling reactions of disulfides with iodobenzenes using POCSP–Ni(II) pincer complex **86**. The reactions showed good catalytic activity and selectivity with various functional groups (Scheme 80).[57]

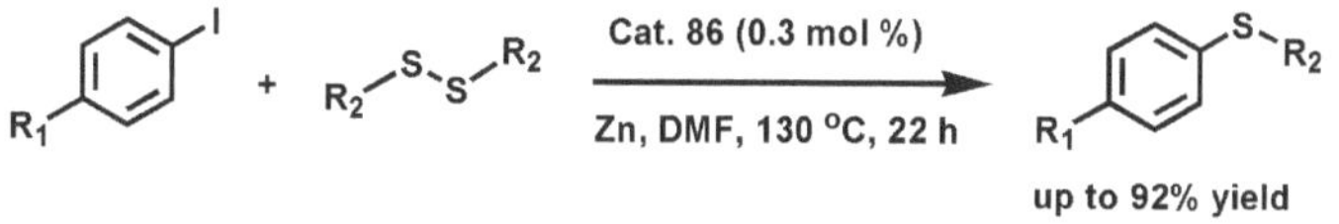

Scheme 80 Ni complex-catalyzed C–S coupling reactions.[57]

Yamaguchi and coworkers reported Ni–NNP pincer compounds and their applications in C–C bond forming reactions. NNP pincer-based Ni complex **98** was efficiently employed in the C–C coupling of a series of

ArF compounds with various aryl magnesium bromide Grignard reagents (Scheme 81).[61]

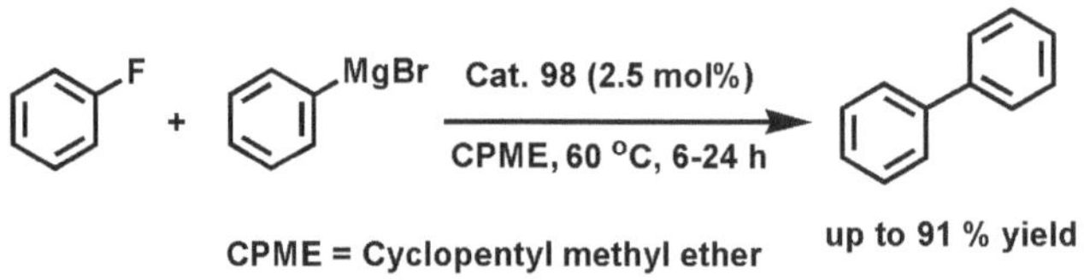

Scheme 81 Cross-coupling reactions catalyzed by NNP-based Ni pincer complex.

Joshi and coworkers synthesized highly stable Pd–pincer complexes, which were utilized as catalysts in coupling reactions of two heteroarenes. These catalysts were successfully applied to the C–H activation of heterocycles, including imidazole, benzimidazole, imidazopyridine, benzothiazole, furan, and thiophene, with low catalyst loadings (Scheme 82).[70]

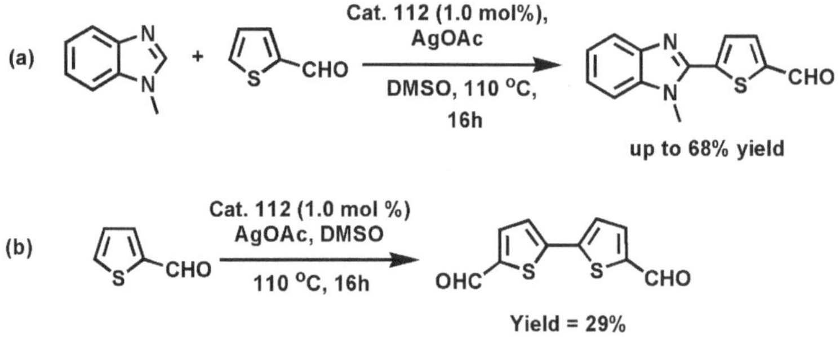

Scheme 82 Dehydrogenative coupling reactions of heteroarenes.

Ghosh and coworkers screened the catalytic reactivity of palladacycle compounds **133–136**. The authors reported direct C(sp^2)–H bond arylation using aryl halides with a catalyst loading of just 0.05 mmol% (Scheme 83).[76]

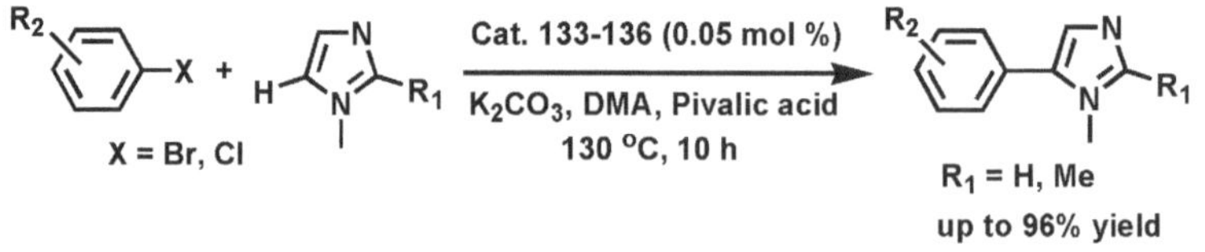

Scheme 83 C–H activation for arylation of imidazole by aryl halides.

Yamaguchi and coworkers reported the hydroboration of vinylarenes using bis(pinacolato)diboron with Markovnikov selectivity, which was

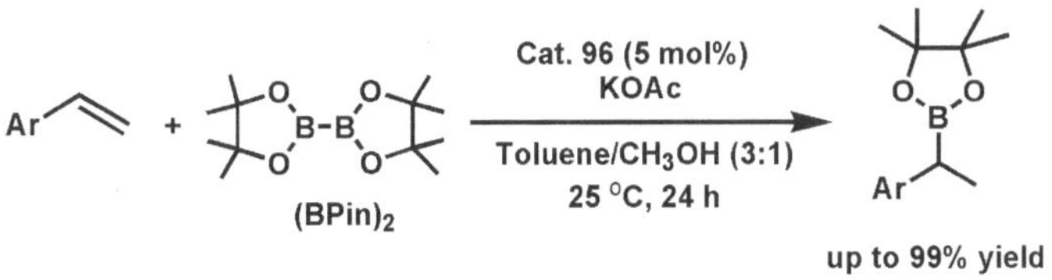

Scheme 84 Selective hydroboration of vinylarenes with (BPin)$_2$.

catalyzed by a β-aminoketonato ONP-based Ni pincer complex (Scheme 84). The authors performed the hydroboration reaction under normal Markovnikov reaction conditions, obtaining the desired products in high yields.[173]

Yamaguchi and coworkers also reported the coupling reaction of allylic ethers with magnesium bromide reagents catalyzed by Ni pincer metal complexes (Scheme 85).[174]

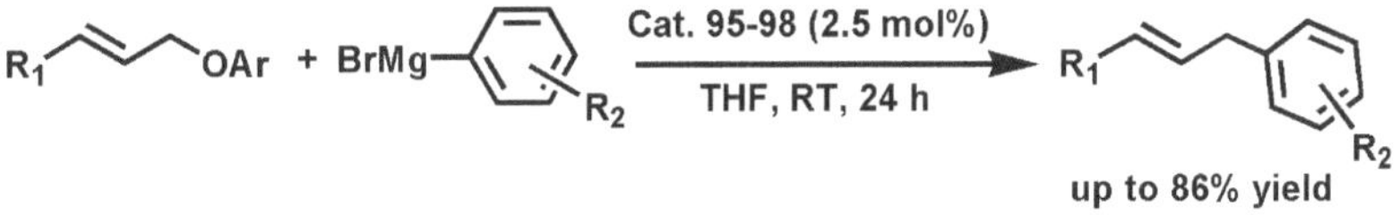

Scheme 85 Ni-catalyzed cross coupling reaction.

Bhanage and coworkers described carbonylative Sonogashira and Suzuki–Miyaura cross-coupling reactions with improved reactivity and high TONs using an aminophosphine Pd pincer catalyst (Scheme 86). The carbonylative Sonogashira (CS) cross-coupling was performed with a low catalyst loading, while the carbonylative Suzuki–Miyaura (CSM) reaction was conducted with a much lower catalyst loading, affording TONs of 10^5 and 10^7, respectively.[86]

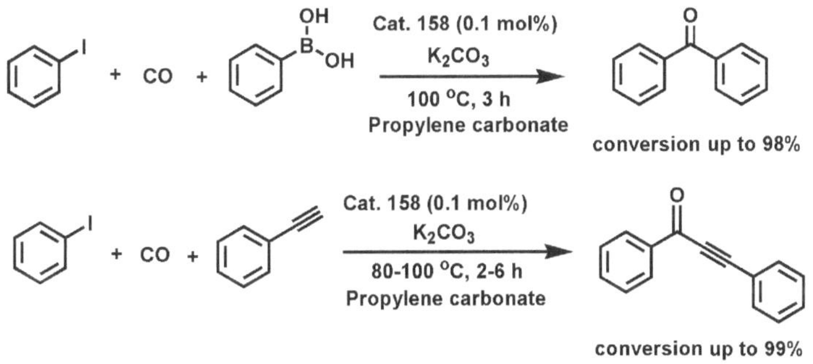

Scheme 86 Palladium catalyzed carbonylative Suzuki and Sonogashira coupling reactions.

6. Conclusions and perspectives

Group 10 metal–pincer chemistry has received increasing interest owing to the development of new functionalized stable pincer metal complexes with structural diversity and promising applications. Current research on potential pincer metal chemistry and its contribution to numerous organic conversions has resulted in progress toward further effective pincer metal-based catalysts, which has improved the understanding and altered the previous vision regarding various organic transformations. Consequently, the future of group 10 metal-pincer complexes as organometallic catalysts for organic transformations seems positive. Advancements in this field are evidenced by applications to important coupling reactions and other potential catalytic reactions.

Acknowledgments

This work was supported by the National Natural Science Foundation of China (21702038), Shenzhen Fundamental Research Project (JCYJ20170811161344569), the Natural Scientific Research Innovation Foundation in Harbin Institute of Technology (Grant No. HIT.NSRIF.2020061), and a startup grant from the Harbin Institute of Technology, Shenzhen.

References

1. Moulton CJ, Shaw BL. Transition metal−carbon bonds. Part XLII. Complexes of nickel, palladium, platinum, rhodium and iridium with the tridentate ligand 2,6-Bis [(di-T-Butylphosphino)- methyl] phenyl. *J Chem Soc Dalton Trans*. 1976;11:1 020–1024.
2. Albrecht M, van Koten G. Platinum group organometallics based on "pincer" complexes: sensors, switches, and catalysts. *Angew Chem Int Ed*. 2001;40:3750–3781.
3. Peris E, Crabtree RH. Key factors in pincer ligand design. *Chem Soc Rev*. 2018;47:1959–1968.
4. van Koten G. Tuning the reactivity of metals Held in a rigid ligand environment. *Pure Appl Chem*. 1989;61:1681–1694.
5. van der Vlugt JI, JNH R. Neutral tridentate PNP ligands and their hybrid analogues: versatile non-innocent scaffolds for homogeneous catalysis. *Angew Chem Int Ed*. 2009;48:8832–8846.
6. Valdés H, García-Eleno MA, Canseco-Gonzalez D, Morales-Morales D. Recent advances in catalysis with transition- metal pincer compounds. *ChemCatChem*. 2018;10:3136–3172.
7. van Koten G, Milstein D. *Organometallic Pincer Chemistry*. Springer; 2013.
8. van Koten G, Gossage RA. *The Privileged Pincer-Metal Platform: Coordination Chemistry & Applications*. Springer; 2016.
9. Morales-Morales D. *Pincer Compounds*. Elsevier; 2018.
10. O'Reilly ME, Veige AS. Trianionic pincer and pincer-type metal complexes and catalysts. *Chem Soc Rev*. 2014;43:6325–6369.

11. Crabtree RH. Homogeneous transition metal catalysis of acceptorless dehydrogenative alcohol oxidation: applications in hydrogen storage and to heterocycle synthesis. *Chem Rev*. 2017;117:9228–9246.
12. Gunanathan C, Milstein D. Bond activation and catalysis by ruthenium pincer complexes. *Chem Rev*. 2014;114:12024–12087.
13. Schneider S, Meiners J, Askevold B. Cooperative aliphatic PNP amido pincer ligands—versatile building blocks for coordination chemistry and catalysis. *Eur J Inorg Chem*. 2012;2012:412–429.
14. Choi J, MacArthur AHR, Brookhart M, Goldman AS. Dehydrogenation and related reactions catalyzed by iridium pincer complexes. *Chem Rev*. 2011;111:1761–1779.
15. Luca OR, Crabtree RH. Redox-active ligands in catalysis. *Chem Soc Rev*. 2013;42:1440–1459.
16. Lavoie CM, Stradiotto M. Bisphosphines: a prominent ancillary ligand class for application in nickel-catalyzed C–N cross-coupling. *ACS Catal*. 2018;8:7228–7250.
17. Budnikova YH, Vicic DA, Klein A. Exploring mechanisms in Ni terpyridine catalyzed C–C cross-coupling reactions–a review. *Inorganics*. 2018;6:18. https://doi.org/10.3390/inorganics6010018.
18. Shi R, Zhang Z, Hu X. Nickamine and analogous nickel pincer catalysts for cross-coupling of alkyl halides and hydrosilylation of alkenes. *Acc Chem Res*. 2019;52:1471–1483.
19. Cornella J, Zarate C, Martin R. Metal-catalyzed activation of ethers via C–O bond cleavage: a new strategy for molecular diversity. *Chem Soc Rev*. 2014;43:8081–8097.
20. Borjesson M, Moragas T, Gallego D, Martin R. Metal-catalyzed carboxylation of organic (pseudo)halides with CO_2. *ACS Catal*. 2016;6:6739–6749.
21. Baba S, Negishi E. A novel stereospecific alkenyl–alkenyl cross-coupling by a palladium- or nickel-catalyzed reaction of alkenylalanes with alkenyl halides. *J Am Chem Soc*. 1976;98:6729–6731.
22. Negishi E-I, King AO, Okukado N. Selective carbon–carbon bond formation via transition metal catalysis. A highly selective synthesis of unsymmetrical biaryls and diarylmethanes by the nickel- or palladium-catalyzed reaction of aryl- and Benzylzinc derivatives with aryl halides. *J Org Chem*. 1977;42:1821–1823.
23. Hu X. Nickel-catalyzed cross coupling of non-activated alkyl halides: a mechanistic perspective. *Chem Sci*. 2011;2:1867–1886.
24. Tasker SZ, Standley EA, Jamison TF. Recent advances in homogeneous nickel catalysis. *Nature*. 2014;509:299–309.
25. Ananikov VP. Nickel: the "spirited horse" of transition metal catalysis. *ACS Catal*. 2015;5:1964–1971.
26. Aihara Y, Chatani N. Nickel-catalyzed direct arylation of C(sp^3)–H bonds in aliphatic amides via bidentate-chelation assistance. *J Am Chem Soc*. 2014;136:898–901.
27. Uemura T, Yamaguchi M, Chatani N. Phenyltrimethylammonium salts as methylation reagents in the nickel-catalyzed methylation of C–H bonds. *Angew Chem Int Ed*. 2016;55:3162–3165.
28. Ruan Z, Lackner S, Ackermann L. A general strategy for the nickel-catalyzed C–H alkylation of anilines. *Angew Chem Int Ed*. 2016;55:3153–3157.
29. Zhang S-K, Samanta RC, Sauermann N, Ackermann L. Nickel-catalyzed electrooxidative C–H amination: support for nickel(IV). *Chem A Eur J*. 2018;24:19166–19170.
30. Xu J, Qiao L, Shen J, Chai K, Shen C, Zhang P. Nickel(II)-catalyzed site-selective C–H bond trifluoromethylation of arylamine in water through a coordinating activation strategy. *Org Lett*. 2017;19:5661–5664.
31. Liu X, Mao G, Qiao J, et al. Nickel-catalyzed C–H bond trifluoromethylation of 8-aminoquinoline derivatives by acyl-directed functionalization. *Org Chem Front*. 2019;6:1189–1193.

32. Selander N, Szabo KJ. Catalysis by palladium pincer complexes. *Chem Rev.* 2011;111:2048–2076.
33. Morales-Morales D, Jensen CM. *The Chemistry of Pincer Compounds.* Amsterdam: Elsevier; 2007.
34. Tsuji J. *Palladium Reagents and Catalysts. New Perspectives for the 21st Century.* Chichester: Wiley; 2004.
35. Sebastian LG, Morales-Morales D. Cross-coupling reactions catalysed by palladium pincer complexes. A review of recent advances. *J Organomet Chem.* 2019;893:39–51.
36. Bender CF, Widenhoefer RA. Platinum-catalyzed intramolecular hydroamination of unactivated olefins with secondary alkylamines. *J Am Chem Soc.* 2005;127:1070–1071.
37. Cucciolito ME, D'Amora A, Vitagliano A. Catalytic coupling of ethylene and internal olefins by dicationic palladium(II) and platinum(II) complexes: switching from hydrovinylation to cyclopropane ring formation. *Organometallics.* 2005;24:3359–3361.
38. Feducia JA, Campbell AN, Doherty MQ, Gagne MR. Modular catalysts for diene cycloisomerization: rapid and enantioselective variants for bicyclopropane synthesis. *J Am Chem Soc.* 2006;128:13290–13297.
39. Gibbons SK, Xu Z, Hughes RP, Glueck DS, Rheingold AL. Chiral bis(phospholane) PCP pincer complexes: synthesis, structure, and nickel-catalyzed asymmetric phosphine alkylation. *Organometallics.* 2018;37(13):2159–2166.
40. Patel UN, Pandey DK, Gonnade RG, Punji B. Synthesis of quinoline-based NNN-pincer nickel(II) complexes: a robust and improved catalyst system for C–H bond alkylation of azoles with alkyl halides. *Organometallics.* 2016;35(11):1785–1793.
41. Mandapati P, Braun JD, Sidhu BK, Wilson G, Herbert DE. Catalytic C–H bond alkylation of azoles with alkyl halides mediated by nickel(II) complexes of phenanthridine-based N^N^N pincer ligands. *Organometallics.* 2020;39(10):1989–1997.
42. Mazaud L TM, Bourcier S, Cordier M, Gandon V, Auffrant A. Tridentate NNN ligand associating amidoquinoline and iminophosphorane: synthesis and coordination to Pd and Ni centers. *Organometallics.* 2020;39:719–728.
43. Ghannam J, Sun Z, Cundari TR, et al. Intramolecular C–H functionalization followed by [2σ + 2π] addition via an intermediate nickel–nitridyl complex. *Inorg Chem.* 2019;58:7131–7135.
44. Chang MC, Jesse KA, Filatova AS, Anderson JS. Reversible homolytic activation of water via metal–ligand cooperativity in a T-shaped Ni(ii) complex. *Chem Sci.* 2019;10:1360–1367.
45. Wang Z, Li X, Sun H, Fuhr O, Fenske D. Synthesis of NHC pincer hydrido nickel complexes and their catalytic applications in hydrodehalogenation. *Organometallics.* 2018;37(4):539–544.
46. Lapointe S, Khaskin E, Fayzullin RR, Khusnutdinova JR. Stable nickel(I) complexes with electron-rich, sterically-hindered, innocent PNP pincer ligands. *Organometallics.* 2019;38(7):1581–1594.
47. Lapointe S, Khaskin E, Fayzullin RR, Khusnutdinova JR. Nickel(II) complexes with electron-rich, sterically hindered PNP pincer ligands enable uncommon modes of ligand dearomatization. *Organometallics.* 2019;38(22):4433–4447.
48. Daw P, Kumar A, Oren D, et al. Redox noninnocent nature of acridine-based pincer complexes of 3d metals and C–C bond formation. *Organometallics.* 2020;39(2):279–285.
49. Yoo C, Lee Y. A T-shaped nickel(I) metalloradical species. *Angew Chem Int Ed.* 2017;341:9502–9506.
50. Tamizmani M, Sivasankar C. Synthesis, characterization and catalytic application of some novel PNP-Ni(II) complexes: regio-selective [2+2+2] cycloaddition reaction of alkyne. *J Organomet Chem.* 2017;845:82–89.

51. Mousa AH, Bendix J, Wendt OF. Synthesis, characterization, and reactivity of PCN pincer nickel complexes. *Organometallics*. 2018;37(15):2581–2593.
52. Mousa AH, Polukeev AV, Hansson J, Wendt OF. Carboxylation of the Ni−Me bond in an electron-rich unsymmetrical PCN pincer nickel complex. *Organometallics*. 2020; 39(9):1553–1560.
53. Scheerder AR, Lutz M, Broere DLJ. Unexpected reactivity of a PONNOP 'expanded pincer' ligand. *Chem Commun*. 2020;56:8198–8201.
54. Valderrama-García BX, Rufino-Felipe E, Valdés H, Hernandez-Ortega S, Aguilar-Castillo BA, Morales-Morales D. Novel and facile procedure for the synthesis of Ni(II) and Pd(II) PSCOP pincer complexes. Evaluation of their catalytic activity on C-S, C-Se and C-C cross coupling reactions. *Inorg Chim Acta*. 2020;502:119283.
55. Gafurov ZN, Bekmukhamedov GE, Kagilev AA, et al. Unsymmetrical pyrazole-based PCN pincer Ni^{II} halides: reactivity and catalytic activity in ethylene oligomerization. *J Organomet Chem*. 2020;912:121163.
56. Ando S, Nakano N, Matsunaga H, Ishizuka T. Synthesis and catalytic activities of Ni complexes bearing a novel N–C–N pincer ligand containing NHC with a bicyclic motif. *J Organomet Chem*. 2020;913:121200.
57. Serrano-Becerra JM, Valdés H, Canseco-González D, Gómez-Benítez V, Hernández-Ortega S, Morales-Morales D. C-S cross-coupling reactions catalyzed by a non-symmetric phosphinitothiophosphinito PSCOP-Ni(II) pincer complex. *Tetrahedron Lett*. 2018;59:3377–3380.
58. Hasche P, Joksch M, Vlachopoulou G, Agarwala H, Spannenberg A, Beweries T. Synthesis of symmetric and non-symmetric Ni(II) thiophosphinitoPECSP (E = S, O) pincer complexes and applications in Kumada coupling under mild conditions. *Eur J Inorg Chem*. 2018;2018:676–680.
59. Kaur-Ghumaan S, Hasche P, Spannenberga A, Beweries T. Nickel(II) PE^{1}CE^{2}P pincer complexes (E = O, S) for electrocatalytic proton reduction. *Dalton Trans*. 2019;48:16322–16329.
60. Asano E, Hatayama Y, Kurisu N, et al. Acetylacetonato-based pincer-type nickel(II) complexes: synthesis and catalysis in cross-couplings of aryl chlorides with aryl grignard reagents. *Dalton Trans*. 2018;47:8003–8012.
61. Kurisu N, Asano E, Hatayama Y, et al. Yamaguchi, a β-diketiminato-based pincer-type nickel(II) complex: synthesis and catalytic performance in the cross-coupling of aryl fluorides with aryl grignard reagents. *Eur J Inorg Chem*. 2019;1:126–133.
62. Arora V, Dutta M, Das K, Das B, Srivastava HK, Kumar A. Solvent-free N-alkylation and dehydrogenative coupling catalyzed by a highly active pincer-nickel complex. *Organometallics*. 2020;39(11):2162–2176.
63. Wang Z, Li X, Xie S, Zheng T, Sun H. Transfer hydrogenation of ketones catalyzed by nickel complexes bearing an NHC [CNN] pincer ligand. *Appl Organomet Chem*. 2019;33, e4932.
64. Sun Y, Li X, Sun H. [CNN]-pincer nickel(II) complexes of N-heterocyclic carbene (NHC): synthesis and catalysis of the Kumada reaction of unactivated C–Cl bonds. *Dalton Trans*. 2014;43:9410–9413.
65. Morales-Espinoza EG, Coronel-García R, Valdés H, et al. Synthesis, characterization and catalytic evaluation of non-symmetric Pd(II)-POCOP pincer compounds derived from 2′,4′-dihydroxyacetophenone. *J Organomet Chem*. 2018;867:155–160.
66. Kozlov VA, Aleksanyan DV, Nelyubina YV, et al. 5,6-Membered palladium pincer complexes of 1-thiophosphoryloxy-3-thiophosphorylbenzenes. Synthesis, X-ray structure, and catalytic activity. *Dalton Trans*. 2009;8657–8666.
67. Yao Q, Kinney EP, Zheng C. Selenium-ligated palladium(II) complexes as highly active catalysts for carbon−carbon coupling reactions: the heck reaction. *Org Lett*. 2004;6:2997–2999.

68. Yao Q, Sheets M. A SeCSe−Pd(II) pincer complex as a highly efficient catalyst for allylation of aldehydes with allyltributyltin. *J Org Chem*. 2006;71:5384–5387.
69. Zhang Y, Song G, Ma G, Zhao J, Pan C-L, Li X. 1,3-Dinitrone pincer complexes of palladium and nickel: synthesis, structural characterizations, and catalysis. *Organometallics*. 2009;28:3233–3238.
70. Shinde VN, Bhuvanesh N, Kumar A, et al. Design and syntheses of palladium complexes of NNN/CNN pincer ligands: catalyst for cross dehydrogenative coupling reaction of heteroarenes. *Organometallics*. 2020;39(2):324–333.
71. Yadav S, Singh A, Rashid N, et al. Phosphine-free bis(pyrrolyl)pyridine based NNN-pincer palladium(II) complexes as efficient catalysts for Suzuki-Miyaura cross-coupling reactions of aryl bromides in aqueous medium. *ChemistrySelect*. 2018;3:9469–9475.
72. McNeece AJ, Chang MC, Filatov AS, Anderson JS. Redox activity, ligand protonation, and variable coordination modes of diimino-pyrrole complexes of palladium. *Inorg Chem*. 2018;57(12):7044–7050.
73. Ding Y, Ma QQ, Kang J, Zhang J, Li S, Chen X. Palladium(ii) complexes supported by PBP and POCOP pincer ligands: a comparison of their structure, properties and catalytic activity. *Dalton Trans*. 2019;48:17633–17643.
74. Therrien JA, Wolf MO, Patrick BO. Synthesis and comparison of nickel, palladium, and platinum bis(N-heterocyclic carbene) pincer complexes for electrocatalytic CO_2 reduction. *Dalton Trans*. 2018;47:1827–1840.
75. Ortega-Gaxiola JI, Valdés H, Rufino-Felipe E, Toscano RA, Morales-Morales D. Synthesis of Pd(II) complexes with P-N-OH ligands derived from 2-(diphenylphosphine)-benzaldehyde and various aminoalcohols and their catalytic evaluation on Suzuki-Miyaura couplings in aqueous media. *Inorg Chim Acta*. 2020;504:119460.
76. Maji A, Singh O, Singh S, Mohanty A, Maji PK, Ghosh K. Palladium-based catalysts supported by unsymmetrical XYC−1 type pincer ligands: C5 Arylation of Imidazoles and synthesis of Octinoxate utilizing the Mizoroki–heck reaction. *Eur J Inorg Chem*. 2020;17:1596–1611.
77. Maji A, Singh O, Rathi S, Singh UP, Ghosh K. Rational design of sterically hindered and unsymmetrical NpyNimOph pincer-type ligands and their palladium(II) complexes: catalytic applications in Suzuki–Miyaura reaction and allylation of aldehydes. *ChemistrySelect*. 2019;4:7246–7259.
78. Jia WG, Gao LL, Wang ZB, Wang JJ, Sheng EH, Han YF. NHC-palladium(II) mononuclear and binuclear complexes containing phenylene-bridged bis(thione) ligands: synthesis, characterization, and catalytic activities. *Organometallics*. 2020;39(10):1790–1798.
79. Mahmood Q, Yue E, Zhang W, Solan GA, Liang T, Sun WH. Bisimino-functionalized dibenzo[a,c]acridines as highly conjugated pincer frameworks for palladium(ii): synthesis, characterization and catalytic performance in Heck coupling. *Org Chem Front*. 2016;3:1668–1679.
80. SanMartin R, Ines B, Moure MJ, Herrero MT, Domínguez E. Mizoroki-Heck and Sonogashira cross-couplings catalyzed by CNC palladium pincer complexes in organic and aqueous media. *Helv Chim Acta*. 2012;95:955–962.
81. Bolliger JL, Frech CM. Highly convenient, clean, fast, and reliable sonogashira coupling reactions promoted by aminophosphine-based pincer complexes of palladium performed under additive- and amine-free reaction conditions. *Adv Synth Catal*. 2009;351:891–902.
82. Zhang J-H, Li P, Hu W-P, Wang H-X. Substituent effect of diimino-palladium (II) pincer complexes on the catalysis of Sonogashira coupling reaction. *Polyhedron*. 2015;96:107–112.

83. Inés B, SanMartin R, Churruca F, Domínguez E, Urtiaga MK, Arriortua MI. A non-symmetric pincer-type palladium catalyst in Suzuki, Sonogashira, and Hiyama couplings in neat water. *Organometallics*. 2008;27:2833–2839.
84. Wang H, Liu J, Deng Y, et al. Pincer thioamide and pincer thioimide palladium complexes catalyze highly efficient Negishi coupling of primary and secondary alkyl zinc reagents at room temperature. *Chem A Eur J*. 2009;15:1499–1507.
85. Erken C, Hindemith C, Weyhermüller T, Hölscher M, Werle C, Leitner W. Hydroamination of aromatic alkynes to imines catalyzed by Pd(II)–anthraphos complexes. *ACS Omega*. 2020;5:8912–8918.
86. Gautam P, Tiwari NJ, Bhanage BM. Aminophosphine palladium pincer-catalyzed carbonylative Sonogashira and Suzuki–Miyaura cross-coupling with high catalytic turnovers. *ACS Omega*. 2019;4:1560–1574.
87. Ranieri AM, Burt LK, Stagni S, et al. Anionic cyclometalated platinum(II) tetrazolato complexes as viable photoredox catalysts. *Organometallics*. 2019;38(5):1108–1117.
88. Mandapati P, Braun JD, Killeen C, Davis RL, Williams JAG, Herbert DE. Luminescent platinum(II) complexes of N̂N−̂N amido ligands with benzannulated N-heterocyclic donor arms: quinolines offer unexpectedly deeper red phosphorescence than phenanthridines. *Inorg Chem*. 2019;58:14808–14817.
89. Li M, Liska T, Swetz A, et al. (Isonitrile)platinum(II) complexes of an amido Bis(N-heterocyclic carbene) pincer ligand. *Organometallics*. 2020;39(10):1667–1671.
90. Liska T, Swetz A, Lai PN, Zeller M, Teets TS, Gray TG. Room-temperature phosphorescent platinum(II) alkynyls with microsecond lifetimes bearing a strong-field pincer ligand. *Chem A Eur J*. 2020;26:8417–8425.
91. Sun CY, To WP, Hung FF, Wang XL, Su ZM, Che CM. Metal–organic framework composites with luminescent pincer platinum(II) complexes: ^{3}MMLCT emission and photoinduced dehydrogenation catalysis. *Chem Sci*. 2018;9:2357–2364.
92. Ruan J, Wang D, Vedernikov AN. CH$_3$−X reductive elimination reactivity of Pt^{IV}Me complexes supported by a sulfonated CNN pincer ligand (X = OH, CF$_3$CO$_2$, PhNMe$_2^+$). *Organometallics*. 2020;39:142–152.
93. Garbe S, Krause M, Klimpel A, et al. Cyclometalated Pt complexes of CNC pincer ligands: luminescence and cytotoxic evaluation. *Organometallics*. 2020;39:746–756.
94. Castonguay A, Zargarian D, Beauchamp AL. Preparation and reactivities of PCP-type pincer complexes of nickel. Impact of different ligand skeletons and phosphine substituents. *Organometallics*. 2008;27:5723–5732.
95. Pandarus V, Zargarian D. New pincer-type diphosphinito (POCOP) complexes of Ni^{II} and Ni^{III}. *Chem Commun*. 2007;978–980.
96. Zhang J, Medley CM, Krause JA, Guan H. Mechanistic insights into C−S cross-coupling reactions catalyzed by nickel bis(phosphinite) pincer complexes. *Organometallics*. 2010;29:6393–6401.
97. Chakraborty S, Zhang J, Krause JA, Guan H. An efficient nickel catalyst for the reduction of carbon dioxide with a borane. *J Am Chem Soc*. 2010;132:8872–8873.
98. Hasche P, Spannenberg A, Beweries T. Study of the reactivity of the [(PE^{1}CE^{2}P) Ni(II)] (E^1, E^2 = O, S) pincer system with acetonitrile and base: formation of cyanomethyl and A midocrotononitrile complexes versus ligand decomposition by P–S bond activation. *Organometallics*. 2019;38:4508–4515.
99. Hao J, Mougang-Soumé B, Vabre B, Zargarian D. On the stability of a POCsp^{3}OP-type pincer ligand in nickel(II) complexes. *Angew Chem Int Ed*. 2014;53:3218–3222.
100. van Koten EG, Milstein D. *Organometallic Pincer Chemistry. Topics in Organometallic Chemistry*. Vol. 40. Heidelberg: Springer; 2013.
101. Grove DM, van Koten G, Zoet R. Unique stable organometallic nickel(III) complexes; syntheses and the molecular structure of [Ni[C$_6$H$_3$(CH$_2$NMe$_2$)$_2$-2,6]I$_2$]. *J Am Chem Soc*. 1983;105:1379–1380.

102. Grove DM, van Koten G, Mul P, et al. Syntheses and characterization of unique rrganometallicnickel(III) aryl species. ESR and electrochemical studies and the X-ray molecular study of square-pyramidal [Ni{C$_6$H$_3$(CH$_2$NMe$_2$)$_2$-o,o'}I$_2$]. Inorg Chem. 1988;27:2466–2473.
103. Patel UN, Jain S, Pandey DK, Gonnade RG, Vanka K, Punji B. Mechanistic aspects of pincer nickel(II)-catalyzed C−H bond alkylation of azoles with alkyl halides. Organometallics. 2018;37:1017–1025.
104. Kruithof CA, Dijkstra HP, Lutz M, Spek AL, RJMK G, van Koten G. X-Ray and NMR study of the structural features of SCS-pincer metal complexes of the group 10 triad. Organometallics. 2008;27:4928.
105. Koizumi T-A, Teratani T, Okamoto K, Yamamoto T, Shimoi Y, Kanbara T. Nickel(II) complexes bearing a pincer ligand containing thioamide units: comparison between SNS- and SCS-pincer ligands. Inorg Chim Acta. 2010;363:2474.
106. Peterson SM, Helm ML, Appel AM. Nickel complexes of a binucleating ligand derived from an SCSPincer. Dalton Trans. 2015;44:747.
107. Xu T, Wodrich MD, Scopelliti R, Corminboeuf C, Hu X. Nickel pincer model of the active site of lactate racemase involves ligand participation in hydride transfer. Proc Natl Acad Sci U S A. 2017;114:1242.
108. Martınez-Prieto LM, Melero C, del Rio D, Palma P, Campora J, Alvarez E. Synthesis and reactivity of nickel and palladium fluoride complexes with PCP pincer ligands NMR-based assessment of electron-donating Properties of fluoride and other monoanionic ligands. Organometallics. 2012;31:1425–1438.
109. Connelly SJ, Chanez AG, Kaminsky W, Heinekey DM. Characterization of a palladium dihydrogen complex. Angew Chem Int Ed. 2015;54:5915–5918.
110. Johansson R, Wendt OF. Synthesis and reactivity of (PCP) palladium hydroxy carbonyl and related complexes toward CO$_2$ and phenylacetylene. Organometallics. 2007;26:2426–2430.
111. Ananthnag GS, Mague JT, Balakrishna MS. Cyclodiphosphazane based pincer ligand, [2,6-{μ-(tBuN)P(tBuHN)PO}$_2$C$_6$H$_3$I]: Ni^{II}, Pd^{II}, Pt^{II} and Cu^{I} complexes and catalytic studies. Dalton Trans. 2015;44:3785–3793.
112. Anderson BG, Spencer JL. The coordination chemistry of pentafluorophenylphosphino pincer ligands to platinum and palladium. Chem A Eur J. 2014;20:6421–6432.
113. Ozerov OV, Guo C, Foxman BM. Missing link: PCP pincer ligands containing P–N bonds and their Pd complexes. J Organomet Chem. 2006;691:4802–4806.
114. Wilson GLO, Abraha M, Krause JA, Guan H. Reactions of phenylacetylene with nickel POCOPPincer hydride complexes resulting in different outcomes from their palladium analogues. Dalton Trans. 2015;44:12128–12136.
115. Serrano-Becerra JM, Hernandez-Ortega S, Morales-Morales D. Synthesis of a novel non-symmetric Pd(II) phosphinito–thiophosphinito PSCOP pincer compound. Inorg Chim Acta. 2010;363:1306–1310.
116. Solano-Prado MA, Estudiante-Negrete F, Morales-Morales D. Group 10 phosphinite POCOP pincer complexes derived from 4-n-dodecylresorcinol: an alternative way to produce non-symmetric pincer compounds. Polyhedron. 2010;29:592–600.
117. Naghipour A, Ghasemi ZH, Morales-Morales D, Serrano-Becerra JM, Jensen CM. Simple protocol for the synthesis of the asymmetric PCP pincer ligand [C$_6$H$_4$-1-(CH$_2$PPh$_2$)-3-(CH(CH$_3$)PPh$_2$)] and its Pd(II) derivative [PdCl{C$_6$H$_3$-2-(CH$_2$PPh$_2$)-6-(CH(CH$_3$)PPh$_2$)}]. Polyhedron. 2008;27:1947–1952.
118. Yoon MS, Ryu D, Kim J, Ahn KH. Palladium pincer complexes with reduced bond angle strain: efficient catalysts for the heck reaction. Organometallics. 2006;25:2409–2411.
119. Luo Q-L, Tan J-P, Li Z-F, Qin Y, Ma L, Xiao D-R. Novel bis(azole) pincer palladium complexes: synthesis, structures and applications in Mizoroki–Heck reactions. Dalton Trans. 2011;40:3601–3609.

120. Xu G, Luo Q, Eibauer S, et al. Palladium(ii)- and platinum(ii) phenyl-2,6-bis(oxazole) pincer complexes: syntheses, crystal structures, and photophysical properties. *Dalton Trans*. 2011;40:8800–8806.
121. Wang T, Hao X-Q, Zhang X-X, Gong J-F, Song M-P. Synthesis, structure and catalytic properties of CNN pincer palladium(ii) and ruthenium(ii) complexes with N-substituted-2-aminomethyl-6-phenylpyridines. *Dalton Trans*. 2011;40:8964–8976.
122. van de Kuil LA, Luitjes H, Grove DM, et al. Electronic tuning of Arylnickel(II) complexes by para substitution of the terdentate monoanionic 2,6-bis[(dimethylamino) methyl]phenyl ligand. *Organometallics*. 1994;13:468–477.
123. Jung IG, Son SU, Park KH, Chung K-C, Lee JW, Chung YK. Synthesis of novel Pd−NCN pincer complexes having additional nitrogen coordination sites and their application as catalysts for the heck reaction. *Organometallics*. 2003;22:4715–4720.
124. Feng J, Lu G, Lv M, Cai C. Palladium catalyzed direct C-2 arylation of indoles. *J Organomet Chem*. 2014;761:28–31.
125. Fossey JS, Richards CJ. Catalysis of aldehyde and imine silylcyanation by platinum and palladium NCN-pincer complexes. *Tetrahedron Lett*. 2003;44:8773–8776.
126. Errington J, McDonald WS, Shaw BL. Cyclopalladation of $C_6H_4(CH_2SBu^t)_2$-1,3 and the crystal structure of $[PdCl\{C_6H_3(CH_2SBu^t)_2$-2,6$\}]$. *J Chem Soc Dalton Trans*. 1980;2312–2314.
127. Tyson GE, Tokmic K, Oian CS, et al. Synthesis, characterization, photophysical properties, and catalytic activity of an SCS Bis(N-heterocyclic thione) (SCS-NHT) Pd pincer complex. *Dalton Trans*. 2015;44:14475–14482.
128. Basauri-Molina M, Hernandez-Ortega S, Morales-Morales D. Microwave-assisted C–C and C–S couplings catalysed by organometallic Pd-SCS or coordination Ni-SNS pincer complexes. *Eur J Inorg Chem*. 2014;2014:4619–4625.
129. da Costa RC, Jurisch M, Gladysz JA. Synthesis of fluorous sulfur/carbon/sulfur pincer ligands and palladium complexes: new catalyst precursors for the heck reaction. *Inorg Chim Acta*. 2008;361:3205–3214.
130. Aleksanyan DV, Kozlov VA, Nelyubina YV, et al. Synthesis, catalytic activity, and photophysical properties of 5,6-membered Pd and PtSCS'-pincer complexes based on thiophosphorylated 3-amino(hydroxy)benzoic acid thioanilides. *Dalton Trans*. 2011;40:1535–1546.
131. Selander N, Szabo KJ. Performance of SCS palladium pincer complexes in borylation of allylic alcohols. Control of the regioselectivity in the one-pot borylation − allylation process. *J Org Chem*. 2009;74:5695–5698.
132. Sarkar S, Carlson AR, Veige MK, Falkowski JM, Abboud KA, Veige AS. Synthesis, characterization, and reactivity of a d^2, Mo(IV) complex supported by a new OCO − trianionic pincer ligand. *J Am Chem Soc*. 2008;130:1116–1117.
133. Adhikary A, Schwartz JR, Meadows LM, Krause JA, Guan H. Interaction of alkynes with palladium POCOP-pincer hydride complexes and its unexpected relation to palladium-catalyzed hydrogenation of alkynes. *Inorg Chem Front*. 2014;1:71–82.
134. Trost BM, Gerusz VJ. Palladium-catalyzed addition of pronucleophiles to allenes. *J Am Chem Soc*. 1995;117:5156–5157.
135. Camacho DH, Nakamura I, Byoung HO, Saito S, Yamamoto Y. Palladium-catalyzed addition of ketones to alkylidenecyclopropanes. *Tetrahedron Lett*. 2002;43:2903–2907.
136. Goddard R, Hopp G, Jolly PW, Kruger C, Mynott R, Wirtz C. The palladium-catalyzed intramolecular cyclization of alkadienyl-substituted 1,3-diketones. *J Organomet Chem*. 1995;486:163–170.
137. Chianese AR, Lee SJ, Gagne MR. Electrophilic activation of alkenes by platinum(II): so much more than a slow version of palladium(II). *Angew Chem Int Ed*. 2007;46:4042–4059.
138. Barone CR, Benedetti M, Vecchio VM, Fanizzi FP, Maresca L, Natile G. New chemistry of olefin complexes of platinum(ii) unravelled by basic conditions: synthesis and properties of elusive cationic species. *Dalton Trans*. 2008;5313–5322.

139. Calmuschi-Cula B, Englert U. Orthoplatination of primary amines. *Organometallics*. 2008;27:3124–3130.
140. Brissy D, Skander M, Retailleau P, Frison G, Marinetti A. Platinum(II) complexes featuring chiral diphosphines and N-heterocyclic carbene ligands: synthesis and evaluation as cycloisomerization catalysts. *Organometallics*. 2009;28:140–151.
141. Hahn C, Cucciolito ME, Vitagliano A. Coordinated olefins as incipient carbocations: catalytic codimerization of ethylene and internal olefins by a dicationic Pt(II)−ethylene complex. *J Am Chem Soc*. 2002;124:9038–9039.
142. Martín M, Sola E. Chapter two—recent advances in the chemistry of group 9-pincer organometallics. *Adv Organomet Chem*. 2020;73:79–193.
143. Jin L, Wei W, Sun N, Hu B, Shen Z, Hu X. Unsymmetrical CNN-palladacycles with geometry-constrained iminopyridyl ligands: an efficient precatalyst in Suzuki coupling for accessing 1,1-diarylalkanes from secondary benzylic bromides. *Org Chem Front*. 2018;5:2484–2491.
144. Majumder P, Paul P, Sengupta P, Bhattacharya S. Formation of organopalladium complexes via C–Br and C–C bond activation. Application in C–C and C–N coupling reactions. *J Organomet Chem*. 2013;736:1–8.
145. Maji A, Singh A, Mohanty A, Maji PK, Ghosh K. Ferrocenyl palladacycles derived from unsymmetrical pincer-type ligands: evidence of Pd(0) nanoparticle generation during the Suzuki–Miyaura reaction and applications in the direct arylation of thiazoles and isoxazoles. *Dalton Trans*. 2019;48:17083–17096.
146. Hamasaka G, Ichii S, Uozumi Y. A palladium NNC-pincer complex as an efficient catalyst precursor for the Mizoroki−Heck reaction. *Adv Synth Catal*. 2018;360: 1833–1840.
147. Broring M, Kleeberg C, Kohler S. Palladium(II) complexes of unsymmetrical CNN pincer ligands. *Inorg Chem*. 2008;47:6404–6412.
148. Olsson D, Nilsson P, Masnaouy ME, Wendt OF. A catalytic and mechanistic investigation of a PCP pincer palladium complex in the Stille reaction. *Dalton Trans*. 2005;1924–1929.
149. Marset X, Gea SD, Guillena GJ, Ramón D. NCN–pincer–Pd complex as catalyst for the hiyama reaction in biomass-derived solvents. *ACS Sustain Chem Eng*. 2018;6: 5743–5748.
150. Eberhard MR. Insights into the heck reaction with PCP pincer palladium(II) complexes. *Org Lett*. 2004;6:2125–2128.
151. Sommer WJ, Yu K, Sears JS, et al. Investigations into the stability of tethered palladium(II) pincer complexes during heck catalysis. *Organometallics*. 2005;24: 4351–4361.
152. Xue L, Lin Z. Theoretical aspects of palladium-catalysed carbon–carbon cross-coupling reactions. *Chem Soc Rev*. 2010;39:1692–1705.
153. Aydin J, Larsson JM, Selander N, Szabo KJ. Pincer complex-catalyzed redox coupling of alkenes with iodonium salts via presumed palladium(IV) intermediates. *Org Lett*. 2009;11:2852–2854.
154. Balakrishna MS. Unusual and rare pincer ligands: synthesis, metallation, reactivity and catalytic studies. *Polyhedron*. 2018;143:2–10.
155. Ohff M, Ohff A, van der Boom ME, Milstein D. Highly active Pd(II) PCP-type catalysts for the heck reaction. *J Am Chem Soc*. 1997;119:11687–11688.
156. Das D, Rao GK, Singh AK. Palladium(II) complexes of the first pincer (Se,N,Se) ligand, 2,6-bis((phenylseleno)methyl)pyridine (L): solvent-dependent formation of [PdCl(L)]Cl and Na[PdCl(L)][PdCl$_4$] and high catalytic activity for the heck reaction. *Organometallics*. 2009;28:6054–6058.
157. Miyaura N, Yamada K, Suzuki A. A new stereospecific cross-coupling by the palladium-catalyzed reaction of 1-alkenylboranes with 1-alkenyl or 1-alkynyl halides. *Tetrahedron Lett*. 1979;20:3437–3440.

158. Chinchilla R, Najera C. Recent advances in Sonogashira reactions. *Chem Soc Rev.* 2011;40:5084–5121.
159. Tamao K, Sumitani K, Kumada M. Selective carbon-carbon bond formation by cross-coupling of Grignard reagents with organic halides. Catalysis by nickel-phosphine complexes. *J Am Chem Soc.* 1972;94(12):4374–4376.
160. Corriu RJP, Masse JP. Activation of grignard reagents by transition-metal complexes. A new and simple synthesis of trans-stilbenes and polyphenyls. *J Chem Soc Chem Commun.* 1972;144a.
161. Frisch AC, Beller M. Catalysts for cross-coupling reactions with non-activated alkyl halides. *Angew Chem Int Ed.* 2005;44:674.
162. Kambe N, Iwasaki T, Terao J. Pd-catalyzed cross-coupling reactions of alkyl halides. *Chem Soc Rev.* 2011;40:4937.
163. Heravi MM, Hajiabbasi P. Recent advances in Kumada-Tamao-Corriu cross-coupling reaction catalyzed by different ligands. *Monatsh Chem.* 2012;143:1575.
164. Phapale VB, Cardenas DJ. Nickel-catalysed Negishi cross-coupling reactions: scope and mechanisms. *Chem Soc Rev.* 2009;38:1598–1607.
165. Haas D, Hammann JM, Greiner R, Knochel P. Recent developments in Negishi cross-coupling reactions. *ACS Catal.* 2016;6:1540–1552.
166. Han FS. Transition-metal-catalyzed Suzuki–Miyaura cross-coupling reactions: a remarkable advance from palladium to nickel catalysts. *Chem Soc Rev.* 2013;42:5270–5298.
167. Maluenda I, Navarro O. Recent developments in the Suzuki-Miyaura reaction: 2010–2014. *Molecules.* 2015;20:7528–7557.
168. Egorova KS, Ananikov VP. Which metals are green for catalysis? Comparison of the toxicities of Ni, Cu, Fe, Pd, Pt, Rh, and Au salts. *Angew Chem Int Ed.* 2016;55: 12150–12162.
169. Liang Y, Fu GC. Stereoconvergent negishi arylations of racemic secondary alkyl electrophiles: differentiating between a CF_3 and an alkyl group. *J Am Chem Soc.* 2015;137:9523.
170. Terao J, Kambe N. Cross-coupling reaction of alkyl halides with grignard reagents catalyzed by Ni, Pd, or Cu complexes with π-carbon ligand(s). *Acc Chem Res.* 2008;41: 1545–1554.
171. Liang Y, Fu GC. Nickel-catalyzed alkyl-alkyl cross-couplings of fluorinated secondary electrophiles: a general approach to the synthesis of compounds having a perfluoroalkyl substituent. *Angew Chem Int Ed.* 2015;54:9047–9051.
172. Alig L, Fritz M, Schneider S. First-row transition metal (De)hydrogenation catalysis based on functional pincer ligands. *Chem Rev.* 2019;119:2681–2751.
173. Hashimoto T, Shiota K, Yamaguchi Y. Selective synthesis of secondary alkylboronates: markovnikov-selective hydroboration of vinylarenes with bis(pinacolato)diboron catalyzed by a nickel pincer complex. *Org Lett.* 2020;22:4033–4037.
174. Hashimoto T, Funatsu K, Ohtani A, Asano E, Yamaguchi Y. Cross-coupling reaction of allylic ethers with aryl Grignard reagents catalyzed by a nickel pincer complex. *Molecules.* 2019;24:2296.